Path-Integral Methods and their Applications

PATH-INTEGRAL METHODS and their APPLICATIONS

D C Khandekar
S V Lawande
K V Bhagwat

*Bhabha Atomic Research Centre,
Bombay, India*

World Scientific
Singapore • New Jersey • London • Hong Kong

Published by

World Scientific Publishing Co. Pte. Ltd.
P O Box 128, Farrer Road, Singapore 912805
USA office: Suite 1B, 1060 Main Street, River Edge, NJ 07661
UK office: 57 Shelton Street, Covent Garden, London WC2H 9HE

British Library Cataloguing-in-Publication Data
A catalogue record for this book is available from the British Library.

First published 1993
First reprint 1998

PATH-INTEGRAL METHODS AND THEIR APPLICATIONS

Copyright © 1993 by World Scientific Publishing Co. Pte. Ltd.

All rights reserved. This book, or parts thereof, may not be reproduced in any form or by any means, electronic or mechanical, including photocopying, recording or any information storage and retrieval system now known or to be invented, without written permission from the Publisher.

For photocopying of material in this volume, please pay a copying fee through the Copyright Clearance Center, Inc., 222 Rosewood Drive, Danvers, MA 01923, USA. In this case permission to photocopy is not required from the publisher.

ISBN 981-02-0563-5

Printed in Singapore.

PREFACE

Quantum mechanics is a probabilistic theory while the classical mechanics is a deterministic one. The connection between the two was eluding for a long time till Feynman blended them through his famous path integral representation of the kernel of the Schrödinger equation. Since then the path integral has come a long way and continues to inspire both physicists and mathematicians. For physicists, it is a very powerful tool, both at the conceptual as well as computational level. For mathematicians, on the other hand, it is a priori a meaningless object and continues to provide a challenge for developing an appropriate framework within which it can be defined, understood and manipulated.

These aspects of Feynman path integrals have been a stimulant to us during our long association with this field. Also during this period of three decades the path integral literature has grown rapidly. The first classic book by Feynman is already more than thirty years old while another book by Schulman is almost a decade old. The third excellent book by Wiegel concentrates primarily on problems on polymer science. Thus a need to write yet another book on the subject was felt by us.

The book attempts to cover the recent developments in the field of path integrals with the emphasis on exact results and approximation schemes along with applications in various areas. A conscious effort has been put to make the book useful to specialists in the field and simultaneously transparent to non-specialists primarily interested in the practical applications of path integration.

The first chapter introduces the concept of Feynman path integral and its relation with allied areas like Brownian motion and Wiener integrals. The second chapter discusses the instantaneous quadratic Lagrangians and prepares the reader for the basic algebraic manipulations inherent in a path integral treatment of a quantal problem. The third chapter deals with the path integration of the so called "two-time quadratic actions". The results derived in this chapter are particularly useful from the point of view of applications. The path integration in curved spaces forms the subject matter of chapter 4. In particular, the path integration in polar coordinates is treated in detail considering its importance in applications. The fifth chapter introduces the recently developed global and local time transformation techniques because of their potential in the context of path integration. The next chapter is concerned with the path integration methods to deal with the problems involving topological constraints. This chapter provides also an account of selected applications of these techniques.

Most of the literature on the subject have not considered the relation between invariants of motion and Feynman propagator and hence it was felt desirable to include a chapter on this theme. Chapter 7 discusses in detail how the knowledge of invariants of motion can be helpful in obtaining Feynman propagator.

The next three chapters are concerned with some approximation techniques frequently used in path integral applications. Notable amongst them is the so-called cumulant approximation method. A separate chapter is devoted to this method and its applications. The chapter on perturbation method has been written with a slightly different perspective. The major question which we pose here concerns the summability of the series rather than the conventional applications which can be found in almost all text books on the subject. The chapter on semiclassical analysis has been added primarily for the sake of completeness. An extensive and excellent discussion of this topic is already available in published form.

A significant departure from the earlier published books is the inclusion of a chapter on numerical methods of computing path integrals. In particular, Monte Carlo methods have been discussed in detail in chapter 11.

The last chapter presents briefly the difficulties encountered in providing a proper mathematical interpretation to Feynman path integral and some of the attempts in resolving them. Most of the discussion has been deliberately kept at a qualitative level to make it more transparent. The more curious reader may find the references listed at the end of chapter 12 useful for more rigorous treatment.

Writing of this book has taken more than two years. Stimulating discussions with several colleagues have helped in bringing the book to its present form. It is a pleasure to thank them all. In particular we acknowledge constant encouragement from Dr. R. Chidambaram all throughout this trying period. Lastly, we must express our sincere thanks to Mrs. Smita Khandekar for her assistance in typing and editing of the present manuscript.

Authors

CONTENTS

Preface   v

Chapter 1   INTRODUCTION TO PATH INTEGRALS   1
- 1.1 Introduction   1
- 1.2 Feynman Path Integral   2
  - 1.2.1 Feynman formulation   4
  - 1.2.2 Relation to other approaches of quantum description   9
  - 1.2.3 Functional calculus   11
- 1.3 Random Walk, Brownian Motion and Wiener integral   14
  - 1.3.1 Random walk and Brownian motion   14
  - 1.3.2 Wiener integral   16
- 1.4 Trotter Product Formula and Alternative Derivation of Path Integral   21
- 1.5 Path Integral Subtleties   24
- Notes and References   26

Chapter 2   PROPAGATORS FOR LOCAL QUADRATIC LAGRANGIANS   29
- 2.1 Introduction   29
- 2.2 Derivation of the Propagator   29
  - 2.2.1 Evaluation of limiting expression for D_N 32
  - 2.2.2 Evaluation of S_{cl} 33
- 2.3 Specific Cases   35
- 2.4 Velocity Dependent Potentials   38
- 2.5 General Quadratic forms   40
- 2.6 Propagator Beyond the First Singularity of VPD   43
- Notes and References   46

Chapter 3	**NON-LOCAL QUADRATIC ACTIONS**	48
	3.1 Elimination of Degrees of Freedom	48
	3.1.1 System coupled to a harmonic oscillator	48
	3.1.2 An electron in a random potential	50
	3.2 Two-time Quadratic Actions	53
	3.3 Extension to More General Non-local Actions	57
	3.4 Examples of Explicit Evaluation	59
	3.4.1 Generalized one-time actions	59
	3.4.2 Translationally invariant two-time actions	61
	3.4.3 Propagator for the polaron kernel	63
	3.5 Applications of Two-time Quadratic Actions	69
	3.5.1 Exactly solvable model of electronic density of states (DOS)	69
	3.5.2 Polymer distribution functions	71
	3.5.3 Propagation of waves in random media	75
	Notes and References	79
Chapter 4	**PATH INTEGRALS IN GENERAL COORDINATE SYSTEMS**	82
	4.1 Introduction	82
	4.2 Path Integrals in Polar Coordinates	83
	4.2.1 Polar coordinates in two dimensions	84
	4.2.2 Polar coordinates in three dimensions	88
	4.2.3 Generalization to d-dimensional polar coordinates	93
	4.3 Examples of Explicit Evaluation	98
	4.3.1 Free particle propagator	98
	4.3.2 Rotor in d-dimensions	100
	4.3.3 Central potentials	102
	4.3.4 Non-spherically symmetric potentials	104
	4.4 Path Integration in General Curved Spaces	107
	4.4.1 Quantization of classical Hamiltonian	107
	4.4.2 Derivation of a path integral	112
	Notes and References	116

Chapter 5 COORDINATE TIME TRANSFORMATIONS IN PATH INTEGRALS — 119

5.1 Introduction — 119
5.2 Local Time Transformations in Classical Mechanics — 120
5.3 Concept of the Promotor — 124
5.4 Coordinate-Time Transformations in Path Integral — 125
5.5 Illustrative Examples — 131
   5.5.1 Coulomb potential — 131
   5.5.2 Morse potential — 134
5.6 Propagators Related to a Rigid Rotor — 136
   5.6.1 Infinite square well — 140
5.7 Coulomb Problem Based on KS Transformation — 142
Notes and References — 149

Chapter 6 CONSTRAINED PATH INTEGRALS — 151

6.1 Examples of Constrained Path Integrals — 151
   6.1.1 Problems in polymer physics — 151
   6.1.2 Aharonov-Bohm effect — 153
6.2 The Constraint As a Functional — 155
6.3 Evaluation of The Path Integral — 156
   6.3.1 A simple entanglement problem — 157
6.4 Evaluation of The Propagator — 161
6.5 Propagators corresponding to More Than One Constraints — 164
6.6 Total Winding Index and Stochastic Area — 167
6.7 Statistical Mechanics of Entangled Polymers — 171
   6.7.1 The properties of entangled polymers — 171
   6.7.2 The properties of entangled polymers: Entanglement with clusters — 174
6.8 Aharonov-Bohm Effect — 179
Notes and References — 183

Chapter 7 TIME DEPENDENT INVARIANTS AND FEYNMAN PROPAGATOR — 186

7.1 Introduction — 186
7.2 Classical and Quantal Invariants — 188

7.2.1 Noether invariants	188
7.2.2 Derivation of invariant based on Hamiltonian description	192
7.2.3 Examples of invariants	194
7.3 Schrödinger Equation and Invariants	198
7.4 Feynman Propagator	202
7.5 Invariants Quadratic in Momentum and the Propagator	205
7.5.1 Illustrative examples	208
7.6 Role Played by the Invariant I	210
7.7 Global Time Transformation in Feynman Path Integral	213
Notes and References	218

Chapter 8 THE CUMULANT APPROXIMATION FOR FEYNMAN PROPAGATORS — 220

8.1 Introduction	220
8.2 The Cumulant Approximation	220
8.3 Spectrum of Positionally Disorderd Systems	225
8.3.1 Basic formulation	225
8.3.2 Evaluation of G	226
8.3.3 The Behaviour of density of states	229
8.3.4 The density of states for Gaussian correlation	231
8.4 The Polaron Problem	236
8.4.1 Free energy and ground state energy	239
8.4.2 The effective mass	243
8.5 The Bi-Polaron Problem	245
8.6 Polymer Distribution Functions	250
Notes and References	252

Chapter 9 THE PERTURBATION APPROACH — 258

9.1 Introduction	258
9.2 The Perturbation Series	258
9.3 One-Dimensional Delta-function Potential	262
9.4 Inverse Square Potential	265
9.5 The Coulomb Potential	267
9.6 General Formulation	270

	9.6.1 Inverse-square potential (1-dimensional)	272
	9.6.2 Harmonic oscillator (1-dimensional)	272
	9.6.3 Coulomb potential	275
	9.6.4 The singular potentials	276
	Notes and References	278

Chapter 10 SEMICLASSICAL PROPAGATOR — 280

    10.1 Introduction — 280
    10.2 Asymptotic Analysis — 281
    10.3 Semiclassical Approximation — 284
    10.4 A Particle in an Inverse Square Potential — 289
    Notes and References — 291

Chapter 11 NUMERICAL METHODS OF SUMMING OVER PATHS — 292

    11.1 Introduction — 292
    11.2 Monte Carlo Method — 293
    11.3 Path Integral by Monte Carlo — 307
    11.4 Deterministic Techniques of Path Summation — 317
    Notes and References — 319

Chapter 12 MATHEMATICAL NATURE OF FEYNMAN PATH-INTEGRAL — 323

    12.1 Introduction — 323
    12.2 Definition through Limiting Procedures — 324
        12.2.1 Trotter product formula — 324
        12.2.2 Generalized Gaussian integrals — 325
    12.3 Definition through White Noise Calculus — 327
        12.3.1 White noise calculus — 327
        12.3.2 Feynman propagator — 334
        12.3.3 A new expansion for the propagator — 339
    Notes and References — 342

CHAPTER 1

INTRODUCTION TO PATH INTEGRALS

1.1. Introduction

This book is on path integrals and their applications to physics at large. Naturally we begin with Feynman, who first introduced the concept to physicists in his new space-time formulation of non-relativistic quantum mechanics published in his classic 1948 paper in Reviews of Modern Physics. Feynman himself established that this third formulation of quantum mechanics, the path integral approach, is equivalent to the usual formulations of Schrödinger as well as that of Heisenberg and Dirac. Incidentally, three independent mathematical disciplines are associated with the three formulations of quantum mechanics. While Heisenberg-Dirac method relies on "algebra", Schrödinger's approach is based on differential equations and hence uses "analysis". Feynman's method on the other hand is based on "geometry". This geometrical way of expressing the quantum superposition principle is intuitively appealing since it allows us to directly visualize the constructive or destructive interference arising from many different paths. Feynman himself attributed this multiplicity of possible descriptions of quantum phenomena to *our having captured key elements in our description of atomic phenomena and is an expression and representation of the simplicity of nature.*

In fact the notion of a path integral usually called a functional integral had been familiar to mathematicians much before Feynman. It was Volterra who used this idea in his work on functional calculus. A functional is to be considered analogous to a function of infinitely many variables. Calculations involving a functional are carried out by assuming it to be a function of a finite number N of variables and subsequently letting N → ∞. The procedure is similar to the "time slicing prescription" used by Feynman. We shall encounter this many times later in the book. Early papers of Daniell deals with some attempts of integrating a functional over a space of functions and are reviewed by Kac. Subsequently Wiener introduced a proper measure-theoretic definition of an integral of a functional over a space of functions. The analogy between Feynman path integral and Wiener integral has been extensively discussed in the literature.

The major motivation of this chapter is to introduce the notion of a path integral from a physicist's point of view. We shall outline the path integral formulation of non-relativistic quantum mechanics in a way that Feynman presented in his 1948 paper and subsequently in the book by Feynman and Hibbs. We shall then discuss the alternative ways of looking at a path integral using the ideas of random walk, Brownian motion and Wiener measure.

1.2. Feynman Path Integral

Before we present Feynman's intuitive arguments for introducing the path integral formulation we may outline the scenario at his time. Quantum mechanics was traditionally based on the Hamiltonian formulation of classical mechanics. The rules governing the transition from classical to quantum description are summarized in Table 1.

Table 1 : Classical vs. Quantum Description

	Classical Mechanics	Quantum Mechanics
1. Variables:	x, p (c-numbers), {x,p} = 1	$\hat{x}, \hat{p}$ (operators), $[\hat{x},\hat{p}]$ = 1
2. Hamiltonian:	H(x, p)	$\hat{H}(\hat{x}, \hat{p})$
3. Dynamical Law:	df(x,p)/dt = {f,H}	a) Heisenberg Eq : $i\hbar\frac{d\hat{f}}{dt} = [\hat{f},\hat{H}]$
	H. J. equation ?	b) Schrödinger Eq : $i\hbar\frac{\partial\psi}{\partial t} = \hat{H}\psi$
		$\hat{H} = H(x, -i\hbar\partial/\partial x)$
4: Lagrangian:	L(x, ẋ)	?

The classical description of a system is through a set of conjugate variables x, p satisfying the Poisson Bracket (PB) relation {x,p} = 1.

The dynamics is described by Hamilton's equations written in the general PB notation in Table 1. In quantum description these conjugate variables are replaced by non-commuting operators $\hat{x}$, $\hat{p}$ and a statement governing the time evolution of these operators. The commutation relation is $[\hat{x},\hat{p}] = i\hbar\hat{I}$, $\hat{I}$ representing the identity operator. Further, the dynamical law is

$$i\hbar \, d\hat{x}/dt = [\hat{x}, \hat{H}] \quad (2.1)$$

in Heisenberg picture which is reminiscent of classical PB theory. Alternatively the dynamical law may take the form of the Schrödinger equation

$$i\hbar \, \partial\psi/\partial t = \hat{H}\psi \quad (2.2)$$

where the quantum Hamiltonian operator $\hat{H}$ is obtained from the classical Hamiltonian $H(x, p)$ by the simple replacement $p \rightarrow -i\hbar\partial/\partial x$. In all this formulation there remained an important gap between quantum and classical mechanics. The Lagrangian formulation, which had preceded the Hamiltonian approach in classical mechanics had practically no role in the quantum formulation. There was, however, one remote connection. This involved the derivation of classical Hamilton-Jacobi (HJ) equation from the Schrödinger equation by means of the transformation

$$\psi \rightarrow C \exp(iS/\hbar) \quad (2.3)$$

where S is the classical action and C is real. In fact the substitution of (2.3) in Schrödinger equation

$$i\hbar \, (\partial/\partial t)\psi = (-\hbar^2/2m) \, (\partial^2/\partial x^2)\psi + V \psi \quad (2.4)$$

for a particle of mass m moving in a one-dimensional potential $V(x)$, yields the HJ equation

$$(\partial S/\partial t) + (1/2m) \, (\partial S/\partial x)^2 + V = 0 \quad (2.5)$$

if terms $O(\hbar/S)$ are neglected. If, however, these terms are retained, we obtain the continuity equation

$$\partial\rho/\partial t + \partial(\rho v)/\partial x = 0 \quad (2.6)$$

where $\rho = |\psi|^2 = C^2$ and $v = (\partial S/\partial x)/m$. Note that (2.3) with S determined from (2.5) represents just the solution of (2.4) in WKB approximation valid for $\hbar \to 0$. This correspondence between Schrödinger equation and HJ equation in the semiclassical limit ($\hbar \to 0$) answers the first question raised in Table 1.

Regarding the second question in Table 1, it was Dirac who had first emphasized the possible importance of Lagrangian in quantum mechanics in one of his early papers and later in his book. Feynman exploited Dirac's remarks to arrive at his so-called Lagrangian formulation of quantum mechanics.

1.2.1. Feynman formulation

In standard quantum mechanics one assigns a complex probability amplitude $\psi(x,t)$ (called the wave function) with the position x of a particle at time t. For simplicity, we shall use notation corresponding to a single dimension, the generalization to more than one being trivial. The probability density for the particle to be found at x at time t is then simply given by $|\psi(x,t)|^2$. Instead, Feynman associated a probability amplitude with the "entire motion of a particle as a function of time" characterized by the path or the trajectory $x(t)$ of the particle. He then extended the quantum superposition principle to apply for paths making use of the following basic distinction between classical and quantum probabilities.

Let $P(a|b)$ denote the conditional probability of an event a given that the event b occurred and similarly for $P(a|c)$ and $P(b|c)$. Then, classically one has the rule

$$P(a|c) = \sum_b P(a|b) P(b|c) \qquad (2.7)$$

where the sum is over all events (or states) b that can occur between c and a. The quantum mechanical rule, however, is somewhat different. Here we have to work with a probability amplitude ϕ_{ab} which like $P(a|b)$ depends on two states and follows a relation similar to (2.7)

$$\phi_{ac} = \sum_b \phi_{ab} \phi_{bc} \qquad (2.8)$$

where as before the sum is over all possible states b. The difference in these two situations is that ϕ is not a probability but amplitude such that $|\phi|^2$ is interpreted as a probability. Equation (2.8) is, in fact, the quantum superposition law as extended to paths and is illustrated in Fig. 1.1 by the double slit experiment.

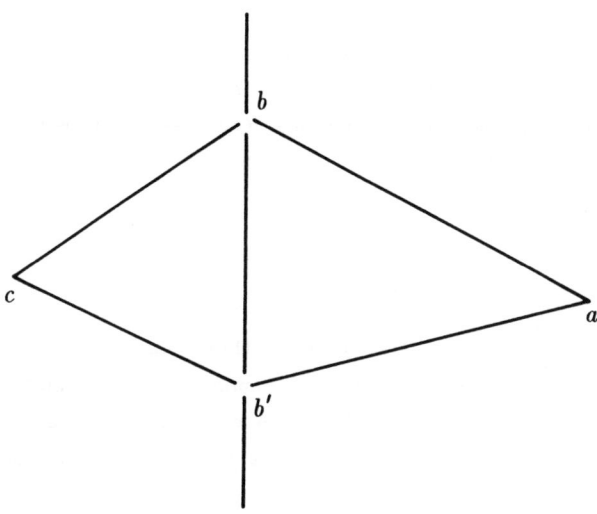

Fig. 1.1. Double slit experiment. *The electron at c can reach the point a on the screen either by taking the path cba or the path cb'a. The amplitude for the path cba is* $\phi_{abc} = \phi_{ab} \cdot \phi_{bc}$ *and that for the path cb'a is* $\phi_{ab'c} = \phi_{ab'} \cdot \phi_{b'c}$. *The total amplitude is thus* $\phi_{ac} = \phi_{abc} + \phi_{ab'c} = \phi_{ab}\phi_{bc} + \phi_{ab'}\phi_{b'c}$. *Here* ϕ_{ab} *is the amplitude for the path ab, etc.*

An electron passes through either (or both) of two slits on its way to the screen. The superposition rule would imply an interference pattern on the screen. Any attempt to verify through which slit the electron went will destroy the interference pattern. Feynman translated this in the Language of "sum over paths" by simply interpreting ϕ_{ab} to be the amplitude for "path" of the electron from b to a and used (2.8) as a postulate to derive his path integral formulation. There is no assertion here that the particle followed definite paths with certain probabilities. One has to obtain the probability amplitudes for various paths and add them to obtain the total amplitude.

In analogy with the double slit experiment the basic quantity in Feynman formulation is the probability amplitude $K(x'',t'';x',t')$ for a non-relativistic particle to go from a space-time point (x',t') to (x'',t''). We must then sum over all intermediate possibilities, that is, our sum must include contributions from each trajectory connecting (x',t') to (x'',t''). If we denote the probability amplitude for a trajectory $x(t)$ $(x(t') = x'$ and $x(t'') = x'')$ by $\phi[x(t)]$, the total amplitude

$$K(x'',t'';x',t') = \sum_{\{x(t)\}} \phi[x(t)] . \qquad (2.9)$$

How do we obtain $\phi[x(t)]$? It is here that connection with Dirac's idea comes in. Dirac noticed that the transformation function K, taking particle from (x',t') to (x'',t''), more commonly known as the propagator is "analogous" to exp $[iS/\hbar]$ where S is the solution of the HJ equation. As mentioned earlier, this is to be expected for small $\hbar$ (WKB approximation). However, Dirac also observed that exp $[iS/\hbar]$ is also a reasonable approximation when the time interval over which K propagates tends to zero. Thus for the simple Lagrangian of a particle of mass m moving in a potential $V(x)$ given by

$$L = \frac{1}{2} m \dot{x}^2 - V(x) , \qquad (2.10)$$

Dirac propagator assumes the form

$$K(x,y;\varepsilon) = K(x,t+\varepsilon;y,t) = \frac{1}{A} \exp\left\{\frac{i\varepsilon}{\hbar}\left[\frac{m}{2}\left(\frac{x-y}{\varepsilon}\right)^2 - V(x)\right]\right\} \qquad (2.11)$$

for small ε. A consequence of the superposition law is the integral equation

$$\psi(x,t+\varepsilon) = \int K(x,t+\varepsilon;y,t) \psi(y,t) dy \qquad (2.12)$$

connecting the wave function at time t to the wave function at time t+ε. Summation over intermediate state b now corresponds to the particle position y at time t. If (2.11) is inserted in (2.12), one obtains

Schrödinger equation

$$i\hbar (\partial/\partial t)\psi = (-\hbar^2/2m)(\partial^2/\partial x^2)\psi + V\psi \qquad (2.13)$$

after one sets the proportionality constant $A^{-1} = (m/2\pi i\hbar\epsilon)^{1/2}$.

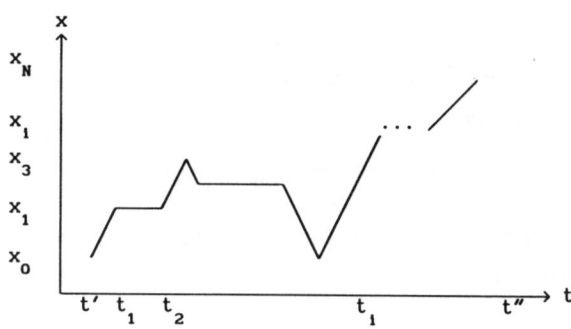

Fig. 1.2. *A typical representation of a path x(t).*

Next, we discretize the path by dividing the time interval $[t',t'']$ into N equal subintervals $[t_{i-1},t_i]$ each of length ϵ with the identification $t_0 = t'$, $t_N = t''$ and denote $x(t_i)$ by x_i. Thus the path $x(t)$ is represented by $\{x', x_1, \ldots, x_{N-1}, x''\}$ as shown in Fig.1.2. The expression (2.9) takes the form

$$K_N = \int \ldots \int \phi(x',x_1,\ldots,x_{N-1},x'') \, dx_1 \ldots dx_{N-1} . \qquad (2.14)$$

Using composition rule (2.8) and after identifying the probability amplitude $\phi(x_i,x_{i-1})$ with $K(x_i,x_{i-1},\epsilon)$, we may write

$$K_N = \int K(x_N,x_{N-1};\epsilon) \prod_{i=1}^{N-1} K(x_i,x_{i-1};\epsilon) \, dx_i \qquad (2.15)$$

and expect that the finite time propagator K to be the limiting form of K_N as $N \to \infty$, that is,

$$K = \lim_{N \to \infty} K_N . \qquad (2.16)$$

Finally inserting the expression (2.11) for $K(x_i,x_{i-1},\epsilon)$, the propagator takes the form

$$K(x'',t'';x',t') = \lim_{\substack{N \to \infty \\ \varepsilon \to 0 \\ (N\varepsilon = t''-t')}} \frac{1}{A} \int \exp\left[\frac{i}{\hbar} \sum_{k=1}^{N} S(x_k, x_{k-1}; \varepsilon)\right] \prod_{k=1}^{N-1} \frac{dx_k}{A} \quad (2.17)$$

where $S(x_k, x_{k-1}, \varepsilon)$, is the action over an infinitesimal time interval $\varepsilon = t_k - t_{k-1}$. Feynman writes this in the compact form as a "sum over paths or histories" as

$$K(x'',t'';x',t') = N \sum_{\{x(t)\}} \exp\left\{\frac{i}{\hbar} S[x(t)]\right\} \quad (2.18a)$$

where N is a normalization constant. An alternative more conventional form is the "path integral"

$$K(x'',t''; x',t') = \int \exp\left\{\frac{i}{\hbar} S[x(t)]\right\} \mathcal{D}[x(t)] \quad (2.18b)$$

where the symbol $\mathcal{D}[x(t)]$ is the "path differential measure" signifying summation or integration over all paths and

$$S = \int_{t'}^{t''} L \, dt$$

is the classical action evaluated between the two end-points. The equation (2.17) is usually referred to as the polygonal or the lattice definition of the path integral of (2.18b).

In summary the physical quantity central to Feynman's approach is the propagator K which represents the probability amplitude for a particle to travel from one space-time point to another. Classically there is only one path or trajectory connecting the two end points over which the particle moves, viz., the path of "minimum" action. Quantum mechanics is, however, more democratic. Here all paths (and not just the classical path) contribute to the amplitude. Moreover all paths contribute equally in amplitude but differently in phases, the phase of each path being the classical action measured in terms of $\hbar$, the unit of quantum action.

1.2.2. Relation to other approaches of quantum description

As remarked earlier, the propagator is the transformation matrix in the algebraic approach of Heisenberg-Dirac or equivalently the matrix element of the evolution operator $\exp[-i(t''-t')\hat{H}/\hbar]$ taken between the two position states:

$$K(x'',t'';x',t') = \langle x'',t''|x',t'\rangle = \langle x''|\exp[-\frac{i}{\hbar}(t''-t')\hat{H}]|x'\rangle. \quad (2.19)$$

Thus K contains the entire dynamics of the system. If one were to use (2.19) as the definition of the propagator one would have to solve Heisenberg equation of motion to obtain K. The path integral (2.18) gives a prescription of obtaining K directly.

Next, if ψ is the wave function described by Schrödinger equation (2.2), K propagates ψ by means of the integral equation

$$\psi(x'',t'') = \int K(x'',t'';x',t')\,\psi(x',t')\,dx', \quad t'' > t'. \quad (2.20)$$

K is, therefore the Green's function of the Schrödinger equation and the equivalence between the geometric approach of Feynman and the analytical approach of Schrödinger is complete.

More precisely, we may introduce the operator form of the Green's function as

$$\hat{G} = \theta(t''-t')\exp[-\frac{i}{\hbar}(t''-t')\hat{H}] \quad (2.21)$$

where $\hat{H}$ is assumed to be independent of time and $\theta(t)$ is a step function which is unity for $t > 0$ and zero otherwise. If we take the matrix element of $\hat{G}$ between the states $|x'\rangle$ and $|x''\rangle$, we have

$$G(x'',t'';x',t') = \langle x''|\hat{G}|x'\rangle = K(x'',t'';x',t')\,\theta(t''-t'). \quad (2.22)$$

It is easy to show that $G(x'',t'';x',t')$ obeys the equation

$$[i\hbar\,\partial/\partial t'' - \hat{H}(x'')]G(x'',t'';x',t') = i\hbar\,\delta(x''-x')\delta(t''-t'). \quad (2.23)$$

In deriving (2.23), the obvious fact that

$$\lim_{t'' \to t'} K(x'',t'';x',t') = \delta(x''-x') \quad (2.24)$$

is used. Also when the Hamiltonian or Lagrangian does not depend on time explicitly, the following stationarity property holds

$$K(x'',t'';x',t') = K(x'',t''-t';x',0) . \qquad (2.25)$$

Another consequence of the superposition principle is the property

$$K(x'',t'';x',t') = \int K(x'',t'';x,t) K(x,t;x',t')dx , \quad t' < t < t'' \qquad (2.26)$$

which is reminiscent of the Chapman-Kolmogorov equation satisfied by the conditional probabilities in a Markov stochastic process.

When the Hamiltonian $\hat{H}$ admits a complete set of eigenfunctions ψ_n corresponding to eigenvalues E_n the following expansion formula

$$K(x'',t'';x',t') = \sum_n \psi_n(x'')\psi_n^*(x') \exp[-\frac{i}{\hbar}(t''-t')E_n], \quad t'' > t' \qquad (2.27)$$

applies. In the case when the spectrum is discrete as well as continuous the summation sign implies also an integration over the continuum. Writing $T = t''-t'$ and taking the Fourier Transform with respect to T, one obtains from (2.22) the energy-dependent Green's function

$$\tilde{G}(x'',x';E) = \frac{1}{i\hbar} \int_{-\infty}^{\infty} dT \exp[iET/\hbar]G(x'',T;x',0)$$

$$= \sum_n \frac{\psi_n(x'') \psi_n^*(x')}{(E - E_n)} \qquad (2.28)$$

where Im E > 0 and in the second step use is made of (2.27). Since the spectrum of $\hat{H}$ is real, G has singularities when Im E = 0. For a discrete spectrum, we identify the poles of $\tilde{G}$ with the location of bound states and the corresponding residues with the bound state wave functions. For the continuous spectrum of $\hat{H}$, $\tilde{G}$ has a cut.

In brief, Feynman path integral has reduced the quantum problem to quadratures. A single compact formula contains the complete quantum dynamics. The formulation also provides a link between the classical mechanics and the quantum superposition principle. An instantaneous correspondence principle is suggested by the form (2.18) since in the limit $\hbar \rightarrow 0$, the important contributions to K arise from the paths obeying the classical variational principle $\delta S = 0$ (Fig.1.3). Thus, on the one hand, contact with classical mechanics is manifest and on the other, the superposition principle of quantum mechanics is built in the path integral. As we shall see later, this has important consequences in applications of path integrals in various fields.

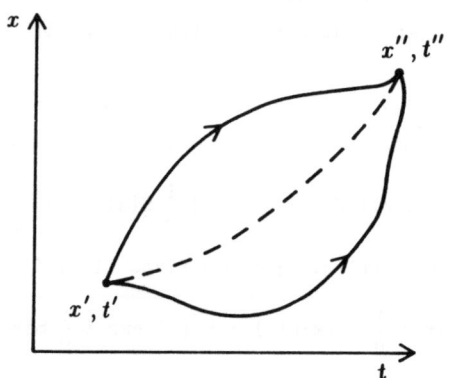

Fig. 1.3. Correspondence Principle. The — line indicates a general trajectory while the -- line indicates the classical trajectory.

1.2.3. Functional calculus

In some respects, a functional is like a function of many variables. For operational purpose, Feynman path integral can be viewed as a functional integral though its rigorous mathematical justification as yet remains to be shown. Therefore, it is useful to derive some quantum mechanical results from this alternative formulation of Feynman.

Let $F[x(\tau)]$ be a functional. A derivative of F is expected to give some idea about difference between two neighbouring functionals $F[x(\tau)+\eta(\tau)] - F[x(\tau)]$, where $\eta(\tau)$ is a small function. If $x(\tau)$ and $\eta(\tau)$ are discretised by $x_i = x(\tau_i)$, $\eta_i = \eta(\tau_i)$, $i = 1, 2, 3, \ldots, N$, we may write

$$F(x_1+\eta_1, x_2+\eta_2, \ldots, x_N+\eta_N) - F(x_1, x_2, \ldots, x_N) = \sum_{i=1}^{N} \left(\frac{\partial F}{\partial x_i}\right) \eta_i \quad (2.29)$$

for sufficiently small η_i's. The definition of the functional derivative is an extension of this to the continuum limit. For "small" η,

$$F[x(\tau)+\eta(\tau)] - F[x(\tau)] = \int \left(\frac{\delta F}{\delta x(s)}\right) \eta(s) \, ds. \quad (2.30)$$

It is clear that the usual rules for differentiation of products, integration by parts, etc., may be derived also for the functional derivatives.

Consider now the "expectation value" of a functional over the Feynman measure. For a functional F, we define the expectation value as

$$<F>_S = \int \mathcal{D}[x(\tau)] \, F[x(\tau)] \exp\left\{\frac{i}{\hbar} S[x(\tau)]\right\}. \quad (2.31)$$

Now, since $\mathcal{D}[x(\tau)] = \mathcal{D}[x(\tau)+\eta(\tau)]$ for a fixed $\eta(\tau)$, it follows that

$$\int \mathcal{D}[x(\tau)] \, [F[x+\eta] \exp\left\{\frac{i}{\hbar} S[x+\eta]\right\} - F[x] \exp\left\{\frac{i}{\hbar} S[x]\right\}]$$

$$= \int \mathcal{D}[x(\tau)] \int ds \left[\left(\frac{\delta F}{\delta x(s)}\right) + \frac{i}{\hbar} F \left(\frac{\delta S}{\delta x(s)}\right)\right] \exp\left\{\frac{i}{\hbar} S[x(\tau)]\right\} \eta(s). \quad (2.32)$$

The L.H.S. of (2.32) is obviously zero and as η is arbitrary the quantity in the square brackets in the integrand on the R.H.S. vanishes. This results in an important relation

$$<\left(\frac{\delta F}{\delta x(s)}\right)>_S = -\frac{i}{\hbar} <F \left(\frac{\delta S}{\delta x(s)}\right)>_S. \quad (2.33)$$

This relation is known as Schwinger's action principle.

The relation (2.33) holds for arbitrary value of the parameters and it is possible to write it at a discrete point x_k corresponding to the value of $x(s)$ at $s = s_k$, in the form

$$<\left(\frac{\partial F}{\partial x_k}\right)>_S = -\frac{i}{\hbar} <F \left(\frac{\partial S}{\partial x_k}\right)>_S. \quad (2.34)$$

We can use the relations (2.33) and (2.34) to derive some interesting results. For simplicity we use the derivative form (2.33). Consider the discretised form of the action

$$S = \sum_{k=1}^{N} \left[\frac{m}{2\varepsilon} (x_{k+1} - x_k)^2 - \varepsilon V(x_k) \right] . \quad (2.35)$$

Selecting a particular time t_k and the associated x_k, we have

$$\frac{\partial S}{\partial x_k} = -\left[m\left(\frac{x_{k+1} - x_k}{\varepsilon} - \frac{x_k - x_{k-1}}{\varepsilon}\right) + \varepsilon V'(x_k) \right] \quad (2.36)$$

and using (2.33)

$$\left\langle \frac{\partial F}{\partial x_k} \right\rangle_S = \frac{i\varepsilon}{\hbar} \left\langle F \left\{ \frac{m(x_{k+1} - 2x_k + x_{k-1})}{\varepsilon^2} + V'(x_k) \right\} \right\rangle_S . \quad (2.37)$$

Now choosing a special form of the function F, viz., letting F = 1, $\partial F/\partial x_k = 0$ and taking the limit $\varepsilon \to 0$, we obtain

$$\langle [m \ddot{x} + V'(x)] \rangle_S = 0 . \quad (2.38)$$

This is just the Ehrenfest's theorem, that is, the quantum mechanical analogue of Newton's equation of motion.

Next, suppose $F = x_k$, so that $\partial F/\partial x_k = 1$. We obtain from (2.37)

$$\frac{\hbar}{i} \langle 1 \rangle_S = \left\langle \left[m\left(\frac{x_{k+1} - x_k}{\varepsilon}\right) x_k - x_k m\left(\frac{x_k - x_{k-1}}{\varepsilon}\right) \right] \right\rangle_S \quad (2.39)$$

where the R.H.S. is written in a specific order. We surmise that in the first term $m(x_{k+1} - x_k)/\varepsilon$ is momentum $p = m\dot{x}$ evaluated at $t = t_k + \varepsilon/2$, while x_k is the position of the particle at $t = t_k$. In the second term, the position is again taken at time t_k but the momentum corresponds to time $t_k - \varepsilon/2$. The relation (2.39) says that these two expectation values are different. In fact, (2.39) is the path integral analogue of the commutation relation $[\hat{x}, \hat{p}] = i\hbar$.

As a further remark we point out that the expectation value of the kinetic energy is not given by $(m/2)\langle [(x_{k+1} - x_k)/\varepsilon]^2 \rangle_S$ since this quantity becomes infinite as ε approaches zero. The correct expression corresponds to $(1/2m)\langle p(t+\varepsilon)p(t) \rangle_S$, that is,

$$\langle K.E. \rangle_S = \left\langle \frac{m}{2} \left(\frac{x_{k+1} - x_k}{\varepsilon}\right) \left(\frac{x_k - x_{k-1}}{\varepsilon}\right) \right\rangle_S . \quad (2.40)$$

This may also be reconciled with Eq. (2.37). Letting $F = (x_{k+1} - x_k)$ and retaining terms of lower order in ε, we obtain

$$\left\langle \frac{m}{2} \left(\frac{x_{k+1} - x_k}{\varepsilon} \right) \left(\frac{x_k - x_{k-1}}{\varepsilon} \right) \right\rangle_S = \left\langle \frac{m}{2} \left(\frac{x_{k+1} - x_k}{\varepsilon} \right)^2 \right\rangle_S + \frac{\hbar}{2i\varepsilon} \langle 1 \rangle_S .$$

Hence the R.H.S. is expected to be finite although the individual terms blow up as $\varepsilon \to 0$.

Feynman calls these expectation values as transition elements. More general transition elements are defined as the matrix elements

$$\langle \chi | F | \psi \rangle_S = \int dy \int dz \int \chi^*(y) F [x(t)] \exp \left\{ \frac{i}{\hbar} S[x(t)] \right\} \psi(z) \, \mathcal{D}[x(t)] .$$

The transition elements can be shown to be related to operator averages in the conventional notation of wave functions. The reader may consult the book by *Feynman and Hibbs* for details.

1.3. Random Walk, Brownian Motion and Wiener Integral

1.3.1. *Random walk and Brownian motion*

In this subsection we discuss a simple model of random walk. For a more general treatment the reader may refer to the classic article of Chandrasekhar. A particle starts from the origin and moves freely in a three-dimensional space taking some discrete steps along directions which are random. The position of the particle at the end of mth step is denoted by $\vec{q}_m$ ($m = 0, 1, 2, \ldots$). Keeping in mind the applications to polymer physics to be discussed in later chapters, we assume that the steps $\Delta \vec{q}_m = (\vec{q}_{m+1} - \vec{q}_m)$ are of equal length $|\Delta \vec{q}_m| = \ell$.

We are interested in the probability $P(\vec{q}, N) \, d\vec{q}$ of finding the particle in the vicinity of $d\vec{q}$ around the point $\vec{q}$ at the end of N steps. It is clear from the assumptions that the single-step probability distribution (conditional probability)

$$p(\vec{q}|\vec{q}') = (1/4\pi \ell^2) \, \delta(|\vec{q} - \vec{q}'| - \ell) \tag{3.1}$$

while $P(\vec{q}, N)$ takes the form of the convolution integral

$$P(\vec{q}, N) = \int d\vec{q}' \, p(\vec{q} | \vec{q}') \, P(\vec{q}', N-1) \tag{3.2}$$

with $P(\vec{q}, 0) = \delta(\vec{q})$.

The interpretation of (3.2) is very simple. The probability that the particle is at q' after (N-1) steps is $P(\vec{q}', N-1)$. A single step then takes the particle from $\vec{q}'$ to $\vec{q}$ with the conditional probability $p(\vec{q}|\vec{q}')$. Since $\vec{q}'$ is arbitrary, integration over all values of $\vec{q}'$ yields $P(\vec{q}, N)$.

In order to obtain the solution of (3.2) it is best to take the Fourier transform with respect to $\vec{q}$. Writing

$$\tilde{P}(\vec{k}, N) = \int \exp[i\vec{k}\cdot\vec{q}] \, P(\vec{q}, N) \, d\vec{q} \tag{3.3}$$

and noting that

$$(1/4\pi\ell^2) \int d\vec{q} \, \delta(|\vec{q}| - \ell) \exp[i\vec{k}\cdot\vec{q}] = \frac{\sin k\ell}{k\ell} \tag{3.4}$$

we obtain

$$\tilde{P}(\vec{k}, N) = \frac{\sin k\ell}{k\ell} \tilde{P}(\vec{k}, N-1) = \left(\frac{\sin k\ell}{k\ell}\right)^N. \tag{3.5}$$

It follows from (3.5) that the inverse Fourier transform $\tilde{P}$ is

$$P(\vec{q}, N) = \left(\frac{1}{2\pi}\right)^3 \int d\vec{k} \left(\frac{\sin k\ell}{k\ell}\right)^N e^{-i\vec{k}\cdot\vec{q}}. \tag{3.6}$$

This is an exact expression for the probability distribution of a N-step random walk. In order to obtain further insight we assume N to be large and the step size to be sufficiently small in certain sense to be made more precise later. We may then use the approximation

$$\frac{\sin k\ell}{k\ell} \simeq 1 - \frac{(k\ell)^2}{6} \simeq \exp\left(-\frac{(k\ell)^2}{6}\right) \tag{3.7}$$

which when inserted in Eq.(3.6) yields the approximate form

$$P(\vec{q}, N) = \left(\frac{2\pi N\ell^2}{3}\right)^{-3/2} \exp\left[-\frac{3q^2}{2N\ell^2}\right]. \tag{3.8}$$

Thus the probability distribution of a N-step random walk to reach a point $\vec{q}$ tends to be a Gaussian when N becomes sufficiently large. More precise condition for the validity of formula (3.8) is that $\ell \to 0$, $N \to \infty$ but $N\ell^2$ is finite.

We add two modifications in the above treatment. First, if the particle is initially at the position $\vec{q}_0$ instead of at the origin, the relation (3.8) gets modified to

$$P(\vec{q},N;\vec{q}_0,0) = \left(\frac{2\pi N\ell^2}{3}\right)^{-3/2} \exp\left[-\frac{3(\vec{q}-\vec{q}_0)^2}{2N\ell^2}\right] \qquad (3.9)$$

where we have modified the notation to indicate the dependence on the initial position $\vec{q}_0$. When there is no room for confusion, we may use the notation $P(\vec{q}-\vec{q}_0,N)$ as well.

The other modification involves introduction of the time variable. We may associate with each step a time ε required to take the step; the total time for N steps is simply $t = N\varepsilon$. We may thus write (3.9) in the form

$$P(\vec{q},t;\vec{q}_0,0) = \left(\frac{2\pi t\ell^2}{3\varepsilon}\right)^{-3/2} \exp\left[-\frac{3\varepsilon(\vec{q}-\vec{q}_0)^2}{2t\ell^2}\right] . \qquad (3.10)$$

For a finite value of t, $N \to \infty$ implies $\varepsilon \to 0$. It is clear now that as $\varepsilon \to 0$, $\ell \to 0$, ℓ^2/ε is required to have a finite value if P has to be meaningful which is possible only if ℓ^2 is proportional to ε. This is precisely the concept of Brownian motion where the square of the displacement is proportional to the time interval during which such a displacement takes place. In Brownian motion the size of the displacement usually depends on the properties of the medium in which the particle moves and it is useful to introduce another quantity D, called the diffusion constant through the relation $D = \ell^2/6\varepsilon$. The expression (3.10) then takes the familiar form

$$P(\vec{q},t;\vec{q}_0,0) = \left(\frac{1}{4\pi Dt}\right)^{3/2} \exp\left[-\frac{(\vec{q}-\vec{q}_0)^2}{4Dt}\right] . \qquad (3.11)$$

It is a simple matter to show by actual differentiation w.r.t. t and $\vec{q}$ that P satisfies the diffusion equation

$$\frac{\partial P}{\partial t} - D \nabla_q^2 P = \delta(\vec{q}-\vec{q}_0) . \qquad (3.12)$$

1.3.2. Wiener integral

The basic property of the distribution (3.11) is the particular manner in which the time t enters in the formula. The distribution depends only on the length of the time interval and the displacement $\vec{\Delta q}$ between the initial and the final position. Moreover, since

$$\left(\frac{1}{4\pi Dt_1}\right)^{3/2}\left(\frac{1}{4\pi Dt_2}\right)^{3/2}\int_{-\infty}^{\infty} d\vec{q}_1 \exp\left[-\frac{(\vec{q}_2-\vec{q}_1)^2}{4Dt_2}\right]\exp\left[-\frac{(\vec{q}_1-\vec{q}_0)^2}{4Dt_1}\right]$$

$$=\left(\frac{1}{4\pi D(t_1+t_2)}\right)^{3/2}\exp\left[-\frac{(\vec{q}_2-\vec{q}_0)^2}{4D(t_1+t_2)}\right],$$

we have

$$P(\vec{q}_2,t_2+t_1;\vec{q}_0,0) = \int d\vec{q}_1\, P(\vec{q}_2,t_2+t_1;\vec{q}_1,t_1)\, P(\vec{q}_1,t_1;\vec{q}_0,0). \qquad (3.13)$$

This fundamental identity implies that the probability that $\vec{q}$ lies between $\vec{q}_2$ and $\vec{q}_2+d\vec{q}_2$ after a time t_1+t_2 is the total compound probability that the change of $\vec{q}$ over time t_1 may be anything but that it should move in a subsequent interval of length t_2 to a position between $\vec{q}_2$ and $\vec{q}_2+d\vec{q}_2$. This property of Gaussian distribution or randomness makes the introduction of a measure possible. Indeed, we can now introduce the conditional probability that the particle beginning its motion at $\vec{q}_0$ ($t=0$) ends up at $\vec{q}$ at time t such that at any intermediate time t_k, its position $\vec{q}_k = \vec{q}(t_k)$ lies within a closed interval $[\vec{q}_{k1},\vec{q}_{k2}]$. This conditional probability is given by

$$\mu(\vec{q},t;\vec{q}_0,0) = \int_{\vec{q}_{N-1,1}}^{\vec{q}_{N-1,2}}\cdots\int_{\vec{q}_{11}}^{\vec{q}_{12}} P(\vec{q},t;\vec{q}_{N-1},t_{N-1})\prod_{k=1}^{N-1} P(\vec{q}_k,t_k;\vec{q}_{k-1},t_{k-1})\, d\vec{q}_k$$

$$(3.14)$$

where $P(\vec{q}_k,t_k;\vec{q}_{k-1},t_{k-1})$ is defined by Eq. (3.11).

One can visualize the appearance of "paths" here. Let us imagine a continuous function of $\vec{q}(t)$ taking values $\vec{q}_k$ on a partition of time interval $[0,t]$ denoted by $[0, t_1, \ldots t_{N-1}, t_N = t]$. P is then the probability for the particle to go through the windows of "width" $|\vec{q}_{k2}-\vec{q}_{k1}|$ at successive times t_k. As the widths of the windows $|\vec{q}_{k2}-\vec{q}_{k1}|$ become smaller and smaller the particle goes through positions which lie in narrower intervals. In the limit the particle is pinned down to go along a single trajectory $\vec{q}(t)$ which takes values $\vec{q}_k$ on the partition

t_k. Thus as $N \to \infty$, we arrive at a measure known as the Wiener measure. With this measure we define the integral over the space of continuous functions $\vec{q}(t)$ which assume the fixed end-point values $\vec{q}(0) = \vec{q}_0$ and $\vec{q}(t) = \vec{q}$. For a more precise mathematical justification of this procedure the reader may consult Wiener's book.

For our purpose, as can be seen from (3.14), the conditional Wiener measure can be defined as

$$d\mu_W[\vec{q}(t)] = \left(\frac{4\pi Dt}{N}\right)^{-3N/2} \prod_{k=1}^{N} \exp\left\{-\frac{N}{4Dt}(\vec{q}_k - \vec{q}_{k-1})^2\right\} \prod_{k=1}^{N-1} d\vec{q}_k \qquad (3.15)$$

and in this notation we may write the conditional probability

$$P(\vec{q},t\,;\vec{q}_0,0) = \int d\mu_W[\vec{q}(t)] \qquad (3.16)$$

where the integration now extends over all possible "paths" $\vec{q}(t)$ connecting the end-points $\vec{q}_0$ and $\vec{q}$.

Having defined the conditional Wiener measure, it is possible to introduce the expectation value of a functional $Q[\vec{q}(t)]$ over this measure. In compact notation

$$E(Q\,[\vec{q}(t)]) = \int Q\,[\vec{q}(\tau)]\,d\mu_W[\vec{q}(\tau)]\,. \qquad (3.17)$$

This has the following operational meaning

$$E(Q\,[\vec{q}(t)]\,) = \lim_{N \to \infty} \int\cdots\int \left(\frac{4\pi Dt}{N}\right)^{-3N/2} Q(\vec{q}_1,\,\vec{q}_2,\,\vec{q}_N)$$

$$\times \prod_{k=1}^{N} \exp\left[-\frac{N}{4Dt}(\vec{q}_k - \vec{q}_{k-1})^2\right] \prod_{k=1}^{N-1} d\vec{q}_k\,. \qquad (3.18)$$

Note that this expectation value is not normalized. In fact

$$E(1) = P(\,\vec{q},t\,;\vec{q}_0,0)\,. \qquad (3.19)$$

In particular, if we choose

$$Q = \exp\left[-\int_0^t d\tau\, U[\vec{q}(\tau)]\right] \qquad (3.20)$$

we may write

$$\rho_U(\vec{q},\vec{q}_0,t) = E\left(\exp\left[-\int_0^t d\tau\, U\,[\vec{q}(\tau)]\right]\right) = \sum_k (-1)^k \rho_U^k(\vec{q},\vec{q}_0,t) \qquad (3.21)$$

where ρ_U^k is given by

$$\rho_U^k(\vec{q},\vec{q}_0,t) = E\left(\frac{1}{k!}\left\{\int_0^t d\tau\, U\,[\vec{q}(\tau)]\right\}^k\right). \tag{3.22}$$

It is left as an exercise to verify that

$$\rho_U^0(\vec{q},\vec{q}_0,t) = P(\vec{q}-\vec{q}_0,\,t), \tag{3.23a}$$

$$\rho_U^1(\vec{q},\vec{q}_0,t) = \int_0^t dt_1 \int d\vec{q}_1\, P(\vec{q}-\vec{q}_1,\,t-t_1)\, U(\vec{q}_1)\, \rho_U^0(\vec{q}_1,\,\vec{q}_0,t_1), \tag{3.23b}$$

$$\rho_U^k(\vec{q},\vec{q}_0,t) = \int_0^t dt_1 \int d\vec{q}_1\, P(\vec{q}-\vec{q}_1,\,t-t_1)\, U(\vec{q}_1)\, \rho_U^{k-1}(\vec{q}_1,\,\vec{q}_0,t_1). \tag{3.23c}$$

We may now write Eq.(3.21) in the form

$$\rho_U(\vec{q},\vec{q}_0,t) = P(\vec{q}-\vec{q}_0,t) - \int_0^t dt_1 \int d\vec{q}_1 P(\vec{q}-\vec{q}_1,t-t_1) U(\vec{q}_1)\rho_U(\vec{q}_1,\vec{q}_0,t_1). \tag{3.24}$$

Applying the operator $\left(\frac{\partial}{\partial t} - D\nabla_q^2\right)$ on the left of (3.24) and making use of the fact that P satisfies the diffusion equation (3.12), we find that ρ_U obeys the modified diffusion equation

$$\frac{\partial \rho_U}{\partial t} = D\nabla_q^2 \rho_U - U(\vec{q})\rho_U + \delta(\vec{q}-\vec{q}_0). \tag{3.25}$$

Equation (3.25) describes the motion of Brownian particles in a medium in which the particles are being annihilated at a rate $U(\vec{q})$ per unit time.

Analogy between Wiener integral and Feynman integral

The analogy between Wiener integral of Brownian Motion and the path integral of Feynman is apparent when one compares, for instance, equations (2.17) and (3.18) (with Q as defined in (3.20)). Alternatively we may compare the diffusion equation (3.25) with the Schrödinger equation (2.23). If we make the correspondence

$$D \longrightarrow i\hbar/m,\ U = iV/\hbar,\ \rho_U \longrightarrow K, \tag{3.26}$$

then the diffusion equation goes over into Schrödinger equation and the conditional probability $\rho_u(\vec{q},\vec{q}_0,t)$ becomes formally identical to the Feynman propagator $K(\vec{q},t;\vec{q}_0,0)$. However, Feynman path integral is an oscillatory integral and hence precludes the development of a bonafide measure-theoretic argument to render it as a rigorous functional integral. However, there have been several attempts towards providing the mathematical justification of Feynman path integral. For details the reader may consult various references given at the end of this chapter.

Despite this lack of rigour, physicists have been able to apply path integral representation as a technique of solving a large class of problems. For problems in statistical mechanics the partition function may be represented as a path integral which is in fact like a Wiener integral. In other problems, one takes a recourse to letting $t \to i\tau$ and reduce the Feynman integral to the Wiener integral. This is like assuming Feynman integral to be an analytic continuation of Wiener integral in some sense. There are, however, a number of physical problems where the presence of the ubiquitous imaginary number "i" makes the Feynman integral a legitimate tool to use. An example of this is the so-called semiclassical approximation and its applications. This and other applications will be treated in later chapters.

The Partition function as a Wiener integral

The statistical mechanical and consequently the thermodynamic properties of a system are described by the partition function Z which is defined as

$$Z = \text{Tr}\,[\exp(-\beta \hat{H})] \qquad (3.27)$$

where β is related to the inverse of temperature, $\beta = 1/kT$. In the coordinate representation Z can be rewritten as

$$Z = \int <x\,|\,\exp(-\beta \hat{H})\,|\,x_0> \delta(x-x_0)\,dx\,dx_0 . \qquad (3.28)$$

Comparing the representation in Eq.(6.2) with Eq.(2.19) the expression for Z takes the form

$$Z = \int \rho(x,x_0;\beta)\,\delta(x-x_0)\,dx\,dx_0 \qquad (3.29)$$

where $\rho(x,x_0;\beta)$ is called the two-point density matrix and can be represented as a Wiener integral

$$\rho(x,x_0;\beta) = \lim_{N\to\infty} \left(\frac{m}{2\pi\varepsilon}\right)^{N/2} \int \exp\left(-\sum_{j=1}^{N} \frac{(x_j-x_{j-1})^2}{2\varepsilon} - \varepsilon V(x_{j-1})\right) \prod_{j=1}^{N-1} dx_j \tag{3.30}$$

where $\varepsilon = \beta/N$. Formally this expression may also be written as

$$\rho(x,x_0;\beta) = \int \exp\left[-\int_0^\beta H(s)\,ds\right] \mathcal{D}[x(s)], \tag{3.31}$$

H being the classical Hamiltonian of the system under consideration.

1.4. Trotter Product Formula and Alternative Derivation of Path Integral

In this section we make use of the fact that a Feynman propagator is the matrix element of the evolution operator $\exp[-it\hat{H}/\hbar]$ between two Schrödinger states. We have

$$K(x,t;x_0,0) = \langle x|\exp[-\frac{i}{\hbar}t\hat{H}]|x_0\rangle. \tag{4.1}$$

The procedure of converting this into a path integral amounts to insertion of extra degrees of freedom. If we divide the time interval $[0,t]$ into N equal subintervals

$$\exp[-\frac{i}{\hbar}t\hat{H}] = \{\exp[-\frac{it}{\hbar N}\hat{H}]\}^N. \tag{4.2}$$

Consider the Hamiltonian operator of the form

$$\hat{H} = \hat{p}^2/2m + \hat{V}(\hat{x}) \equiv \hat{A} + \hat{B} \tag{4.3}$$

where the operators $\hat{A} = \hat{p}^2/2m$ and $\hat{B} = V(\hat{x})$ do not commute. However,

$$\exp[-\frac{it}{\hbar N}(\hat{A}+\hat{B})] = \exp[-\frac{it}{\hbar N}\hat{B}]\exp[-\frac{it}{\hbar N}\hat{A}] + O((t/N)^2). \tag{4.4}$$

We may, therefore, write

$$K(x,t;x_0,0) = \lim_{N\to\infty} \langle x|\left[\exp\left(-\frac{it}{\hbar N}\hat{B}\right)\exp\left(-\frac{it}{\hbar N}\hat{A}\right)\right]^N|x_0\rangle$$

$$= \lim_{N\to\infty} \int \prod_{j=1}^{N} \langle x_j|\exp\left(-\frac{it}{\hbar N}\hat{B}\right)\exp\left(-\frac{it}{\hbar N}\hat{A}\right)|x_{j-1}\rangle \prod_{j=1}^{N-1} dx_j \tag{4.5}$$

where $x_N = x$ and in the second step we have inserted the identity operator $\int |x_j\rangle \, dx_j \, \langle x_j| = 1$ between every two terms of the product. Consider now, one such term

$$Q_{j,j-1} \equiv \langle x_j| \exp[-\frac{it}{\hbar N}\hat{B}]\exp[-\frac{it}{\hbar N}\hat{A}]|x_{j-1}\rangle$$

which can be explicitly written as

$$Q_{j,j-1} = \exp[-\frac{it}{\hbar N}V(x_j)] \langle x_j| \exp[-\frac{it}{\hbar N}\hat{A}]|x_{j-1}\rangle . \tag{4.6}$$

The second factor in (4.6) may be evaluated by making use of the identity operator $\int |p\rangle \, dp \, \langle p| = 1$ in the momentum space where p is the c-number eigenvalue of the momentum operator $\hat{p}$. We also use the momentum eigenfunctions $\langle p|x\rangle = (2\pi\hbar)^{-1/2}\exp[-ipx/\hbar]$. The result is

$$\langle x_j| \exp(-\frac{it}{\hbar N}\hat{A})|x_{j-1}\rangle = \frac{1}{2\pi\hbar}\int_{-\infty}^{\infty} \exp[-\frac{i}{\hbar}\{\frac{t}{N}\frac{p_j^2}{2m} - p_j(x_j - x_{j-1})\}] \, dp_j$$

$$= \left[\frac{m}{2\pi i\hbar \, (t/N)}\right]^{1/2} \exp\left[\frac{im}{2\hbar}\frac{N}{t}(x_j - x_{j-1})^2\right]. \tag{4.7}$$

Inserting (4.7) in (4.6) and writing $\varepsilon = t/N$

$$Q_{j,j-1} = \left(\frac{m}{2\pi i\hbar\varepsilon}\right)^{1/2} \exp\left\{\frac{i}{\hbar}\left[\frac{m}{2\varepsilon}(x_j-x_{j-1})^2 - \varepsilon V(x_j)\right]\right\} . \tag{4.8}$$

Comparing with (2.11), we identify $Q_{j,j-1}$ with Dirac's transformation function. It is clear from (4.5) and (4.8) that the propagator

$$K(x,t;x_0,0) = \lim_{N\to\infty} \left(\frac{m}{2\pi i\hbar\varepsilon}\right)^{N/2} \int \exp\left[\frac{i}{\hbar}\sum_{j=1}^{N}\left\{\frac{m}{2\varepsilon}(x_j-x_{j-1})^2 - \varepsilon V(x_j)\right\}\right] \prod_{j=1}^{N-1} dx_j$$

where $x_j = x(t_j)$ and $t_j = jt/N$ ($j = 1, 2, \ldots, N$), which is the familiar path integral obtained by Feynman.

Hamiltonian path integral

An alternative formulation may also be derived from (4.5). We go back to the step preceding (4.7) and do not carry out the integration over the momentum coordinate and insert the expression as it is in (4.6). The result is

$$Q_{j,j-1} = \frac{1}{2\pi\hbar} \int \exp\{\frac{i}{\hbar} [p_j(x_j - x_{j-1}) - \varepsilon(\frac{p_j^2}{2m} + V(x_j))]\} \, dp_j \ .$$

When this expression is inserted in (4.5), we arrive at the Hamiltonian path integral

$K(x,t;x_0,0)$

$$= \lim_{N\to\infty} \left(\frac{1}{2\pi\hbar}\right)^N \int \cdots \int \exp\{\frac{i}{\hbar} \sum_{j=1}^{N} [p_j(x_j - x_{j-1}) - \varepsilon H(x_j, p_j)]\} \prod_{j=1}^{N-1} dx_j \prod_{j=1}^{N} dp_j \ .$$

(4.9)

This is the formal expression for the path integral. It is clear that the exponent in the integrand is similar to the action integral written in terms of both coordinate and momentum. In the limit $N \to \infty$ ($\varepsilon \to 0$, $N\varepsilon = t$) this exponent can be written as

$$I = \int_0^t (p\dot{x} - H(x,p)) \, dt \ . \tag{4.10}$$

Also in a formal manner we can rewrite the Eq.(4.9) as

$$K(x,t;x_0,0) = \iint \exp\{\frac{i}{\hbar} \int_0^t (p\dot{x} - H(x,p))dt\} \mathcal{D}[x(t)] \, \mathcal{D}[p(t)] \quad (4.11)$$

the multiple differential products in (4.9) being denoted by $\mathcal{D}[x(t)]$ and $\mathcal{D}[p(t)]$ respectively. Note that Eq.(4.10) is strictly valid for continuous histories ($p_j \to p(t)$, $x_j \to x(t)$). Expression (4.11) is also formal since two limiting operations have been interchanged in deriving it from (4.9), viz., the integration over the 2N-1 variables of (4.9) and the operation "limit" wherein $N \to \infty$ were interchanged. On the other hand, in the discrete form of (4.9) the function $x(t)$ is approximated by a piece-wise linear function going from x_{j-1} to x_j in the time interval (t_{j-1}, t_j) while the function $p(t)$ in the same time interval is approximated by a constant p_j. This piece-wise constant but discontinuous representation for the momentum is consistent with the fact that the velocity associated with our approximation to $\dot{x}(t)$ is piece-wise constant but discontinuous.

A difficulty with the Hamiltonian formulation, as seen from (4.9) is that the integration over p_j involves one more variable than the corresponding integration over x_j. The integration measure is thus not invariant under a canonical transformation. Nevertheless most physicists use Eq.(4.9) as an operational definition of a Hamiltonian path integral

The question of "what paths contribute to the sum ?" in Eq.(4.11) has always plagued the Hamiltonian form of path integration. Formally we may invoke the fact that the paths which contribute most to the integral are the ones which satisfy the variational principle $\delta I = 0$. These are the phase-space trajectories $x(t)$, $p(t)$ satisfying the initial conditions $x(t) = x$ and $x(0) = x_0$ with no restrictions on the initial and final values of momentum or energy. The variational problem $\delta I = 0$ singles out the classical trajectories which obey the Hamilton's equations of motion with the specified initial and final conditions.

1.5. Path Integral Subtleties

Midpoint rule

From the very beginning Feynman had realized an important mathematical feature of the paths that enter the sum. Let $x(t)$ be each path and let $x(t+\Delta t) - x(t) = \Delta x$. then the path must satisfy the requirement

$$(\Delta x)^2 \sim \Delta t . \qquad (5.1)$$

The paths are continuous but the velocity associated with them (that is, the limit $\Delta x/\Delta t$ as $\Delta t \rightarrow 0$) is infinite since $\Delta x/\Delta t = (\sqrt{\Delta t})/\Delta t \rightarrow \infty$. This sort of situation is familiar from the theory of Brownian motion (see Sec. 1.3.1) where we have the relation

$$(dx)^2 = D\, dt \qquad (5.2)$$

with D as the diffusion coefficient for the process.

This subtle nature of the paths may be demonstrated when magnetic field is present. For correct quantization of the Lagrangian

$$L = \frac{m}{2} \dot{x}^2 + \frac{e}{c} \vec{A} \cdot \dot{\vec{x}} \qquad (5.3)$$

the finite N (discretised) form of the contribution due to the vector potential must be written as

$$\int \vec{A} \cdot \dot{\vec{x}} \, dt \approx \sum_j \vec{A} \, [(\vec{x}_j + \vec{x}_{j-1})/2] \cdot (\vec{x}_j - \vec{x}_{j-1}) \tag{5.4}$$

with $\vec{A}$ evaluated at the mid-point of the interval. An alternative expression

$$\int \vec{A} \cdot \dot{\vec{x}} \, dt \approx \frac{1}{2} \sum_j [\vec{A}(\vec{x}_j) + \vec{A}(\vec{x}_{j-1})] \cdot (\vec{x}_j - \vec{x}_{j-1}) \tag{5.5}$$

may be used too. The main point here concerns the choice of an appropriate point at which the variables are to be evaluated. To see this more clearly, consider the difference

$$\sum_j \vec{A}(\vec{x}_j) \cdot (\vec{x}_j - \vec{x}_{j-1}) - \sum_j \vec{A}(\vec{x}_{j-1}) \cdot (\vec{x}_j - \vec{x}_{j-1}) \approx \sum_j \vec{A}'(\vec{x}_j^*)(\vec{x}_j - \vec{x}_{j-1})^2$$

$$\approx \sum_j \vec{A}_j' \cdot \Delta t \tag{5.6}$$

where, $\vec{x}_{j-1} \leq \vec{x}_j^* \leq \vec{x}_j$ and the prime on $\vec{A}$ denotes its derivative with respect to its argument. Clearly Eq. (5.6) may have a non-zero limit in general implying thereby that the point where $\vec{A}(\vec{x})$ is evaluated indeed matters.

Gauge invariance

An interesting feature of the path integral formulation is that it allows an easy demonstration of gauge invariance of the theory. Let us introduce the change of the gauge

$$\vec{A} \longrightarrow \tilde{\vec{A}} = \vec{A} + \vec{\nabla} \phi \tag{5.7}$$

under which the propagator

$$K(b,t;a,0) \longrightarrow \tilde{K}(b,t;a,0) = \int \exp\left[\frac{i}{\hbar}\left(S + \frac{e}{c}\int \vec{\nabla}\phi \cdot \dot{\vec{x}} \, dt\right)\right] \mathcal{D}\,[x(t)]$$

$$= K(b,t;a,0) \, \exp\left[\frac{ie}{\hbar c} \int_a^b \vec{\nabla}\phi \cdot d\vec{x}\right]. \tag{5.8}$$

We would normally write

$$\int_a^b \vec{\nabla}\phi \cdot d\vec{x} = \phi(b) - \phi(a) \tag{5.9}$$

and the relation between K and $\tilde{K}$ amounts to applying a unitary transformation (multiplication by a suitable factor). The gauge transformation leaves the physics unchanged. The relation (5.9) is, however, incorrect for curves with the property $(\Delta \vec{x})^2 \approx \Delta t$. The problem is essentially the same as shown in Eq. (5.6). The relation (5.9) holds only when one uses the mid-point rule or its equivalent $[\phi(\vec{x}_j) + \phi(\vec{x}_{j-1})]/2$.

For quantum mechanical calculations the mid-point rule is an accepted definition of the path integral as was pointed out by Feynman. The mid-point rule is also used in the theory of Brownian motion and is known as the Stratanovich integral. However, for Brownian motion there is also another choice given by K. Ito. This latter choice happens to be a natural choice for a large number of probabilistic applications. In our context this choice amounts to using a vector $\vec{A}(\vec{x}_{j-1})$ in Eqs. (5.4) and (5.5). This also modifies Eq. (5.9) as

$$\int_a^b \vec{\nabla}\phi \cdot d\vec{x} = \phi(b) - \phi(a) - \frac{i\hbar}{2m} \int_0^t \nabla^2 \phi [x(s)] \, ds . \qquad (5.10)$$

We might also mention here that the Ito choice involves the use of an effective Lagrangian. For example, in the case of the vector potential, the effective Ito Lagrangian has the form

$$L_I = \frac{m}{2} \dot{\vec{x}}^2 + \frac{e}{c} \vec{A} \cdot \dot{\vec{x}} + \frac{i\hbar}{2m} \nabla (e\vec{A}/c) . \qquad (5.11)$$

It is easy to check the gauge invariance of the Ito formulation with the help of Eq. (5.10). The details of the use of Ito formulation in path integrals may be found in Schulman's book.

Notes and References

For physicists a good and lucid introduction to path-integrals is available in Feynman's original paper

R. P. Feynman, Rev. Mod. Phys. **20**, 367 (1948)

and his book

R. P. Feynman and A. R. Hibbs, "Quantum mechanics and Path integrals" (McGraw-Hill, New York, 1965).

A more recent book which gives a readable account of path integrals is

L. S. Schulman, "Techniques and Applications of Path Integration" (John Wiley, New York 1981).

Our treatment of introducing path integrals closely follows that of Feynman. Apparently, in his formulation, Feynman was inspired by some remarks by Dirac about the similarity between the evolution operator of Schrödinger equation in quantum mechanics and the exponential of the classical action functional. An account of this can be found in the book

P. A. M. Dirac, "The Principles of quantum mechanics", fourth edition (Oxford, London 1958) and in an early paper

P. A. M. Dirac, Physicalische der Sowjetunion **3**, 1 (1933).

However the first serious attempts to define a functional integral can be found in the early mathematics literature:

P. J. Daniell, Ann. Math. **19**, 279 (1918); P. J. Daniell, Ann. Math. **20**, 281 (1918).

These attempts did not succeed and one had to wait till Norbert Wiener introduced a measure on the function space. This was really a starting point which made the integration on function spaces a well-defined object. This was also the beginning of the use of the tools of probability theory in analysis in a significant way.

Mark Kac gives a nice review stating the reasons which were responsible for failure of Daniell's approach in his article

M. Kac, Bull. Am. Math. Soc. **72**, 52 (1966).

Norbert Wiener's approach can be found in the following papers:

N. Wiener, Proc. Nat. Acad. Sci.(USA), **7**, 253 (1952);

N. Wiener, Proc. Nat. Acad. Sci.(USA), **7**, 294 (1952).

Our derivation of the conditional Wiener measure is based on a probabilistic approach discussed in

S. Chandrasekhar, Rev. Mod. Phys. **15**, 1 (1943).

A list of readable papers on this topic including Chandrasekhar's above paper is also available in the following book

"Selected Papers on Noise and Stochastic Processes", Ed. N. Wax (Dover, New-York 1954).

A very lucid description of Trotter product formula and derivation of a path integral representation of the propagator can be found in the papers

E. Nelson, J. Math. Phys **5**, 332 (1964); J. B. Keller and H. McLaughlin, Am. Math. Monthly **82**, 451 (1975).

The functional integral representation of partition function can be found in any standard text book on Statistical Mechanics. In particular see

R. P. Feynman, "Statistical Mechanics", (Benjamin, New York, 1972).

Finally, some general references on path integrals discussing different aspects are

E. W. Montroll, Commun. Pure and Appl. Math. **5**, 415 (1952),

I. M. Gelfand and A. M. Yaglom, J. Math. Phys. **1**, 48 (1960),

S. G. Brush, Rev. Mod. Phys. **33**, 79 (1971).

The Hamiltonian path integrals have been discussed in

R. P. Feynman, Phys. Rev. **84**, 108 (1951),

W. Tobocman, Nuovo Cim. **3**, 1213 (1956),

H. Davis, Proc. Camb. Phil. Soc. **59**, 147 (1963),

C. Garrod, Rev. Mod. Phys. **38**, 483 (1966).

The section on functional calculus follows closely the treatment given in Feynman and Hibbs' book.

CHAPTER 2

PROPAGATORS FOR LOCAL QUADRATIC LAGRANGIANS

2.1. Introduction

The simplest class of systems amenable to exact path integration is the one characterized by Lagrangians quadratic in coordinates and velocities. Hence, they provide good examples for proper understanding of various mathematical steps involved in the explicit computation of path integrals. Moreover, they are also used as a first approximation for studying systems characterized by more complex Lagrangians. It is, therefore, very desirable to have explicit expressions for the propagators associated with certain specific Lagrangians often used in practical computations.

With these goals in mind, we derive the expression for the propagator for a general quadratic action. We give all the mathematical steps involved. Next, we consider specific cases and write down resulting expressions for the propagators. These would serve as ready references for discussion of problems considered in subsequent chapters.

2.2. Derivation of the Propagator

The general quadratic Lagrangian L associated with the motion of a particle in one dimension can be written as

$$L = \frac{1}{2}[a(t)\dot{x}^2 - b(t)x^2] + c(t)x, \qquad (2.1)$$

where x refers to the position of the particle and the dot over x refers to the derivative with respect to time. Further, $a(t)$, $b(t)$ and $c(t)$ are well behaved functions of time t and $a(t)$ is positive.

Physically, the Lagrangian in (2.1) corresponds to the motion of a forced oscillator with a time dependent frequency $\omega(t) = \sqrt{(b/a)}$ subject to an external force $c(t)$. The function $a(t)$ represents a time-varying mass. Specific forms of this Lagrangian have been used in literature to discuss various problems. For example, the choice of $a = \exp(\gamma t)$, $b \exp(-\gamma t) = \omega^2$ and $c = 0$ is used to discuss quantum mechanics of damped systems.

The propagator $K(x'',t'';x',t')$, associated with the Lagrangian (2.1) can be expressed as

$$K(x'',t'';x',t') = \int \exp\left[\frac{i}{\hbar}\int_{t'}^{t''} L(x,\dot{x},t)\, dt\right] \mathcal{D}\,[x(t)] \qquad (2.2)$$

with the condition $x(t') = x'$ and $x(t'') = x''$.

Next we introduce a transformation

$$x(t) = x_{cl}(t) + \eta(t), \qquad (2.3)$$

where $x_{cl}(t)$ is the classical path corresponding to the solution of the Eüler-Lagrange equation

$$\frac{d}{dt}(a(t)\,\dot{x}_{cl}(t)) + b(t)\,x_{cl}(t) = c(t), \qquad (2.4)$$

such that $x_{cl}(t') = x'$ and $x_{cl}(t'') = x''$. Hence it follows that $\eta(t)$ must obey the condition $\eta(t') = \eta(t'') = 0$.

This enables us to write

$$\int_{t'}^{t''} L(x,\dot{x},t)\,dt = S_{cl} + \int_{t'}^{t''} L_0(\eta,\dot{\eta},t)\,dt + \int_{t'}^{t''} [a(t)\dot{x}_{cl}\dot{\eta} - b(t)x_{cl}\eta + c(t)\eta]dt \qquad (2.5a)$$

where

$$L_0 = \frac{1}{2}[\,a(t)\,\dot{\eta}^2 - b(t)\,\eta^2\,] \qquad (2.5b)$$

and S_{cl} is the value of action functional evaluated along the classical path x_{cl}. Performing integration by parts and using the equation of motion (2.4), it is easy to see that the third term in (2.5a) vanishes identically. Thus

$$K(x'',t'';x',t') = \exp\left[\frac{i}{\hbar}S_{cl}\right]\int \exp\left[\frac{i}{\hbar}S_0[\eta(t)]\right] \mathcal{D}\,[\eta(t)], \qquad (2.6a)$$

where the action functional S_0 is given by

$$S_0 = \int_{t'}^{t''} L_0(\eta, \dot\eta, t)\, dt. \tag{2.6b}$$

The expression (2.6a) for the propagator can be written in the polygonal definition as

$$K(x'',t'';x',t') = \exp\left[\frac{i}{\hbar} S_{cl}\right] \lim_{N\to\infty} K_N(0,t'';0,t'), \tag{2.7}$$

where K_N is given by

$$K_N(0,t'';0,t') = A_N \int_{-\infty}^{\infty} \prod_{k=1}^{N-1} d\eta_k \exp\left[\frac{i}{\hbar} S_N\right] \tag{2.8a}$$

with

$$A_N = \prod_{k=1}^{N} \left(\frac{a_k}{2\pi i \hbar \varepsilon}\right)^{1/2}. \tag{2.8b}$$

The discretized form S_N of the action S_0 is given by

$$S_N = \frac{1}{2\varepsilon} \sum_{k=1}^{N} \left[a_k(\eta_k - \eta_{k-1})^2 - \varepsilon^2 b_k \eta_k^2 \right]. \tag{2.9}$$

The symbols a_k, b_k and η_k refer to the values of a, b and η at time $t = t_k$, $k = 1,2,\ldots,N$ with $\eta_0 = \eta_N = 0$. When the expression (2.9) for the discretized action S_N is inserted in (2.8a), we obtain

$$K_N(0,t'';0,t') = A_N \int \exp[i\alpha\, \tilde\eta\, P\, \eta]\, d\eta, \tag{2.10}$$

where $\alpha = (2\hbar\varepsilon)^{-1}$ and $\tilde\eta$ is an (N-1)-dimensional vector $(\eta_1,\eta_2,\ldots,\eta_{N-1})$. Further P is an (N-1)×(N-1) square matrix with elements

$$P_{k,k} = a_k + a_{k+1} - b_k \varepsilon^2 \;;\; P_{k+1,k} = P_{k,k+1} = -a_{k+1}$$

$$P_{j,k} = 0, \quad j \neq k, k+1. \tag{2.11}$$

The integral in (2.10) essentially represents a generalization of a one-dimensional Gaussian integral. Since P is Hermitian, it can be diagonalized by a unitary matrix. Let A be such a matrix. Then the transformation $\eta = A\mu$ will reduce the integral (2.10) to a form

$$K_N(0,t'';0,t') = A_N \int \prod_{k=1}^{N-1} d\mu_k \exp[i\alpha \lambda_k \mu_k^2] \tag{2.12}$$

where λ_k, $k = 1, 2, 3, \ldots, N-1$, are the eigenvalues of P. The integration over each μ_i being independent can now be performed separately. This yields

$$K_N(0,t'';0,t') = \left(\frac{D_N}{i\pi}\right)^{1/2} \quad ; \quad D_N = \left\{\left(\prod_{k=1}^{N} a_k\right) \frac{\alpha}{\det P}\right\}. \quad (2.13)$$

The next task in the evaluation of K_N is to compute the limit of D_N as $N \to \infty$.

2.2.1. Evaluation of limiting expression for D_N

In order to evaluate the limit of D_N as $N \to \infty$, we must have an analytical expression for det P. To this end we note that P is a tridiagonal symmetric matrix. Therefore the kth minor Δ_k of P satisfies the recurrence relation

$$\Delta_k = (a_k + a_{k+1} - b_k\varepsilon^2)\Delta_{k-1} - a_k^2 \Delta_{k-2} \quad ; \quad k \geq 1 \quad (2.14)$$

with $\Delta_0 = 1$, $\Delta_{-1} = 0$. Next, for computational convenience we define a new variable

$$\Delta_k = \frac{1}{\varepsilon}\left(\prod_{j=1}^{k+1} a_j\right)\psi_{k+1}. \quad (2.15)$$

The new variable ψ_k satisfies the equation

$$a_{k+1}\psi_{k+1} - (a_k + a_{k-1})\psi_k + a_k\psi_{k-1} + b_k\varepsilon^2\psi_k = 0 \quad (2.16)$$

and the conditions on Δ_0, Δ_{-1} imply that $\psi_0 = 0$, $\psi_1 = \varepsilon/a_1$.

Now we divide both sides of (2.16) by ε^2 and let $\varepsilon \to 0$. The set of relations (2.16) transform to a differential equation

$$\frac{d}{dt}(a\dot\psi) + b\psi = 0, \quad \psi(t') = \psi' = 0 \;;\; \dot\psi(t') = 1/a. \quad (2.17)$$

By changing the variable $\psi \to v : v = \psi\sqrt{a}$, one can write down the solution of (2.17)

$$\psi(t) = \frac{v_1(t)}{\sqrt{[a(t)a(t')]}\,\dot v_1'} \quad (2.18)$$

where v_1 and v_2 are the two independent solutions of

$$\ddot v + \Omega^2(t)v = 0 \quad (2.19)$$

satisfying the conditions $v_1(t') = v_1' = 0$; $v_2(t'') = v_2'' = 0$ and

$$\Omega^2(t) = \frac{1}{2}\left[\frac{\dot{a}^2}{2a^2} - \frac{\ddot{a}}{a}\right] + \frac{b}{a} . \qquad (2.20)$$

Further, since det $P = \Delta_{N-1}$, we have

$$D \equiv \lim_{N \to \infty} D_N = \lim_{\varepsilon \to 0} \left[\frac{\alpha}{\det P} \prod_{k=1}^{N} a_k\right]^{1/2} = \left(\frac{1}{2\hbar\psi''}\right)^{1/2}$$

$$= \frac{(\dot{v}_1'^2\, a'a'')^{1/4}}{\sqrt{2\hbar v_1''}}$$

and hence the propagator $K(x'',t'';x',t')$ of equation (2.7) is given by

$$K(x'',t'';x',t') = \left[\frac{\dot{v}_1'^2\, a'a''}{2\pi i \hbar v_1''}\right]^{1/4} \exp\left[\frac{i}{\hbar} S_{cl}\right] . \qquad (2.22)$$

2.2.2. Evaluation of S_{cl}

It remains still to find an analytical expression for the action S_{cl} along the classical path. First, we observe that x_{cl} and ψ satisfy the same differential equation but with different boundary conditions. Hence, x_{cl} can be expressed in terms of v_1 and v_2 as

$$x_{cl}(t) = \frac{1}{\sqrt{a}}\left[\frac{\sqrt{a'}x'v_2}{v_2'} + \frac{\sqrt{a''}x''v_1}{v_1''} - \int_{t'}^{t''}\frac{G(t,s)c(s)}{\sqrt{a(s)}} ds\right] \qquad (2.23)$$

where $G(t,s)$ is the Green's function of equation (2.17) satisfying the conditions $G(t,0) = G(0,s) = 0$ and is given by

$$G(t,s) = v_1(t)v_2(s)/Q , \qquad t \leq s \qquad (2.24a)$$

$$= v_1(s)v_2(t)/Q , \qquad t \geq s .$$

Here Q is the Wronskian of solutions of equation (2.19) defined as

$$Q = \dot{v}_1(s)v_2(s) - v_1(s)\dot{v}_2(s) \qquad (2.24b)$$

which can be shown to be a constant independent of s. Hence it follows that $Q = \dot{v}_1'v_2'' = v_1''\dot{v}_2'$. Next we use the solution (2.23) to calculate the action

$$S_{cl} = \int_{t'}^{t''} \frac{1}{2} [a(t)\dot{x}_{cl}^2 - b(t)x_{cl}^2] \, dt + \int_{t'}^{t''} c(t)x_{cl}(t) \, dt \qquad (2.25)$$

and obtain

$$S_{cl} = -\frac{1}{4} \left[\frac{\dot{a}'' X''^2}{a''} + \frac{\dot{a}' X'^2}{a'} \right] + \frac{1}{2} \left[\frac{\dot{v}_1'' X''^2}{v_1''} + \frac{\dot{v}_2' X'^2}{v_2'} - \frac{2X'X''\dot{v}_1'}{v_1''} \right]$$

$$+ \frac{2X'}{v_2'} \int_{t'}^{t''} ds \, c(s) v_2(s)/\sqrt{a(s)} + \frac{2X''}{v_1''} \int_{t'}^{t''} ds \, c(s) v_1(s)/\sqrt{a(s)}$$

$$- \int_{t'}^{t''} dt \int_{t'}^{t''} ds \, \frac{c(t)G(t,s)c(s)}{\sqrt{a(t)a(s)}} \qquad (2.26)$$

with $X = x \sqrt{a}$. A straightforward differentiation of S_{cl} in (2.26) yields

$$\left| \frac{\partial^2 S_{cl}}{\partial x' \partial x''} \right| = \left| \frac{\sqrt{a'a''} \, \dot{v}_1'}{v_1''} \right|. \qquad (2.27)$$

From equation (2.22) it is clear that we can rewrite the expression for the propagator K as

$$K(x'',t'';x',t') = \left[\frac{1}{2\pi i \hbar} \left| \frac{\partial^2 S_{cl}}{\partial x' \partial x''} \right| \right]^{1/2} \exp\left[\frac{i}{\hbar} S_{cl} \right]. \qquad (2.28)$$

The result expressed through Eq.(2.28) is the famous Van Vleck-Pauli formula for the propagator. The prefactor before the exponent denotes the normalization constant, which we shall denote by N is often referred to as the Van Vleck Pauli determinant (VPD). This result is of fundamental importance and implies that the quantum dynamics of systems characterized by quadratic Lagrangians is completely specified by classical trajectory.

The propagator (2.28) has been derived by several people in many ways. Moreover, there are several papers in literature which derive the expression for the propagator with specific choices of the parameters a, b and c. In the following section we shall consider some specific Lagrangians, write down the expressions for the classical action S_{cl} and

2.3. Specific Cases

A Free particle

For a free particle the Lagrangian $L = (m/2)\,\dot{x}^2$ corresponds to the choice $a = m$, and $b = c = 0$. For this choice

$$S_{cl} = m(x'' - x')^2/2T \quad , \quad N = (m/2\pi i\hbar T)^{1/2} \,. \tag{3.1}$$

Here and subsequently the symbol T denotes the time difference $t'' - t'$.

A free particle perturbed by a time-dependent force

The Lagrangian for this case, $L = (m/2)\,\dot{x}^2 + f(t)\,x$, corresponds to the choice $a = m$, and $b = 0$, $c = f(t)$. For this choice

$$S_{cl} = \frac{m(x'' - x')^2}{2T} + x'' \int_{t'}^{t''} dt\, f(t)\, \frac{(t - t')}{T} + x' \int_{t'}^{t''} dt\, f(t)\, \frac{(t'' - t)}{T}$$

$$- \int_{t'}^{t''} dt \int_{t'}^{t''} ds\, \frac{f(t)\,f(s)\,G(t,s)}{2mT} \tag{3.2a}$$

where

$$G(t,s) = (t - t')(t'' - s)/T \qquad t \le s \,,$$

$$= (t'' - t)(s - t')/T \qquad t \ge s \,;$$

$$N = (m/2\pi i\hbar T)^{1/2} \,. \tag{3.2b}$$

In the case when the force $f(t)$ does not depend on t, S_{cl} of (3.2a) simplifies to

$$S_{cl} = \frac{m(x'' - x')^2}{2T} + \frac{f\,T(x'' + x')}{2} - \frac{f^2\,T^3}{24m} \,. \tag{3.2c}$$

A case of particular interest which is frequently used while evaluating the propagator within first cumulant approximation corresponds to the choice $c = f\,\delta(t-\tau)$. For this case

36 *Path-Integral Methods and their Applications*

$$S_{cl} = \frac{m(x''-x')^2}{2T} + \frac{f\,[\tau(x''-x') - x''t' + x't'']}{T} - \frac{f^2(\tau-t')(t''-\tau)}{2mT}$$

(3.2d)

and N is the same as given in equation (3.2b).

An oscillator with constant frequency

The Lagrangian for this case corresponds to the choice $a = m$, $b = m\,\omega^2$, $c = 0$, ω being the frequency of the oscillator. Hence

$$S_{cl} = \frac{m\omega}{2\sin\omega T}\,[(x'^2 + x''^2)\cos\omega T - 2x'x''],$$

(3.3a)

$$N = \left[\frac{m\omega}{2\pi i\hbar\sin\omega T}\right]^{1/2}.$$

(3.3b)

A harmonic oscillator perturbed by a time-dependent force

For this case the parameters are chosen as $a = m$, $b = m\omega^2$ and $c = f(t)$, the time dependent force.

$$S_{cl} = \frac{m\omega}{2\sin\omega T}\,[(x'^2 + x''^2)\cos\omega T - 2x'x''] + \frac{x''}{\sin\omega T}\int_{t'}^{t''} dt\, f(t)\sin\omega(t-t')$$

$$+ \frac{x'}{\sin\omega T}\int_{t'}^{t''} f(t)\sin\omega(t''-t)\,dt - \frac{2}{m\omega}\int_{t'}^{t''} dt\int_{t'}^{t''} ds\, f(t)\,f(s)\,G^1(t,s)$$

where

$$G^1(t,s) = \frac{\sin\omega(t-t')\sin\omega(t''-s)}{\sin\omega(t''-t')} \qquad t < s$$

$$= \frac{\sin\omega(s-t')\sin\omega(t''-t)}{\sin\omega(t''-t')} \qquad t > s$$

(3.4)

and N is given by Eq. (3.3b).

When the force $f(t)$ does not depend on time S_{cl} of Eq. (3.4) simplifies to

$$S_{cl} = \frac{m\omega}{2 \sin \omega T} [(x'^2 + x''^2)\cos \omega T - 2x'x''] + f(x'' + x')\frac{1 - \cos \omega T}{\omega \sin \omega T}$$

$$- \frac{[1 - \cos \omega T - (\omega T/2)\sin \omega T]}{m\omega^3 \sin \omega T} f^2 . \qquad (3.5)$$

For the particular case when the force $f(t) = f \delta(t-\tau)$, the expression (3.4) for S_{cl} simplifies to

$$S_{cl} = \frac{m\omega}{2 \sin \omega T} [(x'^2 + x''^2) \cos \omega T - 2x'x'']$$

$$+ \frac{f}{\sin \omega T} [x'' \sin \omega(\tau - t') + x' \sin \omega(t'' - \tau)] - \frac{2f^2}{m\omega} G^1(\tau,\tau) \qquad (3.6)$$

where G^1 is the same as in Eq. (3.4) and N is given by Eq. (3.3b).

A damped oscillator

A damped oscillator is represented by the choice of the parameters $a = me^{\gamma t}$, $b = m\omega^2 e^{\gamma t}$ and $c = 0$. In this case we have

$$S_{cl} = (\mu/\sin \mu T)[(Q'^2 + Q''^2)\cos \mu T - 2 Q'Q''] \qquad (3.7a)$$

where Q' and Q'' refer to the values of $Q(x) = [(m/2\hbar)e^{\gamma t}]^{1/2} x$ at $x = x'$ and $x = x''$ respectively. Further, $\mu^2 = \omega^2 - \gamma^2/4$, while the normalization constant is given by

$$N = \left[\frac{m\mu \exp [\gamma/2(t'+t'')]}{2\pi i\hbar \sin \mu T}\right]^{1/2} . \qquad (3.7b)$$

Runaway oscillator

It can be easily seen that the choice of the parameters $a = me^{\gamma t}$, $b = m\omega^2 e^{-\gamma t}$, $c = 0$ corresponds to a runaway oscillator. For this case

$$S_{cl} = \frac{m\omega}{2 \sin \mu T} [(x'^2 + x''^2) \cos \mu T - 2x'x''], \quad \mu = \frac{\omega(1 - e^{\gamma T})}{\gamma T},$$

$$N = \left(\frac{m\omega}{2\pi i\hbar \sin \mu T}\right)^{1/2} . \qquad (3.8)$$

It may be noted that the above results can be generalized to two and three dimensions in a straightforward way provided the Lagrangian is of the form (2.1). A similar calculation reveals that the expression for S_{cl} remains unchanged with the obvious modification that coordinate x is replaced by a vector $\vec{q}$ in d dimensions and the products x'x", x'f and x"f would now be interpreted as the scalar products $\vec{q}'.\vec{q}''$, $\vec{q}'.\vec{f}$ and $\vec{q}''.\vec{f}$ respectively. However, N now involves the determinant of the matrix D with matrix elements $d_{jk} = \partial^2 S/\partial x'_j \partial x''_k$, where x_j, j = 1, 2, 3 refer to respectively to x, y, and z components of $\vec{q}$. For this reason D is also called as the Van Vleck Pauli determinant (VPD).

2.4. Velocity dependent Potentials

In the foregoing discussion the Lagrangian did not have any term involving a product of coordinates and velocities. In one dimension the product term for a quadratic Lagrangian has to be of the form $x\dot{x}$. This term contributes only an additive constant term in the action. This is due to the fact that $x\dot{x}$ can be expressed as the total time derivative of $x^2/2$. This merely changes the propagator by a constant phase.

However, in higher space dimensions such a coupling introduces non-trivial changes in the propagator even when the associated Lagrangian is quadratic. Whereas the terms of the form $\vec{q}.\dot{\vec{q}}$ still contribute a constant phase to the propagator, the terms involving the product $\vec{q}\times\dot{\vec{q}}$ do introduce non-trivial changes in the propagator. The simplest example of direct relevance to physics where such a situation arises pertains to the motion of a charged particle in a magnetic field. The action functional for such a system can be written as

$$S = S_0 + \frac{e}{c} \int_{t'}^{t''} \vec{A} . \dot{\vec{q}} \, dt \qquad (4.1)$$

where S_0 is the action in the absence of the magnetic field, e refers to the charge on the particle and $\vec{A}$ is the vector potential. It has been pointed out in Chapter 1 that such actions must be discretized using the mid-point rule. This implies that the second term in Eq.(4.1) must be written as

$$\int_{t'}^{t''} \vec{A}\cdot\dot{\vec{q}}\, dt = \epsilon \sum_{j} [\vec{A}(\vec{q}_j) + \vec{A}(\vec{q}_{j-1})] \frac{1}{2\epsilon} [\vec{q}_j - \vec{q}_{j-1}] \, . \qquad (4.2)$$

When this form of action is inserted in the polygonal definition (2.8) the various integrations over $\vec{q}_j$ can be performed as before to arrive at the expression for the propagator. As an example consider the particular case of a charged particle in a uniform magnetic field of strength B in the z direction. The action functional for the particle can be written as

$$S = \int_{t'}^{t''} \left[\frac{m}{2} \dot{\vec{q}}^2 + \frac{e}{c} \vec{A}\cdot\dot{\vec{q}} \right] dt \, . \qquad (4.3a)$$

The vector potential can be chosen as $(e/c)\vec{A} = \omega(-y, x, 0)$, ω being the cyclotron frequency, viz., $\omega = eB/mc$. The discretized form of the action using the mid-point rule reads as

$$S_N = \sum_{j=1}^{N} \frac{m}{2\epsilon} (\vec{q}_j - \vec{q}_{j-1})^2 + \omega (y_j x_{j-1} - x_j y_{j-1}) \, . \qquad (4.3b)$$

With this form of discretized action S_N one can follow the derivation given in Sec. 2.2 to arrive at the following expression for the propagator:

$$K(\vec{q}'',t'';\vec{q}',t') = K_0(\vec{q}'',t'';\vec{q}',t')\, K_1(z'',t'';z',t') \, . \qquad (4.4a)$$

The propagator $K_0(\vec{q}'',t'';\vec{q}',t')$ has the following expression

$$K_0 = \mathcal{N}_0 \exp\left[\frac{im\omega \cot(\omega T/2)}{4\hbar}\left((x''-x')^2 + (y''-y')^2 + (x'y'' - x''y')\right)\right]$$

where the normalization constant $\mathcal{N}_0$ is given by

$$\mathcal{N}_0 = \left(\frac{m}{2\pi i\hbar T}\right) \frac{\omega T}{2\sin(\omega T/2)} \qquad (4.4b)$$

and $K_1(z'',t'';z',t')$ denotes the usual one-dimensional free particle propagator. We might mention in passing that the mid-point rule guarantees uniqueness (up to an additive constant) of the discretized form of action (4.3b) for every admissible choice of the vector potential $\vec{A}$. This implies that the propagator changes by a constant phase which does not affect the time evolution of the system.

Another interesting example involving the motion of a charged particle in a magnetic field has been considered by Wiegel. The system is characterized by the Lagrangian

$$L = \frac{m}{2}\left[\dot{x}^2 + \dot{y}^2 + \gamma^2(x^2-y^2) + \omega(\dot{x}y - \dot{y}x)\right] . \quad (4.5)$$

The propagator corresponding to this case is given by Van-Vleck-Pauli formula. The expression for S_{cl} reads as

$$S_{cl} = \frac{m}{4}\tau_1\tau_2\left\{(\lambda^2+\mu^2)\left[\sinh(\lambda T/2)\sin(\mu T/2)\left(\frac{x_+^2}{\tau_2} - \frac{y_+^2}{\tau_1}\right) + \cosh(\lambda T/2)\cos(\mu T/2)\left(\frac{x_-^2}{\tau_1} + \frac{y_-^2}{\tau_2}\right) - \omega\left(\frac{x_+y_-}{\tau_2^2} - \frac{x_-y_+}{\tau_1^2}\right)\right]\right\},$$

where we use the following abbreviations

$$x_\pm = x' \pm x'' \quad ; \quad y_\pm = y' \pm y'' \quad ;$$

$$\mu^2 = (\omega^2/2) + (\gamma^4 + \omega^4/4)^{1/2} \quad ; \quad \lambda^2 = -(\omega^2/2) + (\gamma^4 + \omega^4/4)^{1/2}.$$

Further the variables τ_1 and τ_2 are defined by

$$\tau_1^{-1} = (\lambda + \mu)\cosh(\lambda T/2)\sin(\mu T/2) + (\lambda - \mu)\sinh(\lambda T/2)\cos(\mu T/2),$$

$$\tau_2^{-1} = (\mu - \lambda)\cosh(\lambda T/2)\sin(\mu T/2) + (\lambda + \mu)\sinh(\lambda T/2)\cos(\mu T/2).$$

Finally the normalization constant N reads as

$$N = \frac{(\lambda^2+\mu^2)}{4\pi i\hbar} m (\tau_1\tau_2)^{1/2} .$$

2.5. General Quadratic forms

The most general quadratic Lagrangian L in three dimension can be written as

$$L = \frac{1}{2}\sum_{i,j=1}^{3} m_{ij}(t)\,\dot{\vec{q}}_i\cdot\dot{\vec{q}}_j - C_{ij}(t)\,\vec{q}_i\cdot\vec{q}_j + \lambda_{ij}(t)\,\vec{q}_i\cdot\dot{\vec{q}}_j + \vec{f}_i\cdot\vec{q}_i . \quad (5.1)$$

Note that (5.1) does not contain any linear term in $\vec{q}_i$. Such a term can be easily eliminated by redefining $\vec{f}_i$. Equation (5.1) can be rewritten in a compact form

$$L = \frac{1}{2} [\dot{q}^+ M(t) \dot{q} - q^+ \cdot C(t) \cdot q + \dot{q}^+ \cdot Z \cdot q) + f^+ \cdot q] , \qquad (5.2)$$

where $M = \{m_{ij}\}$ and $Z = \{\lambda_{ij}\}$ are 3×3 matrices and f denotes a time-dependent force. Without loss of generality we can assume that M and C are symmetric. Proceeding as in Sec. 2.1 we can write the propagator associated with (5.1) as

$$K(\vec{q}'',t'';\vec{q}',t') = F(t'',t') \exp\left[\frac{i}{\hbar} S_{cl}\right] , \qquad (5.3)$$

where S_{cl} is the action functional evaluated along the classical trajectory $\vec{q}_{cl}$ satisfying Euler-Lagrange equations. The function $F(t'',t')$ is the path integral

$$F(t'',t') = \int_{h(t')=0}^{h(t'')=0} \exp\left[\frac{i}{\hbar} S_0\right] \mathcal{D}[h(t)] \qquad (5.4)$$

with S_0 as

$$S_0 = \frac{1}{2} [\dot{h} M \dot{h} - h C h + \dot{h} Z h] . \qquad (5.5)$$

Using the polygonal approach, we write

$$F = \lim_{N \to \infty} F_N \qquad (5.6)$$

and express the Nth approximation to F, viz., F_N as a multiple integral

$$F_N = \int \prod_{j=1}^{N-1} dh_j \left[\prod_{j=0}^{N-1} \det\left(\frac{M_j}{2\pi i \hbar \varepsilon}\right)^{1/2} \exp\left(\frac{i}{\hbar} S_N\right)\right] . \qquad (5.7)$$

The discretized form of the action S_N is given by

$$S_N = \frac{1}{2\varepsilon} \sum_{j=1}^{N} [h_j^+ - h_{j-1}^+) M_j (h_j - h_{j-1}) - \varepsilon^2 h_j^+ C_j h_j$$

$$+ (h_j^+ - h_{j-1}^+)(Z_j h_j + Z_{j-1} h_{j-1})] , \qquad (5.8)$$

Note that the third term in action S_0 of (5.5) involves a coupling of coordinates h and velocities dh/dt. Hence we have used the mid-point rule for discretization.

The appearance of determinants in the normalization factor can be understood as follows. Recall that for one-dimensional Lagrangian (2.5) the normalization factor contained a term $\prod_{k=1}^{n} (a_k/2\pi i\hbar\varepsilon)^{1/2}$. Now, if M is a diagonal matrix with elements $M_{ij} = M_i \delta_{ij}$ then the kinetic energy term will read as $\sum M_i (dh_i/dt)^2$. Each of these would contribute a term $\sqrt{M_i}$ to the normalization factor. Hence the normalization factor will contain the term $\prod \sqrt{M_i}$ which is nothing but the square root of the determinant of M. Next if M is not diagonal we could use a transformation to go over to its diagonal representation and arrive at the same result.

To evaluate the multiple integral, we rewrite (5.8) in the form

$$S_N = \frac{1}{2\varepsilon} \sum_{k=1}^{N-1} (h_k^+ - h_{k+1}^+ Q_k^+) A_k (h_k - Q_k h_{k+1}) \qquad (5.9)$$

where A_k and Q_k are 3×3 matrices. The matrix elements can be written by comparing Eqs. (5.8) and (5.9)

$$A_1 = (M_0 + M_1) - \varepsilon^2 C_1 \;;\; A_{k+1} + Q_k^+ A_k Q_k = (M_k + M_{k+1}) - \varepsilon^2 C_{k+1},$$

$$A_k Q_k = M_k + \frac{\varepsilon}{2} (Z_{k+1} - Z_k) \;;\; Q_k A_k = M_k + \frac{1}{2} (Z_{k+1}^+ - Z_k) \;. \qquad (5.10)$$

Next, we introduce a transformation

$$y_k = h_k - q_k h_{k+1} \qquad (5.11a)$$

and write S_N as

$$S_N = (1/2\varepsilon) \sum_{k=1}^{N} y_k^+ A_k y_k \;. \qquad (5.11b)$$

Since (5.11) is of the same form as (2.10), we can now follow all the subsequent steps and complete the analysis. Thus, we obtain

$$F_N = \left[\frac{\det M_0}{2\pi i\hbar} \right]^{1/2} [\det(\varepsilon \prod_{k=1}^{N-1} A_k M_k^{-1})]^{-1/2} = \left[\frac{\det(M(t')D_{N-1}^{-1})}{2\pi i\hbar} \right]^{1/2} \qquad (5.12)$$

where we have identified M_0 with $M(t')$ and

$$D_k = \varepsilon \, A_1 M_1^{-1} A_2 M_2^{-1} \ldots A_k M_k^{-1} \quad , \quad k = 1, 2, \ldots, N-1 \quad . \tag{5.13}$$

The next task is to generate the recursion relation for D_k, as was done for Δ_k in (2.14), and then obtain a differential equation in the limit $\varepsilon \to 0$. This leads to

$$\frac{d}{dt}\left(\frac{dD}{dt} M\right) + D\frac{dZ^s}{dt} + DC = \frac{dD}{dt} Z^a - DZ^a M^{-1} D^{-1} \frac{dD}{dt} + DZ^a M^{-1} Z^a \tag{5.14}$$

where Z^s and Z^a refer to the symmetric and anti-symmetric part of the matrix Z. The initial condition to be satisfied by $D(t)$ is

$$D(t') = 0; \quad \left(\frac{dD}{dt}\right)_{t=t'} = I \tag{5.15}$$

and hence

$$F(t'',t') = \left[\det\left(\frac{M(t')}{2\pi i \hbar D(t'')}\right)\right]^{1/2} \tag{5.16}$$

which completes the derivation.

The general result obtained above can be used to derive exact expression for the propagator in a number of cases. Consider for example, the case of a three-imensional damped free particle. For this case the matrix $M = me^{\gamma t} I$, so that (5.14) reduces to

$$\frac{d}{dt}\left(\frac{dD}{dt} M\right) = 0 \tag{5.17}$$

which can be easily solved to obtain

$$D(t'') = (m\gamma)^{-1}[1 - \exp(-\gamma T)]I \quad , \quad T = t'' - t'$$

and hence

$$F(t'',t') = \left[\det\left(\frac{M(t')}{2\pi i \hbar D(t'')}\right)\right]^{1/2} = \left\{\frac{m\gamma}{[2\pi i \hbar (e^{-\gamma t'} - e^{-\gamma t''})]}\right\}^{3/2} . \tag{5.19}$$

2.6. Propagator Beyond the First Singularity Of VPD

Note that in all our earlier derivations we had implicitly assumed that the VPD considered as a function of the time t'' had no singularity. Thus the earlier expressions for the propagators are in fact valid up to the time t" before we encounter the first singularity of the VPD. We shall now attempt to relax this restriction and obtain the expression for the propagator for all values of time t".

To fix the ideas let us consider the example of a simple harmonic oscillator. The singularities of VPD are given by the zeros of the function $\sin \omega T$. Therefore the expression (3.3) for the propagator is valid only if $\omega T < \pi$. In this interval the propagator admits the expansion

$$K(x'',x';T) = \sum_{n=0}^{\infty} \exp[-i(n+\tfrac{1}{2})\omega T]\, \psi_n(x'')\, \psi_n(x') \qquad (6.1)$$

where $\psi_n(x)$ are the eigenfunctions of the oscillator corresponding to eigenvalues $(n+\tfrac{1}{2})\hbar\omega$. Next we analytically continue the expression for the propagator through the expansion (6.1) for all values of the time interval T and substitute $T = k\pi/\omega + \tau$, $k = 1, 2, \ldots,$. The expression for the propagator in the interval $k\pi/\omega < T < (k+1)\pi/\omega$ can be written using Eq.(6.1) as

$$K(x'',x';T) = e^{-k\pi i/2} \sum_{n=0}^{\infty} \exp[-i(n+\tfrac{1}{2})\omega\tau]\, e^{-ikn\pi}\, \psi_n(x'')\, \psi_n(x'). \qquad (6.2)$$

Since the eigenfunctions of the harmonic oscillator are of a definite parity we can identify $e^{-ikn\pi}\psi_n(x')$ as $\psi_n((-1)^k x')$. The sum in Eq.(6.2) can be performed to obtain

$$K(x'',x';T) = e^{-k\pi i/2} K(x'',(-1)^k x';\tau). \qquad (6.3)$$

Further as $\omega T \to k\pi$, Eq.(6.3) implies

$$K(x'',x';T) = e^{-k\pi i/2} \delta(x'' - (-1)^k x'). \qquad (6.4)$$

Similar ideas can be used to obtain the expression for the propagator associated with the general quadratic Lagrangian beyond the first singularity of the VPD. In this case the singularities of the VPD are determined by the zeros of the function $v_1(t)$ defined in Eq.(2.19). The functions $v_1(t)$ and $v_2(t)$ of (2.19) can be cast in the form

$$v_1(t) = \rho(t) \sin[\gamma(t) - \gamma(t')] \qquad (6.5a)$$

Propagator for Local Quadratic Lagrangians 45

$$v_2(t) = \rho(t) \sin [\gamma(t'') - \gamma(t)] \tag{6.5b}$$

where the quantities $\rho(t)$ and $\gamma(t)$ obey the differential equations

$$\ddot{\rho} - \rho^{-3} + \Omega^2(t) \rho = 0 \quad ; \quad \rho^2 \dot{\gamma} = 1 \: . \tag{6.6}$$

Moreover, in terms of the functions $\rho(t)$ and $\gamma(t)$ the VPD has the expression $(a'a'')^{1/2} \rho' \rho''^{-1} \dot{\gamma}' \sin[\phi(t'',t')]$.

Thus for the general case the function $\phi(t'',t') = \gamma(t'') - \gamma(t')$ plays the same role as that of ωT for the simple harmonic oscillator. Further, from Eq.(6.6) it can be inferred easily that $\gamma(t)$ is an increasing function of time t. The next step consists of obtaining an expansion as in Eq.(6.1) for the propagator corresponding to the general case. This expansion is given by

$$K(x'',x';T) = \sum_{n=0}^{\infty} e^{-i[\alpha_n(t'') - \alpha_n(t')]} \psi_n(x'',t'') \: \psi_n(x',t') \tag{6.7}$$

where the functions $\alpha_n(t)$ and $\psi_n(x,t)$ have the expressions

$$\alpha_n(t) = -(n+\tfrac{1}{2})\gamma(t) - F(t) \: , \tag{6.8a}$$

$$\psi_n(x,t) = [(\pi\rho)^{1/2} 2^n n!]^{-1/2} \: e^{[-(P^2+iR)/2]} H_n(\tfrac{x}{\rho} + V) \: . \tag{6.8b}$$

The quantities F, P, R and V are functions of time and are related to the solutions $\rho(t)$ and $\gamma(t)$ as

$$P = \rho^{-1} + V \quad ; \quad R = [\: x^2 \tfrac{\dot{\rho}}{\rho} + U \tfrac{x}{\rho}\:] \quad ; \quad F = \int^t \frac{U^2(\tau) - V^2(\tau)}{2\rho} d\tau$$

and

$$U = \int^t \rho(t) c(\tau) \cos \phi(\tau,t) \: d\tau \quad ; \quad V = \int^t \rho(t) c(\tau) \sin \phi(\tau,t) \: d\tau \: . \tag{6.9}$$

Following our earlier arguments the expression for the propagator, in the interval $k\pi < \phi(t'',t') < (k+1)\pi$, $k = 1, 2, \ldots$, can be easily obtained and is given by

$$K(x'',x';T) = e^{-k\pi i/2} K(x'', (-1)^k x'; \tau) \tag{6.10}$$

where τ is defined as $\phi(t'',t') = k\pi + \phi(\tau,t')$. Further it can be readily seen from Eq. (6.7), that in the limit $\phi(t'',t') \to k\pi$, the propagator takes the form

$$K(x'',x';T) = e^{-k\pi i/2} \delta(x''/\rho'' - (-1)^k x'/\rho' + V(t'') - (-1)^k V(t'))$$

$$\times e^{-i [F(t'') - F(t')]} \qquad (6.11)$$

which completes the derivation.

Notes and References

The derivation of the propagator for quadratic systems can be found in several places. For example, see:

R. P. Feynman and A. R. Hibbs, "Quantum Mechanics and Path Integrals", (McGraw Hill New York 1965).

L. S. Schulman, "Techniques and Applications of Path Integration" (John Wiley, New York 1981).

These books as well as several other papers consider the Lagrangians which are not explicitly time-dependent. As far as we know the explicitly time-dependent Lagrangians in the context of path integration were first treated by

D. C. Khandekar and S. V. Lawande, J. Math. Pays. **16**, 384 (1975).

In this paper they derive the expression for the propagator associated with an oscillator with explicitly time-dependent frequency. Later they also generalized this to include a model of quantum decay. This can be found in

D. C. Khandekar and S. V. Lawande, J. Math. Phys **20**, 1870 (1979).

A complete and detailed description including references for Feynman propagators associated with quadratic Lagrangians can be found in a review article

D. C. Khandekar and S. V. Lawande, Phys. Rep. **137**, 115 (1986).

In particular the runaway oscillator used in several applications has been considered by

L. F. Landovitz, A. M. Levine and W. M. Schreiber, Phys. Rev. **20A**, 1162 (1979).

The evaluation of the propagator beyond the first singularity of the Van Vleck-Pauli determinant is usually referred to as the propagator on and beyond caustics. This is due to the fact that the singularities also happen to be related to the solutions of the associated Jacobi field which determines the conjugate points or caustics in classical dynamics. The derivation of the propagator on and beyond caustics was first considered by

P. J. Horvathy, Int. J. Theo. Phys. **18**, 245 (1979).

Our presentation of the derivation of the propagator on and beyond caustics closely follows Hovrathy's work but is different in details. For quadratic systems, the derivation of the propagator on and beyond caustics can also be found in

J. Rezende, J. Math. Phys. **25**, 3264 (1984) ,

B. K. Cheng, J. Phys. **17A**, 2475 (1984).

Finally the propagator associated with the general quadratic Lagrangian has also been considered by

G. J. Papadopoulos, Phys. Rev. **D11**, 2870 (1975).

CHAPTER 3

NON-LOCAL QUADRATIC ACTIONS

3.1. Elimination of Degrees of Freedom

The path integral formulation has the merit that it is possible in many cases to eliminate the variables which are not of interest. The reduced problem is, however, characterized by the occurrence of a non-local action. We give some explicit examples of this below.

3.1.1. *System coupled to a harmonic oscillator*

Consider the case of a system X interacting with a harmonic oscillator Y. The Lagrangian consists of the three terms

$$L = L_X + L_Y + L_I \qquad (1.1)$$

representing the system, the oscillator and the coupling between the system and the oscillator. Let us assume that the coupling is linear in the oscillator coordinate y, that is, $L_I = y\, f(x)$, where $f(x)$ denotes an arbitrary function of the system coordinate x. The path integral for the entire problem is

$$K(x,y,T;x_0,y_0,0) = \int \mathcal{D}[x(t)]\mathcal{D}[y(t)]\exp\left[\frac{i}{\hbar}\left(S_X + S_Y + S_I\right)\right]. \qquad (1.2)$$

We want to eliminate the oscillator motion from the problem. For this purpose we write

$$S_{YI} = S_Y + S_I = \int_0^T dt\, [\tfrac{1}{2}(\dot{y}^2 - \omega^2 y^2) + y\gamma(t)] \qquad (1.3)$$

where $\gamma(t) = f(x(t))$ and we set the mass of the oscillator, $m = 1$ for simplicity. We know from chapter 2, Eq.(2.28) that the propagator for the action S_{YI} is given by the Van Vleck-Pauli formula

$$K_{YI}(y,T;y_0,0) = \left(\frac{\omega}{2\pi i\hbar\sin\omega T}\right)^{1/2} \exp\left[\frac{i}{\hbar} S_{YI}[y_{cl}(t)]\right] \tag{1.4}$$

where $y_{cl}(t)$ is the classical path connecting y_0 and y. The expression for the classical action can be obtained explicitly from Eqs.(3.4) of chapter 2 with the identification $f \to \gamma$, $x' \to y_0$ and $x'' \to y$. Even after the expression (1.4) is inserted in (1.3), the dependence on the end points (y_0,y) of the oscillator coordinates will remain. We can eliminate this dependence by multiplying (1.2) by a suitable distribution $\Phi(y,y_0)$ of the oscillator coordinates and integrating over both y and y_0. For the purpose of illustration we choose

$$\Phi(y,y_0) = \frac{\delta(y-y_0)}{\int dy \, K_{HO}(y,T;y,0)} \tag{1.5}$$

where $K_{HO}(y,T;y_0,0)$ is the propagator for a free harmonic oscillator. The result is an average propagator defined by

$$\bar{K}(x,T;x_0,0) = \frac{\int \mathcal{D}[x(t)] \exp\{\frac{i}{\hbar} S_x\} \int dy \, K_{YI}(y,T;y,0)}{\int dy \, K_{HO}(y,T;y,0)}. \tag{1.6}$$

The integrations over y in the numerator and the denominator involve only Gaussian integrals and can be easily performed. The average propagator (1.6) then takes the form

$$\bar{K}(x,T;x_0,0) = \int \mathcal{D}[x(t)] \exp\left[\frac{i}{\hbar} \bar{S}[x(t)]\right] \tag{1.7}$$

where $\bar{S}[x(t)]$ has the form

$$\bar{S}[x(t)] = S_x[x(t)] + \frac{1}{4\omega} \int_0^T dt \int_0^T ds \, G(t,s) \, f[x(t)] \, f[x(s)] \tag{1.8}$$

with the kernel $G(t,s)$ given by

$$G(t,s) = \frac{\cos\omega(T/2 - |t-s|)}{\sin\omega T/2}. \tag{1.9}$$

The important point here is that the elimination of the oscillator motion has resulted in a non-local action which is typical of memory effects.

3.1.2. An Electron in a random potential

Another example is that of the motion of an electron in a random potential. Edwards and Gulyaev proposed a model in which an electron was assumed to move in a potential of N randomly distributed impurity ions. Denoting the electronic position by $\vec{q}$ and that of an ion α (located in a volume V) by $\vec{Q}_\alpha$ the electron propagator, averaged over the coordinates of the ions, takes the form

$$\bar{K}(\vec{q},T;\vec{q}_0,0) = \int_V \prod_\alpha \left(\frac{d\vec{Q}_\alpha}{V}\right) \int_{\vec{q}(0)=\vec{q}_0}^{\vec{q}(T)=\vec{q}} D[\vec{q}(t)] \exp\left[\frac{i}{\hbar}\int_0^T \left\{\frac{m}{2}\dot{\vec{q}}^2 - \sum_\alpha U(\vec{q}-\vec{Q}_\alpha)\right\} dt\right]$$

(1.10)

where U is the electron-ion potential. Consider now the term

$$F = \int \prod_\alpha \left(\frac{d\vec{Q}_\alpha}{V}\right) \int \exp\left[-\frac{i}{\hbar}\sum_\alpha \int_0^T U(\vec{q}-\vec{Q}_\alpha)dt\right]$$

$$= \left\{\int_V \exp\left[-\frac{i}{\hbar}\int_0^T U(\vec{q}-\vec{Q})dt\right] \frac{d\vec{Q}}{V}\right\}^N$$

$$= \left\{1 + \frac{\rho}{N}\int_V\left[\exp\left(-\frac{i}{\hbar}\int_0^T U(\vec{q}-\vec{Q})dt\right)-1\right]d\vec{Q}\right\}^N \quad (1.11)$$

where in the last step we have introduced the density of ions ρ = N/V. If we now take the thermodynamic limit, that is, let N $\to \infty$, V $\to \infty$ but keep N/V = ρ (finite), we find that

$$F \to \exp\left\{\rho\int_V d\vec{Q}\left(\exp\left[-\frac{i}{\hbar}\int_0^T U(\vec{q}-\vec{Q})dt\right]-1\right)\right\}. \quad (1.12)$$

For dense and weak scatterers, we may expand the exponential under the integral sign and retain first three terms. Also, without loss of generality we may set the average potential energy

$$\int_V U(\vec{q} - \vec{Q}) \, d\vec{Q} = 0 \qquad (1.13)$$

and for convenience introduce a parameter η measuring the strength of the potential. With this

$$F = \exp\left[-\frac{\eta^2}{2\hbar^2} \int_0^T dt \int_0^T ds \, W(\vec{q}(t) - \vec{q}(s))\right] \qquad (1.14)$$

where

$$W(\vec{q}(t) - \vec{q}(s)) \equiv \rho \int d\vec{Q} \, U(\vec{q}(t) - \vec{Q}) \, U(\vec{q}(s) - \vec{Q}) \qquad (1.15)$$

is the correlation function of the potential.

The electron propagator then takes the closed form

$$\bar{K}(\vec{q}, T; \vec{q}_0, 0) = \int \mathcal{D}[\vec{q}(t)] \exp\left[\frac{i}{\hbar} \bar{S}\right] \qquad (1.16)$$

where

$$\bar{S} = \int_0^T \frac{m}{2} \dot{\vec{q}}^2 \, dt + \frac{i\eta^2}{2\hbar} \int_0^T dt \int_0^T ds \, W(\vec{q}(t) - \vec{q}(s)) . \qquad (1.17)$$

The path integral formulation has allowed us to take the average over the ensemble of impurity ions right at the beginning. This is in contrast with the standard perturbation approach for the Green's function where the ensemble average has to be carried out term by term. Also the fluctuating potential which plays an important role in disordered systems is automatically built into the formalism.

Another approach to the problem is to consider the electron to be in a random potential, the origin of randomness being immaterial. The propagator for the electron is now written as

$$K(\vec{q}, T; \vec{q}_0, 0) = \int_{\vec{q}(0) = \vec{q}_0}^{\vec{q}(T) = \vec{q}} \mathcal{D}[\vec{q}(t)] \exp\left[\frac{i}{\hbar} \int_0^T \left\{\frac{m}{2} \dot{\vec{q}}^2 + \lambda V(\vec{q})\right\} dt\right]. \qquad (1.18)$$

We are interested in the average propagator obtained by taking the ensemble average over the random variable V

$$\langle K(\vec{q},T;\vec{q}_0,0)\rangle_V = \int_{\vec{q}(0)=\vec{q}_0}^{\vec{q}(T)=\vec{q}} \mathcal{D}[\vec{q}(t)] \exp\left[\frac{i}{\hbar}\int_0^T \frac{m}{2}\dot{\vec{q}}^2 dt\right] \langle \exp[\frac{i\lambda}{\hbar}\int_0^T V(\vec{q})dt]\rangle.$$

For a Gaussian random variable with mean $\langle V \rangle = 0$, we have

$$\langle \exp[\frac{i\lambda}{\hbar}\int_0^T V(\vec{q})dt]\rangle = \exp\left\{-\frac{\lambda^2}{2\hbar^2}\int_0^T dt \int_0^T ds\, W(\vec{q}(t)-\vec{q}(s))\right\} \quad (1.19)$$

where we have introduced the autocorrelation function

$$W(\vec{q}(t)-\vec{q}(s)) = \langle V(\vec{q}(t))V(\vec{q}(s))\rangle_V . \quad (1.20)$$

Thus, the average propagator for the electron finally reads as

$$\langle K \rangle_V = \int \mathcal{D}[\vec{q}(t)] \exp[\frac{i}{\hbar}\bar{S}_V] \quad (1.21a)$$

where

$$\bar{S}_V = \int_0^T \frac{m}{2}\dot{\vec{q}}^2 dt + \frac{i\lambda^2}{2\hbar}\int_0^T dt \int_0^T ds\, W(\vec{q}(t)-\vec{q}(s)) \quad (1.21b)$$

which is of the same form as in (1.16). If we assume that the origin of randomness is due to the presence of infinitely many impurity ions, then the two approaches are equivalent as a consequence of the central limit theorem. The correlation W, which characterizes the statistical properties of disordered systems, generally contains arguments defined at two different times. Formally, W could be interpreted as a two body interaction in the sense of many-body theory. The actual form of W, in principle, must be determined from the potential V. For instance, if V is a screened Coulomb potential, then W will be an exponential function. Another standard form for W is Gaussian.

The average propagator of Eq.(1.16) or Eq. (1.21) is used in literature to discuss many physical properties of disordered systems such as density of states which is just the Fourier transform of its diagonal part and also to describe the localization problem and wave propagation in random media.

Not all forms of non-local actions can be path integrated. In the subsequent section we derive the propagator for a general quadratic action. This propagator can be used along with some suitable approximation schemes to solve problems involving non-local actions.

3.2. Two-time Quadratic Actions

Consider the general quadratic non-local action

$$S = \frac{1}{2}\int_0^T \dot{x}^2 dt - \frac{1}{2}\int_0^T dt \int_0^T ds\, G(t,s)\, x(t)\, x(s) + \int_0^T f(t) x(t)\, dt \quad (2.1)$$

where without loss of generality $G(t,s)$ is assumed to be a symmetric function of t and s. The function $f(t)$ represents a time dependent external force. As we have seen in Sec. 3.1.1, $G(t,s)$ may often be considered as a phenomenological way of characterizing memory effects. Such effects arise in the reduced description of a system interacting with a reservoir when the reservoir variables are eliminated. The reservoir may be the heat bath of phonons (oscillators) as in the polaron problem or that of photons in quantum electrodynamics or that of a collection of randomly distributed scattering centres.

The most important point here is that we cannot define a Lagrangian or a Hamiltonian in this case (barring the case when $G(t,s) = \delta(t-s)$). Hence the propagator may not have the usual meaning of being the Green's function of a Schrödinger equation. In particular, it does not satisfy the transitivity property. Nevertheless, path integration of such an action is of interest in its own right as well as for applications. Curiously enough, various forms of $G(t,s)$ yield a variety of new propagators and some old ones including those for harmonic oscillator and free particle as special cases. Integration turns out to be no more difficult than that for a local quadratic action treated in chapter 2.

The first step in deriving the propagator is to write

$$x(t) = x_{cl} + y \quad (2.2)$$

where x_{cl} is the classical path connecting the end points $x_{cl}(0) = x_0$ and $x_{cl}(T) = x$ for which $\delta S = 0$ and $y(0) = y(T) = 0$. Hence, x_{cl} obeys the following classical equation of motion

$$\ddot{x}_{cl} + \int_0^T G(t,s) x_{cl}(s)\, ds = f(t). \qquad (2.3)$$

Next, inserting (2.2) in (2.1), we have

$$S[x(t)] = S[x_{cl}] + S_0[y] \qquad (2.4)$$

where

$$S_0[y] = \frac{1}{2}\int_0^T dt\, \dot{y}^2 - \frac{1}{2}\int_0^T dt \int_0^T ds\, G(t,s)\, y(t)\, y(s). \qquad (2.5)$$

The propagator

$$K(x,T;x_0,0) = \exp\left\{\frac{i}{\hbar} S[x_{cl}]\right\} \int_{y(0)=0}^{y(T)=0} \exp\left\{\frac{i}{\hbar} S_0[y(t)]\right\} \mathcal{D}[y(t)]. \qquad (2.6)$$

The path integral occurring in (2.6) is now a sum over paths $y(t)$, beginning and terminating at $y = 0$ and is therefore purely a function of time T. We denote this function by $K_{00}(T)$. According to the usual time-slicing prescription

$$K_{00}(T) = \lim_{N\to\infty} \left(\frac{1}{2\pi i\hbar\varepsilon}\right)^{N/2} \int \prod_{j=1}^{N-1} dy_j \exp\left[\frac{i}{\hbar} S_{ON}\right] \qquad (2.7)$$

where the discretized action is

$$S_{ON} = \sum_{j=1}^{N} \frac{(y_j - y_{j-1})^2}{2\varepsilon} - \frac{\varepsilon^2}{2} \sum_{j,k=1}^{N} G_{jk}\, y_j\, y_k. \qquad (2.8)$$

In fact, in more compact notation

$$K_{00}(T) = \lim_{N\to\infty} (\alpha/\pi)^{N/2} \int_{-\infty}^{\infty} dY \exp[-\alpha(Y^T P Y)] \qquad (2.9)$$

where $\alpha = (2i\hbar\varepsilon)^{-1}$. Here Y is a $(N-1)$ dimensional column vector with components $(y_1,\ldots,y_{N-1})$ while P is an $(N-1)$ dimensional symmetric square matrix ($P_{ij} = P_{ji}$) with the following structure

$$P_{jj} = 2 - \varepsilon^3 G_{jj}\ ;\quad P_{j,j+1} = -\varepsilon^3 G_{j,j+1} - 1\ ; \qquad (2.10a)$$

$$P_{ij} = -\varepsilon^3 G_{ij}\ ,\quad i \ne j,\ j \pm 1. \qquad (2.10b)$$

The Gaussian integral (2.9) may be readily evaluated to obtain

$$K_{00} = \lim_{N \to \infty} \left[\frac{\alpha}{\pi \det P} \right]^{1/2} . \quad (2.11)$$

In order to evaluate det P, we decompose the matrix P as

$$P = L + V . \quad (2.12)$$

Here L and V may be chosen in any manner but L^{-1} must exist. We choose in particular, L to be symmetric and V such that

$$L_{j,j+1} = -1 \; ; \; L_{jj} = 2 \; ; \; L_{ij} = 0 \; (i \neq j, j \pm 1) \; ; \; V_{ij} = P_{ij} . \quad (2.13)$$

Now, we write

$$\det P = (\det L) \det(I + L^{-1}V) \quad (2.14)$$

and use the general result

$$\det (I + \lambda M) = \exp [\text{Tr} \int_0^\lambda d\mu \, R(\mu)] \quad (2.15)$$

where M is a finite dimensional square matrix and the matrix $R(\mu)$ satisfies the equation for the resolvant

$$R = M - \mu M R . \quad (2.16)$$

Writing $\lambda = 1$, $M = L^{-1} V$, the equation for the resolvant becomes

$$L R = V - \mu V R . \quad (2.17)$$

It is convenient to write $\tilde{R} = R/\varepsilon$ and express (2.17) in the component form

$$2 \tilde{R}_{jk} - \tilde{R}_{j-1,k} - \tilde{R}_{j+1,k} = -\varepsilon^2 G_{jk} + \mu \varepsilon^2 \sum_l G_{jl} \tilde{R}_{lk} \quad (2.18)$$

by defining $\tilde{R}_{0k} = \tilde{R}_{N,k} = 0$ for all k. After rearranging the terms in (2.18), dividing throughout by ε^2 and finally taking the limit $\varepsilon \to 0$, we find that the difference equation goes over to the integro-differential equation

$$\frac{d^2}{dt^2} \tilde{R}(t,s) + \mu \int_0^T G(t,r) \tilde{R}(r,s) \, dr = G(t,s) \tag{2.19}$$

with the end-point conditions $\tilde{R}(0,s) = \tilde{R}(T,s) = 0$, $(0 \le s \le T)$. It follows that

$$\det (I + L^{-1}V) = \exp \left[\int_0^1 d\mu \sum_{j=1}^{N-1} \epsilon \tilde{R}_{jj} \right]$$

and upon taking the limit as $\epsilon \to 0$, we have

$$\lim_{N \to \infty} \det (I + L^{-1}V) = \exp \left[\int_0^T dt \int_0^1 d\mu \, \tilde{R}(t,t;\mu) \right]. \tag{2.20}$$

The evaluation of det L is carried out by noting that the kth minor of det L obeys the recurrence relation

$$\Delta_k = 2\Delta_{k-1} - \Delta_{k-2}, \quad k \ge 1 \tag{2.21}$$

with $\Delta_0 = 1$, $\Delta_{-1} = 0$. This has a solution $\Delta_k = k+1$ implying thereby that

$$\det L = \Delta_{N-1} = N = T/\epsilon. \tag{2.22}$$

Using (2.20) and (2.21) in (2.11) and inserting the value of α,

$$K_{00} = \lim_{N \to \infty} \left[\frac{\alpha}{\pi \det P} \right]^{1/2} = (2\pi i \hbar T)^{-1/2} \exp \left[-\frac{1}{2} \int_0^T dt \int_0^1 d\mu \, \tilde{R}(t,t;\mu) \right]. \tag{2.23}$$

Also $S[x_{cl}]$ can be written as

$$S[x_{cl}] = \frac{1}{2} \left\{ x \dot{x}_{cl}(T) - x_0 \dot{x}_{cl}(0) + \int_0^T dt \, f(t) x_{cl}(t) \right\}. \tag{2.24}$$

The propagator for action (2.1) takes the form

$$K(x,T; x_0,0) = (2\pi i \hbar T)^{-1/2} \exp \left[-\frac{1}{2} \int_0^T dt \int_0^1 d\mu \, \tilde{R}(t,t;\mu) \right]$$

$$\times \exp \left[\frac{i}{2\hbar} \left\{ x \dot{x}_{cl}(T) - x_0 \dot{x}_{cl}(0) + \int_0^T dt \, f(t) x_{cl}(t) \right\} \right]. \tag{2.25}$$

3.3. Extension to More General Non-local Actions

It is possible to path integrate more general non-local action

$$S_F = \frac{1}{2}\int_0^T \dot{x}^2 \, dt - \frac{1}{2}\int_0^T dt \int_0^T ds \, G(t,s)x(t)\,x(s) + F\left(\int_0^T f(t)x(t)dt\right) \quad (3.1)$$

where $F(z)$ is an arbitrary function of its argument z. The propagator for the generalized action may also be derived very simply by means of the following trick.

The path integral is first carried out by summing the contributions of all paths that keep the integral

$$\int_0^T f(t)\,x(t)\,dt = u \quad (3.2)$$

a fixed value and then integrating over all possible values of u. Thus

$$K_F = \int_{-\infty}^{\infty} du \int_{x(0)=x_0}^{x(T)=x} \mathcal{D}[x(t)] \exp\left[\frac{i}{\hbar}\{S + F(u)\}\right] \delta\left(u - \int_0^T dt\, f(t)x(t)\right) \quad (3.3)$$

where S has the form (2.5). Inserting the Fourier integral representation of the δ-function in (3.3), we may write

$$K_F = \frac{1}{2\pi\hbar}\int_{-\infty}^{\infty} dq \, \exp\left[-\frac{i}{\hbar}qu\right]\int_{-\infty}^{\infty} du \, \exp\left[\frac{i}{\hbar}F(u)\right] K_q(x,T;x_0,0). \quad (3.4)$$

Here K_q is the propagator (2.25) with $f(t)$ replaced by $qf(t)$. Note, however, that the classical path $x_{cl}(t)$ now depends on q. In order to bring out this dependence on q explicitly, consider the classical equation

$$\frac{d^2 x_{cl}}{dt^2} + \int_0^T G(t,s)\, x_{cl}(s)\, ds = q\, f(t) \quad (3.5)$$

with the conditions $x_{cl}(T) = x$, $x_{cl}(0) = x_0$. The solution of this equation may be written in the form

$$x_{cl}(t) = \psi(t) - q\int_0^T ds\, Q(t,s)f(s) \quad (3.6a)$$

where $\psi(t)$ is the solution of the homogeneous equation

$$\ddot{\psi} + \int_0^T ds\, G(t,s)\, \psi(s) = 0 \qquad (3.6b)$$

with the end-point conditions $\psi(0) = x_0$ and $\psi(T) = x$. The kernel $Q(t,s)$ obeys the integro-differential equation

$$\frac{d^2 Q(t,s)}{dt^2} + \int_0^T dr\, G(t,r)\, Q(r,s) = -\delta(t-s) \qquad (3.7)$$

with $Q(0,s) = Q(T,s) = 0$. The solution (3.6a) is then inserted in the expression for S_{cl} so that

$$S_{cl} = \frac{1}{2}[x\dot{\psi}(T) - x_0\dot{\psi}(0)] + q\int_0^T dt\, f(t)\psi(t) - \frac{q^2}{2}\int_0^T dt \int_0^T ds\, f(t) Q(t,s) f(s).$$

In deriving this expression, the relation

$$[\psi(T)\, \dot{Q}(T,s) - \psi(0)\, \dot{Q}(0,s)] = -\psi(s)$$

is used. This relation is derived by multiplying (3.7) by $\psi(t)$ and integrating over t. The expression (3.4) now takes the form

$$K_F = \frac{K_{00}}{2\pi\hbar} \exp\{\tfrac{i}{\hbar} S_\psi\} \int_{-\infty}^{\infty} du\, e^{iF(u)/\hbar} \int_{-\infty}^{\infty} dq\, \exp\{-\tfrac{1}{2\hbar}[\beta q^2 + 2(u-\alpha)q]\} \qquad (3.8)$$

where for brevity we have introduced the definitions

$$S_\psi = (x\dot{\psi}(T) - x_0\dot{\psi}(0))/2 \;;\; \alpha = \int_0^T f(t)\psi(t)\, dt, \qquad (3.9a)$$

$$\beta = \int_0^T dt \int_0^T ds\, f(t)\, Q(t,s)\, f(s). \qquad (3.9b)$$

The Gaussian integral over q in (3.8) is performed to yield the final result

$$K_F = \frac{K_{00}}{\sqrt{2\pi i\hbar\beta}} \exp[\tfrac{i}{\hbar}(x\dot{\psi}(T) - x_0\dot{\psi}(0))] \int_{-\infty}^{\infty} du\, \exp\left[\tfrac{i}{\hbar}\left(F(u) + \frac{(u-\alpha)^2}{2\beta}\right)\right]. \qquad (3.10)$$

This general formula encompasses all integrable cases of one- and two-time quadratic actions. Actual details of derivation involve the solution of the corresponding classical equations of motion and also some results from the theory of integral equations.

3.4. Examples of Explicit Evaluation

3.4.1. Generalized one-time actions

These actions fall under the common requirement on the kernel $G(t,s)$, viz.

$$G(t,s) = \Omega^2(t)\, \delta(t-s) \tag{4.1}$$

leading to the action functional

$$S[x(t)] = \frac{1}{2} \int_0^T [\dot{x}^2 - \Omega^2(t) x^2]\, dt + F\!\left(\int_0^T f(t)\, x(t)\, dt\right). \tag{4.2}$$

The resulting propagator following from (3.10) can be shown to have the form

$$K(x,T;x_0,0) = \frac{K_{TDHO}(x,T;x_0 0)}{\sqrt{2\pi i \hbar \beta}} \int_{-\infty}^{\infty} du\, \exp\!\left[\frac{i}{\hbar}\left(F(u) + \frac{(u-\alpha)^2}{2\beta}\right)\right]. \tag{4.3}$$

When $F(u) = 0$, we recover the expression for K_{TDHO}, the propagator for the time-dependent harmonic oscillator. If $F(u) = u$, the Gaussian integral in (4.3) can be evaluated to obtain the propagator K_{TDFHO} for a time-dependent forced oscillator.

A new case is obtained if we set $F(u) = u^2$. Here the expression (4.3) reduces to

$$K(x,T;x_0,0) = K_{TDHO}(x,T;x_0 0)\, (2\beta + 1)^{-1/2} \exp\!\left(\frac{i\alpha^2}{\hbar\,(2\beta + 1)}\right). \tag{4.4}$$

An interesting special case of (4.3) is obtained if we let

$$\Omega^2(t) = \Omega_0^2, \quad f(t) = \Omega_0 / (2T)^{1/2}. \tag{4.5}$$

In this case, the action (4.2) takes the form

$$S[x(t)] = \frac{1}{2}\int_0^T \dot{x}^2 \, dt - \frac{\Omega_0^2}{4T}\int_0^T dt \int_0^T ds \, [x(t) - x(s)]^2 \ . \qquad (4.6)$$

This gives the apparent look of a two-time action. It was first introduced by Bezak in his theory of an electron moving in an environment of random scatterers. Denoting the propagator by K_B, it follows from (4.4) that

$$K_B(x,T;x_0,0) = K_{HO}(2\beta + 1)^{-1/2} \exp[\frac{i\alpha^2}{\hbar}(2\beta + 1)]. \qquad (4.7)$$

In order to simplify the result, we need the values of α and β defined in Eqs. (3.9a) and (3.9b). The quantities $\psi(t)$ and $Q(t,s)$ are pertaining to a harmonic oscillator of constant frequency Ω_0. It is easily seen that

$$\alpha = \frac{(x + x_0)(1 - \cos\Omega_0 T)}{\sqrt{2T}} \ , \qquad (4.8a)$$

$$\beta = \frac{(1 - \cos\Omega_0 T)}{(\Omega_0 T \sin\Omega_0 T)} - \frac{1}{2} \ . \qquad (4.8b)$$

Expression for K_{HO} is well known and insertion of (4.8) in (4.7) leads to the Bezak propagator

$$K_B(x,T;x_0,0) = \left(\frac{1}{2\pi i\hbar T}\right)^{1/2} \frac{\Omega_0 T}{2\sin(\Omega_0 T/2)} \exp\left\{\frac{i\Omega_0 \cot(\Omega_0 T/2)}{4\hbar}(x-x_0)^2\right\}. \qquad (4.9)$$

Obviously as $\Omega_0 \to 0$, $K_B \to K_f$, the free particle propagator. Bezak's action is an example of translationally invariant action. For such actions the propagator resembles that for a free particle with an effective mass $m^* = (\Omega_0 T/2)\cot(\Omega_0 T/2)$. There is, however, a correction factor in the normalization arising from the non-local character of action.

Problem 4.1

Use the δ-function technique to derive the following relation between K_B and K_{HO}:

$$K_B(x,T;x_0,0) = \left(\frac{iT\Omega_0^2}{2\pi\hbar} \right)^{1/2} \int_{-\infty}^{\infty} dy \; K_{HO}(x+y,T;x_0+y,0) \; . \quad (4.10)$$

Interpret the result physically.

Problem 4.2

Apply the δ-function technique to path integrate the action

$$S = \frac{1}{2} \int_0^T \dot{x}^2 \, dt - \frac{\omega^2}{2T} \left(\int_0^T x(t) \, dt \right)^2 \quad (4.11)$$

where ω is a constant.

Hint: The action (4.11) corresponds to the case where the kernel $G(t,s) = \omega^2/T$ and $f(t) = 0$. The propagator has the expression

$$K(x,T;x_0,0) = (2\pi i\hbar\beta T)^{-1/2} \exp\left[\frac{i}{2\hbar T} \left\{ (x-x_0)^2 - \frac{\omega^2 T^2}{\beta} (x+x_0)^2 \right\} \right] \quad (4.12)$$

where $\beta = 1-(\omega^2 T^2/12)$. It can also be evaluated from the general expression (2.25).

3.4.2. Translationally invariant two-time actions

In a large class of physical problems the form of $G(t,s)$ is dictated by the translational invariance of K. The general form of $G(t,s)$ in this case reads as

$$G(t,s) = \Omega^2(t) \, \delta(t-s) - G_0(t,s) \; ; \quad \Omega^2(t) = \int_0^T G_0(t,s) \, ds. \quad (4.13)$$

It is clear that

$$\int_0^T dt \int_0^T ds \, G(t,s) \, x(t) \, x(s) = \frac{1}{2} \int_0^T dt \int_0^T ds \, G_0(t,s) \, [x(t) - x(s)]^2. \quad (4.14)$$

Examples of $G_0(t,s)$ are

Bezak's kernel

$$G_0(t,s) = \Omega_0^2/T \; , \quad \Omega_0^2 = \text{const.} \quad (4.15)$$

Feynman's kernel for polaron problem

62 Path-Integral Methods and their Applications

$$G_0(t,s) = \omega \Omega_0^2 \frac{\cos[\omega(T/2 - |t-s|)]}{2\sin(\omega T/2)} . \quad (4.16)$$

Note also that as $\omega \to 0$, polaron kernel goes over to Bezak's kernel.

In order to see how the general formula (3.10) gets modified for the kernel of the form (4.13), we consider first the classical equation of motion (3.6b). This equation now takes the form

$$\ddot{\psi} + \int_0^T G_0(t,s)[\psi(t) - \psi(s)] ds = 0 \quad (4.17)$$

with $\psi(0) = x_0$ and $\psi(T) = x$. Integration of both sides of (4.17) over time t from 0 to T, leads to $\dot{\psi}(0) = \dot{\psi}(T)$ and the general solution of (4.17) reads as

$$\psi(t) = \frac{[x_0 v(T) - x v(0)] + (x-x_0)v(t)}{(v(T) - v(0))} \quad (4.18)$$

where $v(t)$ is a non-constant solution of (4.17). Going over to the formula (3.10), we have

$$\frac{i}{2\hbar}[x\dot{\psi}(T) - x_0\dot{\psi}(0)] = \frac{i\,\dot{v}(T)\,(x-x_0)^2}{2\hbar(v(T) - v(0))} \quad (4.19)$$

and the propagator reads as

$$K_F = K_0 \frac{1}{\sqrt{2\pi i\hbar\beta}} \int_{-\infty}^{\infty} du\, \exp\left\{\frac{i}{\hbar}\left[F(u) + \frac{(u-\alpha)^2}{2\beta}\right]\right\}. \quad (4.20)$$

Here the expression for K_0 has the form

$$K_0 = \frac{1}{\sqrt{2\pi i\hbar T}} \exp\left[-\frac{1}{2}\int_0^T dt \int_0^1 d\mu\, \tilde{R}(t,t;\mu)\right] \exp\left[\frac{i\,\dot{v}(T)\,(x-x_0)^2}{2\hbar(v(T) - v(0))}\right]. \quad (4.21)$$

Note that K_0 is the propagator corresponding to the translationally invariant action

$$S_0[x(t)] = \frac{1}{2}\int_0^T \dot{x}^2 dt - \frac{1}{4}\int_0^T dt \int_0^T ds\, G_0(t,s)[x(t)-x(s)]^2. \quad (4.22)$$

The second factor multiplying K_0 in (4.20) contains the effects due to generalized force term (cf. Eq.(3.1)).

We may note the resemblance of K_0 to K_f the free particle propagator of mass

$$m^* = \frac{T\,\dot{v}(T)}{v(T)-v(0)} = T\,\frac{\partial^2 S_{cl}}{\partial x \partial x_0} \qquad (4.23)$$

and express the equation (4.21) as

$$K_0(x,T;x_0,0) = C\left[\frac{1}{(2\pi i\hbar)}\left|\frac{\partial^2 S_{cl}}{\partial x \partial x_0}\right|\right]^{1/2} \exp\left[\frac{i}{\hbar}S_{cl}\right]. \qquad (4.24)$$

This is like the standard Van Vleck-Pauli formula apart from the correction factor

$$C = (T/m^*)^{1/2} \exp\left[-\frac{1}{2}\int_0^T dt \int_0^1 d\mu\, \tilde{R}(t,t;\mu)\right]. \qquad (4.25)$$

3.4.3. Propagator for the polaron kernel

It may be worthwhile to illustrate the above theory by means of an explicit calculation of the propagator for the polaron kernel (4.16). There are several steps in the derivation. We first use the short hand notation

$$G_0(t,s) = \Omega_0^2 \phi(t,s)\,;\ \phi(t,s) = \frac{\omega\,\cos[\omega(T/2-|t-s|)]}{2\,\sin(\omega T/2)}. \qquad (4.26)$$

We now make use of the following properties of $\phi(t,s)$. First $\phi(t,s)$ is symmetric in t and s, and is normalized such that

$$\int_0^T ds\,\phi(t,s) = 1. \qquad (4.27a)$$

Secondly $\phi(t,s)$ obeys the differential equation

$$(D^2 + \omega^2)\phi(t,s) = \omega^2 \delta(t-s) \qquad (4.27b)$$

where $D \equiv \partial/\partial t$, with the end-point conditions

$$\phi(0,s) = \phi(T,s) = \frac{\omega\,\cos[\omega(T/2-s)]}{2\,\sin(\omega T/2)}, \qquad (4.27c)$$

$$\dot{\phi}(0,s) = \dot{\phi}(T,s) = -\frac{\omega^2 \sin[\omega(T/2 - s)]}{2 \sin(\omega T/2)} \quad . \tag{4.27d}$$

Lastly, if $f(t)$ is a solution of the differential equation

$$(D^2 + \mu^2) f(t) = 0 \, , \tag{4.28}$$

Then it is easy to verify the following identity

$$\int_0^T ds\, \phi(t,s) f(s) = \frac{[\omega^2 f(t) - \dot{\phi}(0,t)\{f(T)-f(0)\} - \phi(0,t)\{\dot{f}(T)-\dot{f}(0)\}]}{(\omega^2 - \mu^2)} \, . \tag{4.29}$$

The evaluation of the propagator proceeds in three steps:

a. *Solution of the classical equation*

The classical equation of motion now reads as

$$(D^2 + \Omega_0^2)\psi(t) - \Omega_0^2 \int_0^T ds\, \phi(t,s)\, \psi(s)\, ds = 0. \tag{4.30}$$

Applying the operator $(D^2+\omega^2)$ and using the property (4.27b), equation (4.30) is converted into an ordinary differential equation

$$D^2 (D^2 + \nu^2)\, \psi(t) = 0 \tag{4.31}$$

where $\nu^2 = \Omega_0^2 + \omega^2$. The solution is readily obtained as

$$\psi(t) = A \cos \nu t + B \sin \nu t + Ct + D \tag{4.32}$$

The constants A, B, C, and D are determined by demanding that (4.32) indeed satisfies the original integro-differential equation (4.30). Inserting (4.32) in (4.30) and using the identity (4.29) we determine the constants B and C in terms of A:

$$B = -A \frac{\sin \nu T}{1 - \cos \nu T} \quad ; \quad C = -\frac{2 A \omega^2}{\Omega_0^2 T} \, . \tag{4.33}$$

With these values of B and C,

$$\psi(t) = \frac{A \ [\cos \nu t - \cos \nu (T-t)]}{1 - \cos \nu T} - \frac{2 A \omega^2 t}{\Omega_0^2 \ T} + D \ . \tag{4.34}$$

The end-point conditions $\psi(0) = x_0$ and $\psi(T) = x$ determine the remaining constants A and D:

$$A = - (x-x_0)\Omega_0^2 /2\nu^2 \ , \quad D = x_0 - A \ . \tag{4.35}$$

We use this solution to derive

$$x \ \dot\psi(T) - x_0 \dot\psi(0) = \Omega_0^2 \left(\nu \cot (\tfrac{1}{2} \nu T) + \frac{2\omega^2}{\Omega_0^2 \ T} \right) \frac{(x-x_0)^2}{2\nu^2} \ . \tag{4.36}$$

b. *Green's Function* $Q(t,s)$

The equation (3.7) for the Green's function $Q(t,s)$ takes the form

$$(D^2 + \Omega_0^2) \ Q(t,s) - \Omega_0^2 \int_0^T \phi(t,r) \ Q(r,s) \ dr = - \delta(t-s) \tag{4.37}$$

with the end-point conditions $Q(0,s) = Q(T,s)$. Once again, applying the operator $(D^2 + \omega^2)$ on both sides of this equation and using the property (4.27b) we obtain

$$(D^2 + \nu^2) \ Q(t,s) = - \delta(t-s) + \omega^2 \ Q_0(t,s) \tag{4.38}$$

where $Q_0(t,s)$ is the Green's function of equation

$$D^2 Q_0(t,s) = - \delta(t-s) \ . \tag{4.39}$$

The solution of (4.38) has the form

$$Q(t,s) = Q_1(t,s) - \omega^2 \int_0^T Q_1(t,r) \ Q_0(r,s) \ dr \tag{4.40}$$

where $Q_1(t,s)$ is the Green's function satisfying the equation

$$(D^2 + \nu^2) \ Q_1(t,s) = - \delta(t-s) \tag{4.41}$$

along with the end-point conditions $Q_1(0,s) = Q_1(T,s) = 0$. The corresponding conditions satisfied by $Q_0(t,s)$ are to be determined by consistency, that is, by demanding that the solution (4.40) indeed satisfies the original equation. For this purpose, we use the result (which follows from (4.29))

$$\int_0^T Q_1(t,r) Q_0(r,s) \, dr = \nu^{-2} [Q_1(t,s) - Q_0(t,s) + A(t,s)] , \quad (4.42)$$

$$A(t,s) = Q_0(0,s) \dot{Q}_1(0,t) - Q_0(T,s) \dot{Q}_1(T,t) . \quad (4.43)$$

Using (4.42), we write (4.40) in the form

$$Q(t,s) = \nu^{-2} [\Omega_0^2 Q_1(t,s) + \omega^2 Q_0(t,s) - \omega^2 A(t,s)] . \quad (4.44)$$

Note that the Green's function of $Q_1(t,s)$ is given by

$$Q_1(t,s) = \sin \nu t \sin \nu(T-s)/(\nu \sin \nu T) \qquad t < s$$

$$= \sin \nu s \sin \nu(T-t)/(\nu \sin \nu T) \qquad t > s \quad (4.45)$$

and the Green's function $Q_0(t,s)$ reads as

$$Q_0(t,s) = Q_0(0,s) + t(T-s)/T \qquad t < s$$

$$= Q_0(T,s) + s(T-t)/T \qquad t > s . \quad (4.46)$$

Now inserting (4.44) in (4.38), we find that

$$Q_0(0,s) = Q_0(T,s) = - (\Omega_0^2/2\omega^2\nu)[1 - \chi(t)][1 - \chi(s)] \cot(\nu T/2), \quad (4.47)$$

$$\chi(t) = [\sin \nu t + \sin \nu(T-t)]/\sin \nu T . \quad (4.48)$$

Combining these results, the Green's function $Q(t,s)$ takes the form

$$Q(t,s) = [\Omega_0^2 Q_1(t,s) + \omega^2 (\tilde{Q}_0(t,s) + Q_0(T,s))]/\nu^2 \quad (4.49)$$

where $Q_1(t,s)$ and $Q_0(t,s)$ are as defined in Eqs. (4.45) and (4.46) respectively and $\tilde{Q}_0$ is the Green's function of (4.39) given by

$$\tilde{Q}_0(t,s) = t(T-s)/T \qquad t < s$$

$$= s(T-t)/T \qquad t > s \quad (4.50)$$

which satisfies the conditions $\tilde{Q}_0(0,s) = \tilde{Q}_0(T,s) = 0$.

c. *Resolvant* $\tilde{R}(t,s;\mu)$

The equation for the resolvant $\tilde{R}(t,s;\mu)$ is given by (2.19). Inserting the form (4.13) for $G(t,s)$ and that of $G_0(t,s)$ given in (4.26), Eq. (2.19) now reads as

$$(D^2 + \mu\Omega_0^2)\tilde{R} - \mu\Omega_0^2 \int_0^T \phi(t,r) \tilde{R}(r,s;\mu)dr = \Omega_0^2 [-\delta(t-s) - \phi(t-s)] . \quad (4.51)$$

Comparing (4.51) with (4.37) we may write the solution

$$\tilde{R}(t,s;\mu) = - \Omega_0^2 Q(t,s;\mu) + \Omega_0^2 \int_0^T Q(t,r;\mu) \phi(r,s)dr \quad (4.52)$$

where $Q(t,s;\mu)$ is obtained from $Q(t,s)$ of (4.49) by replacing Ω_0^2 by $\mu\Omega_0^2$. The integral on the R.H.S. of (4.52) is evaluated explicitly and the result is:

$$\int_0^T Q(t,r;\mu)\phi(r,s)dr = - [Q_1(t,s;\mu) + \tilde{Q}_0(t,s;\mu)](\omega^2/\nu_1^2)$$

$$- (\Omega_1^2/2\nu_1^3)[1 - \chi_1(t)][1 - \chi_1(s)] \cot(\nu_1 T/2)$$

$$- [1 - \chi_1(t)] \chi_1(s) \cot(\nu_1 T/2)/2\nu_1 \quad (4.53)$$

where χ_1 has the same form as in (4.48) with ν replaced by ν_1 and

$$\nu_1 = (\omega^2 + \Omega_1^2)^{1/2}, \quad \Omega_1^2 = \mu \Omega_0^2. \quad (4.54)$$

Inserting (4.53) and the form of $Q_1(t,s;\mu)$ (see Eq. (4.45)) in (4.52), the resolvant

$$\tilde{R}(t,s;\mu) = -\Omega_0^2 [Q_1(s,t;\mu) + \{1 - \chi_1(t)\} \chi_1(s) \cot(\nu_1 T/2)/2\nu_1]. \quad (4.55)$$

It is now easy to verify that

$$\int_0^T \tilde{R}(t,t;\mu) dt = - \Omega_0^2 [\nu_1 T \cot(\nu_1 T/2) - 2]/2\nu_1^2 , \quad (4.56)$$

with the result

$$\int_0^1 d\mu \int_0^T \tilde{R}(t,t;\mu)\, dt = 2 \ln\left[\frac{\omega \sin(\nu T/2)}{\nu \sin(\omega T/2)}\right]. \tag{4.57}$$

Thus the normalization factor occurring in K_0 is given by

$$\exp\left[-\frac{1}{2}\int_0^1 d\mu \int_0^T dt\, \tilde{R}(t,t;\mu)\right] = \left(\frac{\nu}{\sin(\nu T/2)}\right)\left(\frac{\sin(\omega T/2)}{\omega}\right). \tag{4.58}$$

d. *Propagator*

The propagator K_0 for the polaron kernel now reads as

$$K_0(x,T;x_0,0) = \left(\frac{1}{2\pi i \hbar T}\right)^{1/2}\left(\frac{\nu}{\sin(\nu T/2)}\right)\left(\frac{\sin(\omega T/2)}{\omega}\right)$$

$$\times \exp\left\{\frac{i\Omega_0^2}{2\hbar\nu^2}\left(\nu \cot(\nu T/2) + \frac{2\omega^2}{\Omega^2 T}\right)(x-x_0)^2\right\}. \tag{4.59}$$

Note that in the limit $\omega \to 0$, K_0 reduces to Bezak's propagator (4.9).

The general propagator K_F which contains the contribution due to term F depending on the external force $f(t)$, has the same form as in (4.20) with K_0 given by (4.59). We might be interested later in the explicit form of K_F when

$$F(u) = u = \int_0^T x(t)\, f(t)\, dt, \tag{4.60}$$

This can be readily obtained after carrying out the Gaussian integration in (4.20) and reads as

$$K_F(x,T;x_0,0) = K_0(x,T;x_0,0)\, \exp\left[\frac{i}{2\hbar}\int_0^T dt \int_0^T ds\, f(t) f(s)\, Q(t,s)\right]$$

$$\times \exp\left[\frac{i}{\hbar}\int_0^T dt\, f(t)\left\{\left[\omega^2(2t-1) + \frac{\Omega^2 \sin(\nu(t-T)/2)}{\sin(\nu T/2)}\right]\frac{\Delta}{2} + \bar{x}\right\}\right] \tag{4.61}$$

where, $\Delta = x - x_0$, $\bar{x} = (x + x_0)/2$.

As a final comment, we may add that it is easy to allow a slight generalization in the kernel, $G(t,s) = a\,\delta(t-s) - b\,\phi(t,s)$, where a and b are constants while $\phi(t,s)$ is given by (4.26). The case evaluated so far in (4.59) is for $a = b = \Omega_0^2$. However, in applications a and b may be chosen appropriately to suit, for example, a variational principle.

Applications of the exact results derived in these sections will be discussed in chapter 8.

3.5. Applications of Two-time Quadratic Actions

3.5.1. *Exactly solvable model of electronic density of states (DOS)*

Here the starting point is the formulation due to Edwards and Gulyaev outlined in Sec. 3.1.2. The average propagator for the electron moving in the field of a dense cloud of randomly distributed ions is given by (1.26a) corresponding to the non-local action given by (1.27b). The next question is : What is the form of $W(\vec{q}(t) - \vec{q}(s))$? It has been shown by Halperin and Lax that for dense and weak scatterers, W may be chosen as the Gaussian distribution (unnormalized)

$$W = \exp[-(\vec{q}(t) - \vec{q}(s))^2/L^2] \qquad (5.1)$$

with the correlation length L. Physically, L is a measure of ion-electron coupling. In disordered systems, a quantity of interest is the electronic DOS defined as

$$n(E) = \frac{1}{2\pi\hbar V} \int_{-\infty}^{\infty} dT \exp[iET/\hbar]\, \mathrm{Tr}\, \bar{K}, \qquad (5.2)$$

where V is the volume of the system. Now, if the correlation length is large, a reasonable assumption in many physical situations such as polycrystalline conductors is to approximate the correlation function W by the first two terms of its expansion:

$$W = 1 - [\vec{q}(t) - \vec{q}(s)]^2/L^2. \qquad (5.3)$$

Consequently the propagator in (1.20a) takes the form

$$\bar{K} = \exp\left[-\frac{\eta^2 T^2}{2\hbar^2}\right] \int \mathcal{D}[\vec{q}(t)] \exp\left[\frac{i}{\hbar}\left\{\int_0^T dt \, \frac{m\dot{\vec{q}}^2}{2} - \frac{m\Omega^2}{4T}\int_0^T dt \int_0^T ds (\vec{q}(t)-\vec{q}(s))^2\right\}\right]$$

where Ω is a parameter depending on L defined as $\Omega^2 = 2i\eta^2 T/m\hbar L^2$.

The propagator in question has been essentially given by (4.9) except for an obvious generalization to three dimensions. We have then

$$\bar{K} = \exp\left[-\frac{\eta^2 T^2}{2\hbar^2}\right] \left(\frac{m}{2\pi i\hbar T}\right)^{3/2} \left(\frac{\Omega T}{2\sin(\Omega T/2)}\right)^3 \exp\left[\frac{im\Omega \cot(\Omega T/2)}{4\hbar}(\vec{q}-\vec{q}_0)^2\right].$$

In order to correlate the density of states all we have to do is to take the Fourier transform defined in (5.2). Explicitly, n(E) is given by the integral

$$n(E) = \frac{1}{2\pi\hbar} \int_{-\infty}^{\infty} dT \exp\left[\frac{iET}{\hbar}\right] \left(\frac{m}{2\pi i\hbar T}\right)^{3/2} \left(\frac{\Omega T}{2\sin(\Omega T/2)}\right)^3 \exp\left[-\frac{\eta^2 T^2}{2\hbar^2}\right] \quad (5.4)$$

This integral cannot be evaluated exactly. However the cases which are of interest correspond to E in the band tail ($E \to -\infty$) region. When E is large the methods of asymptotic analysis can be applied. We first note that the maximum contribution to the integral comes from the points in the neighborhood of T = 0. A good approximation is to set the factor $[\Omega T/2\sin(\Omega T/2)]^3 \approx 1$ when $T \approx 0$ and write (5.4) as

$$n(E) = \frac{1}{2\pi} \left(\frac{m}{2\pi i\hbar^2}\right)^{3/2} \int_{-\infty}^{\infty} du \, u^{-3/2} \exp[iEu - \eta u^2/2] \quad (5.5)$$

where $u = T/\hbar$. Using the standard methods of asymptotic analysis, it can be shown that

$$n(E) \approx (2/\sqrt{\pi}) \, (m/2\pi\hbar^2)^{3/2} \, E^{1/2} \quad (E > \eta) \quad (5.6a)$$

$$n(E) \approx (1/\sqrt{\pi}) \, (m/2\pi\hbar^2)^{3/2} \, \frac{\eta^{3/2}}{|E|} \exp(-E^2/4\eta^2) \quad (E << -\eta). \quad (5.6b)$$

Equation (5.6a) corresponds to the well known free electron DOS while Eq. (5.6b) gives the DOS in the band "tail". The exponential behavior agrees with the result obtained by Kane. The result (5.6) was obtained by Bezak starting from an approximate expression for the propagator for the action (4.6). The propagator derived by Bezak contained the exponential term present in (4.9) but did not have the complete normalization factor. However, since the contribution to the integral n(E) comes mostly from the neighbourhood of T = 0 where (5.5) applies, the leading order in E in the expression for n(E) does not change though the exact expression for n(E) may be quite different in details.

3.5.2. Polymer distribution functions

The polymeric system consists of large number of macromolecules. The macromolecule itself is made up of several repetitive units called monomers. A typical real macromolecule (like polyethylene) contains around 25 000 monomers. The linear dimension of a monomer is about 30 μm. Thus a macromolecule extends up to around 0.1 mm which is equal to the linear dimension of a living biological cell. These monomers are very tightly bound to each other by chemical bonds and hence the mutual distance between these monomers is more or less constant. However these monomers can be rotated with respect to each other and the molecule as a whole is very flexible. These characteristics of polymeric substances suggest that the physical properties of these materials should be closer to those of gases. The elastic coefficient of polymeric substances is closer to isothermal bulk modulus of gases. A substance like rubber is heated up when adiabatically compressed.

The similarity between the properties of polymeric substances and gases are exploited to build a model for studying the bulk properties of polymers. The model considers a large ensemble of polymers with one of the monomers labelled by s' is fixed at some point $\vec{q}'$. In this ensemble the polymers will be distributed in various configurations. Just as an ideal gas is thought of as a collection of Brownian particles, we imagine each configuration of the polymeric macromolecule to be a random walk of N steps each of length ℓ. Here N is the total number of monomers in the macromolecule and ℓ represents the separation between successive monomers (imagined as molecules of an ideal gas). Hence for large N, the probability density that the other monomer labelled by s" of the macromolecule will be found at some point $\vec{q}''$ will be given by (3.9) of chapter 1. Since in actual practice the total length of the polymer $N\ell$ is much larger than the distance between monomers we can simplify the treatment by only studying the limit $N \to \infty$, $\ell \to 0$, $N\ell^2$ = Constant. In this case, the probability density $P(\vec{q}'',s'';\vec{q}',s')$ which represents the fraction of the total number of polymers with the above constraints will be given by Eq.(3.15) of chapter 1 where the diffusion constant D is related to ℓ^2 and the role of time t is played by the number N. In view of the stationary nature P is a function of $s''-s'$. Having identified P with the probability density associated with free random walks, one can immediately write down the Wiener integral representation of (3.16) of chapter 1. Wiener integral is formally similar to a Feynman integral. In particular in the present context we identify the mass of the particle with $3\hbar/\ell^2$ and the time interval T with $-iN$. This correspondence will be assumed in all our subsequent discussion of the polymer problem and results derived for Feynman propagators will be often employed.

The situation gets complicated when we consider the interactions. Let $V_1(\vec{q}_j)$ be the potential due to an attractive force per monomer. Further, let $V_2(\vec{q}_i-\vec{q}_j)$ represent the potential between two monomers which binds them. Then for a configuration of the polymer which is characterized by $[\vec{q}_0, \vec{q}_1, \ldots, \vec{q}]$ of N+1 end-points of N monomers, the Boltzman factor B involving the interaction energy can be written in the form

$$B = \exp\left\{ -\beta \left(\sum_{j=0}^{N} V_1(\vec{q}_j) + \sum_{j>k=0}^{N} V_2(\vec{q}_j-\vec{q}_k) \right) \right\}. \quad (5.7)$$

For long polymers we will go to continuous limit and B will be given by the functional of the path $\vec{q}(\nu)$ where ν runs along the polymer configuration. Thus

$$B[\vec{q}(\nu)] = \exp\left[-\beta \int_0^N d\nu\, V_1[\vec{q}(\nu)] - \beta \int_0^N d\nu \int_0^N d\nu'\, V_2[\vec{q}(\nu)-\vec{q}(\nu')] \right]. \quad (5.8)$$

Hence the configuration sum or the fraction of polymer configurations with the positions of monomers s'' and s' fixed at $\vec{q}''$ and $\vec{q}'$ respectively can be expressed as a functional integral

$$Q(\vec{q}'',s'';\vec{q}',s') = \int_{\vec{q}(s')=\vec{q}'}^{\vec{q}(s'')=\vec{q}''} B[\vec{q}(\nu)] d\mu_W[\vec{q}(\nu)]. \quad (5.9)$$

Alternatively, Eq.(5.9) represents a path integral corresponding to the non-local action S given by

$$S = \frac{3}{2\ell^2} \int_0^N d\nu \left(\frac{d\vec{q}}{d\nu}\right)^2 + \beta \int_0^N d\nu\, V_1[\vec{q}(\nu)] + \beta \int_0^N d\nu \int_0^N d\nu'\, V_2[\vec{q}(\nu)-\vec{q}(\nu')].$$
(5.10)

The action S essentially represents the interaction energy of a polymeric molecule. The presence of the nonlocal interaction V_2 destroys the Markovian nature of the configuration sum Q and further is no longer a function of $s''-s'$. Moreover, Q does not obey any differential equation like Eq.(3.25) of chapter 1 and it is precisely at this point that the use of path integral techniques becomes unavoidable. If, however, $V_2 = 0$, one can identify Q of (5.9) as ρ_u of Eq.(3.25) of chapter 1 and the potential $V_1(\vec{q})$ will represent the annihilation rate of Brownian particles.

An exactly solvable problem

The interaction energy of the polymer is eventually the elastic energy of the chain and contains terms arising from the elastic energy of stretching and bending. For a chain containing a finite number of molecular units the contribution to the elastic energy of stretching arises primarily from the nearest neighbor interactions and in the limit of $N \to \infty$ is accounted for by the first term of the action functional of Eq. (5.10). The bending is generally associated with second neighbour interactions. For a one-dimensional configuration of polymers characterized by $[x_0, x_1, \ldots, x_N]$, the interaction energy U has the following form

$$U(\{x_j\}) = \sum_{j=1}^{N-1} \frac{\alpha}{2} \frac{(x_{j+1} - x_j)^2}{\Delta s_j} + \sum_{j=0}^{N-1} \frac{\gamma}{2} \frac{(x_{j+1} - 2x_j + x_{j-1})^2}{(\Delta s_j)^3}$$
(5.11)

where the constants α and γ depend on the mean bond length of the polymer. Further, Δs_j denotes the distance between monomers labelled by s_j and s_{j+1}. In the limit $N \to \infty$, $U(\{x_j\})$ goes over to the form

$$U[x(s)] = \frac{1}{2} \int_0^N \left\{ \alpha \left(\frac{dx}{ds}\right)^2 + \gamma \left(\frac{d^2x}{ds^2}\right)^2 \right\} ds . \qquad (5.12)$$

It is easy to see that (5.11) may be written in the alternative form

$$U[x(s)] = \frac{1}{2} \int_0^N \alpha \left(\frac{dx}{ds}\right)^2 ds + \frac{\gamma}{2} \int_0^N dt \int_0^N ds \frac{\partial^2 \delta(t-s)}{\partial t \partial s} x(s)x(t) . \qquad (5.13)$$

Identifying $\gamma \frac{\partial^2}{\partial t \partial s} \delta(t-s)$ as a kernel, the interaction energy functional of Eq.(5.13) represents a two-time quadratic action. The configuration sum or the fraction of the polymer configurations can be readily obtained by evaluating the path integral of Eq.(5.9) as in Sec. 3.2. The algebra is tedious but straightforward and the resulting expression for P(x,x') (we drop the subscripts s and s' here) reads as

$$P(x'',x') = \left[\frac{\gamma \Omega^3 \beta}{2\pi D(s'',s',N)}\right]^{3/2} \exp\left[\frac{\gamma \Omega^2 \beta}{2D} (x'' - x')^2\right] \qquad (5.14)$$

where

$$D = \Omega|s''-s'| - \{1 + \cosh[\Omega(s''+s')] - \cosh[\Omega(s''-s')] $$

$$- \cosh[\Omega(s''+s')]\cosh[\Omega(s''-s')]\} \tanh(\Omega N) + \sinh[\Omega(s''+s')]$$

$$- \sinh[\Omega|s''-s'|] - \sinh[\Omega(s''+s')]\cosh[\Omega|s''-s'|] \qquad (5.15)$$

and $\Omega^2 = \alpha/\gamma$.

From the general expression (5.14) for the configuration sum it is possible to derive some specific results. For example, let us calculate the so-called end-to-end probability, where we let $s'' \to N$ and $s' \to 0$. This leads to the well known result derived by Freed namely,

$$\tilde{P}(x'',x') = \left[\frac{\gamma \Omega^3 \beta}{2\pi(\Omega N - \tanh \Omega N)}\right]^{3/2} \exp\left[\frac{\gamma \Omega^3 \beta (x''-x')^2}{2\pi(\Omega N - \tanh \Omega N)}\right] . \qquad (5.16)$$

As an application of the result (5.14) for the configuration sum let us obtain the scattering probability $P(\mu)$ which gives the angular distribution of the intensity of electromagnetic radiation scattered from a solution of polymers. The probability $P(\mu)$ is related to $P(x'',x')$ as

$$P(\mu) = N^{-2} \int \cdots \int ds'' \, ds' \, dx'' \, dx' \, P(x'',x') \exp[i\mu(x''-x')] \quad (5.17)$$

where $\mu = |k - k_0|$, k_0 and k being the propagation vectors for the incident and scattered beams. Inserting the expressions for $P(x'',x')$ from (5.15) in (5.17) and performing the integrations over x'' and x', we obtain

$$P(\mu) = N^{-2} \int ds'' \int ds' \, \exp[-\mu^2 D / 2 \delta \Omega^3 \beta]. \quad (5.18)$$

As a final remark we might mention that if the interaction potential consists only of the elastic energy of stretching ($\gamma \rightarrow 0$), the integrations over s'' and s' can also be carried out. In this case, the expression for $P(\mu)$ assumes a simple form

$$P(\mu) = 2 \left\{ \frac{(2\alpha\beta)^2}{\mu^4 N^2} \left[\exp\left(-\frac{\mu^2 N}{2\alpha\beta}\right) - 1 \right] + \frac{2\alpha\beta}{\mu^2 N} \right\} .$$

This may be readily recognized as the usual Debye scattering formula.

3.5.3. *Propagation of waves in random media*

The problem of wave propagation in a random medium arises in many areas of physics. Some examples are atmospheric optics and radio-astronomy. As the wave propagates through the medium its amplitude $A(\vec{q},t)$ at any point $\vec{q}$ and its frequency ω are affected by the fluctuations in the medium. This problem has been extensively studied in literature for quite some time. However, among the earlier methods there does not seem to exist a single approach which can treat arbitrary fluctuations in the medium. For example, the so-called Ryotov method is applicable only when the medium properties deviate very little from their average values. On the other hand, the more recent path integral formulation of the problem is capable of handling arbitrary variations in the medium properties. In the following we illustrate the basic ideas involved.

For simplicity we apply the path integral formulation to study the wave propagation due to a point source. Let the vector $\vec{q}_0 = (x_0, y_0)$ denote the location of a monochromatic point source in the plane $z = 0$. Let $E(\vec{q},z,t)$ be the strength of the signal in the plane $z > 0$. Let R be the total range of propagation. Defining a complete envelope $A(\vec{q},z,t)$ as

$$E(\vec{q},z,t) = \text{Re } [A(\vec{q},z,t) \exp \{i(kz-\omega t)\}] \quad , \quad (5.19)$$

ω being the frequency of the source. It can be shown that under certain conditions the full wave equation for A can be approximated by the parabolic equation

$$[i\partial/\partial z + (1/2k) \nabla^2 - k \mu(\vec{q},z,t)] \vec{A}(\vec{q},z,t) = 0 \quad (5.20)$$

where ∇^2 is the Laplacian in two dimensions and the quantity $\mu(\vec{q},z,t)$ is related to the refractive index $n(\vec{q},z,t)$ of the medium by the relation

$$\mu(\vec{q},z,t) = 1 - n(\vec{q},z,t) \quad , \quad (5.21)$$

Further the symbol k denotes the wave number which is related to the frequency ω of the wave by the usual relation $\omega = ck$, c being the speed of the wave. Equation (5.20) has to be further supplemented by the source distribution at $z = 0$

$$A(\vec{q},\vec{q}_0;0) = \delta(\vec{q}-\vec{q}_0) = \lim_{z \to 0} (4\pi z)^{-1} \exp [\frac{ik}{2z} (\vec{q} - \vec{q}_0)^2] \quad , \quad (5.22)$$

Equation (5.20) is just the Schrödinger equation where the coordinate z plays the role of time t. Therefore, the solution of this equation can be represented as a path integral

$$A(\vec{q},z;\vec{q}_0,0) = (i/2k) \int \mathcal{D} [\vec{q}(z)] \exp (ikS) \quad (5.23)$$

where the quantity S is analogous to the action functional and is given by

$$S = S_0 - \int_0^z \mu (\vec{q}(z'),z',t) \, dz' \quad (5.24a)$$

$$S_0 = \frac{1}{2} \int_0^z (d\vec{q}/dz')^2 \, dz' \quad , \quad (5.24b)$$

Further the integrations are to be performed over all paths $\vec{q}(z)$ connecting the source point $(\vec{q}_0, 0)$ to the receiver at $(\vec{q}_0, z)$. The intensity I of the wave can now be obtained by using the relation between I and A, viz. $I = |A|^2$.

Since the medium properties are randomly fluctuating the quantity μ is described by a probability distribution function. In most of the physical applications it is assumed that μ is a Gaussian random variable. Thus μ is completely described by its mean value and a covariance function $\rho(|\vec{q}-\vec{q}'|)$ defined as

$$< \mu(\vec{q},t)\, \mu(\vec{q}',t')> \;=\; \rho(|\vec{q}-\vec{q}'|, t-t') \quad . \tag{5.25}$$

Equation (5.23) can be used to compute average amplitude and intensity of the wave at any space point (q,z,t). Thus

$$<A> \;=\; (i/2k) \int \mathcal{D}\,[\vec{q}(z)] < \exp(ikS) > \tag{5.26a}$$

$$<I> \;=\; (1/2k)^2 \int \int \mathcal{D}\,[\vec{q}_1(z)]\, \mathcal{D}\,[\vec{q}_2(z')] < \exp[ik(S-S')] > \tag{5.26b}$$

where the action functional S' is the same as S except that $\vec{q}_1(z)$ is replaced by $\vec{q}_2(z')$. If μ is a Gaussian random variable, the averages in equation (5.26) may be carried out yielding the average amplitude

$$<A> \;=\; (i/2k) \int \mathcal{D}\,[\vec{q}(z)]\, \exp[ikS_1] \tag{5.27}$$

and the average intensity

$$<I> \;=\; (1/2k)^2 \int \int \mathcal{D}\,[\vec{q}_1(z_1)]\, \mathcal{D}\,[\vec{q}_2(z_2)]\, \exp[ik(S_1 - S_2)]. \tag{5.28}$$

Here

$$S_1 = S_0\,[\vec{q}_1(z)] - \frac{1}{2} k^2 \int_0^z dz'' \int_0^z dz'\, \rho(|\vec{q}_1(z'') - \vec{q}_1(z')|, t''-t') \tag{5.29}$$

$$S_2 = S_0[\vec{q}_2(z)] - k^2 \int_0^z dz'' \int_0^z dz'\, \rho(|\vec{q}_2(z'') - \vec{q}_1(z')|, t''-t')$$

$$+ \frac{k^2}{2} \int_0^z dz'' \int_0^z dz' \, \rho(|\vec{q}_2(z'') - \vec{q}_2(z')|, t''-t') + S_0[\vec{q}_1(z)] \, . \quad (5.30)$$

where in obtaining the averages in Eqs. (5.29) and (5.20), we have used the fact that for a Gaussian random variable y with mean <y>=0

$$< \exp(iy) > \; = \; \exp[\, -<y^2>/2 \,]. \quad (5.31)$$

A similar identity was previously used in deriving Eq. (1.23).

The path integrals occurring in Eq. (5.26) are the familiar path integrals corresponding to non-local actions. Note that the path integral representation for A and I is valid for any variation of ρ since at no stage any assumption about the strength of the medium fluctuations has been used. Explicit evaluation of these path integrals can better be done with a variational technique similar to the one used in treating polaron problem to be discussed in chapter 8. The exact propagators derived in this chapter are useful in the variational treatment. This is because, to a good extent, the essential physics of the problem can be extracted by using the non-local quadratic forms as trial actions. Of course, in the present case, the appropriate form of the kernel G(t,s) to be used is dictated by the behaviour of the correlation function ρ.

Alternatively, we may approximate the correlation function ρ itself by some simpler forms, so that the resulting expressions for A and I are in closed analytical form. For example in the so-called Markovian approximation ρ may be approximated by the relation

$$\rho(|\vec{q}-\vec{q}'|) \equiv \rho((|\vec{q}(z) - \vec{q}(z')|^2 + |z-z'|^2)^{1/2}) \approx \rho(|z-z'|). \quad (5.32)$$

In the path integral language this approximation simply means that the important contribution to the propagators in (5.26) arises from paths for which $|d\vec{q}/dz|$ is small. In this approximation for ρ, the path integrals in (5.26) yield simply the free particle propagators apart from some constant factors. The closed form expressions for A and I, therefore, can be readily written down as

$$\langle A \rangle = (i/2k) \exp\left[-\frac{1}{2} k^2 \int_0^z dz'' \int_0^z dz' \, \rho(|z''-z'|)\right] \exp\left[\frac{ik}{4} (\vec{q}(0) - \vec{q}(z))^2\right]$$

(5.33)

$$\langle I \rangle = (1/4k^2) \exp\left[-2k^2 \int_0^z dz'' \int_0^z dz' \, \rho(|z-z'|)\right]$$
(5.34)

which are the usual formulae obtained in literature after invoking the Markovian approximation. One can improve upon the approximation (5.32) by including the spatial dependence of ρ to some extent. For example one may use

$$\rho(|\vec{q} - \vec{q}'|) = (|\vec{q}(z) - \vec{q}'(z')|^2 + |z - z'|^2) \, f(|z - z'|)$$
(5.35)

and the resulting path integral in this case corresponds to that of two time quadratic actions. The exact expressions for A and I can therefore be evaluated in a straightforward manner and this task is left as an exercise.

Notes and References

General references for the origin of two-time actions and their applications are:

R. P. Feynman and A. R. Hibbs, "Quantum Mechanics and Path Integrals", McGraw-Hill, New York (1965),

R. P. Feynman, "Statistical Mechanics", Benjamin (1975).

Applications of two-time action to density of states were first suggested by

S. F. Edwards and Y. V. Gulyaev, Proc. R. Soc., **83**, 495 (1964) and later used by

V. Bezak, Proc. R. Soc. (London) **A315**, 339 (1970).

Analytical evaluation of two-time quadratic actions is discussed by several authors: Most of the references can be found in

D. C. Khandekar and S. V. Lawande, Phys. Repts **137**, 115-229 (1986).

We may list a few other references here

D. C. Khandekar, K. V. Bhagwat and S. V. Lawande, J. Phys. **A16**, 4209 (1983),

J. Adamowski, B. Gerlach and H. Leschke, J. Math. Phys. **23**, 243 (1983),

D. P. L. Castrigiano and N. Kokiantonis, Phys. Lett. **96A**, 55 (1983),

D. P. L. Castrigiano and N. Kokiantonis, Phys. Lett. **104A**, 123 (1984),

B. K. Cheng, Phys. Lett. **103A**, 357(1984).

Exact propagators for a large class of two-time quadratic actions have also been obtained by

G. J. Papadopoulos, J. Phys. **A18**, 1945 (1985).

Bezak's action was first path integrated by

G. J. Papadopoulos, J. Phys. **A7**, 183 (1974)

and subsequently by

A. Maheshwari, J. Phys., **A8**, 1019 (1975).

The prodistribution approach used in this derivation is extensively discussed by

C. DeWitt-Morette, A. Maheshwari and B. Nelson, Phys. Repts., **50**, 255 (1979)

and has been also used in an alternative derivation of the exact propagator for a general two-time action in a paper by

D. C. Khandekar, S. Datta and S. V. Lawande, Phys. Rev. **31A**, 3574 (1985).

Path integral technique for treating polymer problems was first used by

S. F. Edwards, Proc. Phys. Soc. **85**, 613 (1965) ;

see also

S. F. Edwards, J. Phys., **A7**, 332 (1974),

S. F. Edwards and K. F. Freed, J. Phys. **C3**, 739 (1970),

K. F. Freed, J. Chem. Phys. **54**, 1453 (1971).

The results presented in Sec. 3.5.2 have also been derived by an alternative approach by

G. J. Papadopoulos and J. Thomchick, J. Phys. **A10**, 1115 (1977).

Previous calculations of $P(\mu)$ reported by Freed (op cit) and by Saito et al in

N. Saito, K. Takahashi and Y. Yunoki, J. Phys. Soc. (Japan) **22**, 219 (1967).

have used the end-to-end probability distribution (Eq.(5.16)) rather than the full two-point distribution function (Eq.(5.14)). The latter is more appropriate to use because of the non-Markovian character of the problem.

A more recent reference for applications of path integrals in polymer physics is a book by

F. W. Wiegel, "Introduction to Path Integral Methods in Physics and Polymer science" (World Scientific, Singapore, 1986).

Applications of two-time actions to discuss the propagation of waves in random media are treated in the following papers:

P. L. Chow, J. Math. Phys. 13, 1224, (1972),

R. Dashen, J. Math. Phys. 23, 894 (1982),

R. Dashen, Optics. Lett. 9, 110 (1984).

CHAPTER 4

PATH INTEGRALS IN GENERAL COORDINATE SYSTEMS

4.1. Introduction

Quantum description of a classical dynamical system is in general complicated. The complications arise both in the standard canonical approach and in the path integral approach. In the canonical approach, the difficulties arise because of the ambiguity in ordering of the operators that do not commute with each other. In writing the quantum version of a classical expression one does not know how to order various factors and a prescription is required. One such prescription for ordering is the well-known Weyl ordering rule.

In the path integral formulation, we deal with only c-number functions and one would expect that this method of quantization would be free from ambiguities associated with operators and commutators. This is far from being the case and path integral formulation does not solve the ordering problem. In fact, operator ordering arbitrariness is now translated into arbitrariness in the definition of the path integral. Indeed, different forms of a path integral result depending on the point where functions are evaluated in the discrete sum over paths. Each prescription corresponds to a particular Hermitian ordering for the quantum Hamiltonian. In particular, the path integral obtained by the mid-point choice yields the same quantum theory as described by the Hamiltonian operator with Weyl ordering rule.

In the subsequent sections we first develop path integrals in polar coordinates. We stress here the importance of mid-point rule while deriving the discretized version of the propagator in polar coordinates starting from its form in cartesian coordinates. In particular, we examine the issue of quantum corrections to the potential which arise rather naturally. We treat in detail the cases of central as well as non-central potentials where exact closed analytical form can be derived.

The role of quantum correction to the potential is emphasized in the treatment of the problem of a particle moving on the surface of a d-dimensional sphere. This case naturally leads to the formulation of path integrals on group manifolds. For lack of space, we refrain from discussing this aspect here and cite only some references for the reader at the end of this chapter.

The development of algorithm for discretizing action for a dynamical system in arbitrary curved spaces is considered in the final section. It is sufficient for all practical purposes to invoke the mid-point prescription and we derive the quantum correction to the potential only on this basis. The reader may easily compare our description with those existing in literature.

4.2. Path Integrals in Polar Coordinates

Many physical systems are invariant under rotations. In dealing with such systems one usually makes a transformation to polar coordinates. Thus, for example, in the case of the Schrödinger equation for spherically symmetric potentials, one reduces the problem to the radial Schrödinger equation, an effective one-dimensional problem. It is natural to ask, whether such a transformation to polar coordinates is possible also with a path integral. The answer to this question is non-trivial, because, as stated in the introduction there is apparently no general prescription for writing a path integral in general coordinates.

In fact, in Feynman's original path integral formulation of quantum mechanics, one necessarily has to start from cartesian coordinates. The stochastic nature of paths generally precludes a direct use of the classical action expressed in curvilinear coordinates. Moreover, even if a path integral is expressed in cartesian coordinates, it is not invariant under a formal change of variables. However, with a little bit of hindsight and care, one can obtain the correction to the Lagrangian which ensures that the transformed path integral describes the same quantum problem as the original one. Our motivation in this section is essentially to bring out these subtle issues. We deal with the path integrals in two and three dimensional polar coordinates at first and draw some lessons for dealing with general curvilinear coordinates.

4.2.1. Polar coordinates in two-dimensions

We start with the Feynman propagator for a particle of mass m moving in two-dimensions under a potential $V(\vec{q})$ with $\vec{q} = (x,y)$. The time-sliced prescription for the propagator is

$$K(\vec{q}'',\vec{q}';T) = \lim_{N\to\infty} A_N \int \int \prod_{j=1}^{N-1} d\vec{q}_j \exp\left[\frac{i}{\hbar} \sum_{j=1}^{N} S_j\right] \quad (2.1)$$

where the short-time action over the isometric time interval $[t_{j-1}, t_j]$ of length ε is

$$S_j = \frac{m}{2\varepsilon} (\vec{q}_j - \vec{q}_{j-1})^2 - \varepsilon V(\vec{q}_j) \quad (2.2)$$

and $\vec{q}_0 = \vec{q}'$, $\vec{q}_N = \vec{q}''$. The normalization constant associated with the Feynman measure is $A_N = (m/2\pi i\hbar\varepsilon)^N$.

There are two prescriptions to be followed in the implementation of the transformation from cartesian (x,y) to polar (r,θ) coordinates : first is the mid-point rule, that is, all quantities must be expressed at the midpoint; the second is that the terms up to order ε must be retained in the action and in the transformation of the measure. The mid-point rule implies the use of Stratonovich calculus. The retention of terms up to order ε requires that in the expansion around the midpoint terms up to fourth-order need to be retained.

Consider first the expression

$$(\Delta\vec{q}_j)^2 = (\vec{q}_j - \vec{q}_{j-1})^2 = (\Delta r_j)^2 + 2r_j r_{j-1}(1-\cos(\Delta\theta_j))$$

$$= (\Delta r_j)^2 + \tilde{r}_j^2 (\Delta\theta_j)^2 - \frac{1}{4}(\Delta r_j)^2(\Delta\theta_j)^2 - \tilde{r}_j^2 (\Delta\theta_j)^4/12 \quad (2.3)$$

where the notation used is

$$\Delta u_j \equiv u_j - u_{j-1}, \quad \tilde{u}_j \equiv \frac{1}{2}(u_j + u_{j-1}) . \quad (2.4)$$

The potential term $V(x_j, y_j)$ may be simply replaced by $\tilde{V}_j = V(\tilde{r}_j, \tilde{\theta}_j)$. The short time action now reads as

$$S_j = \frac{m}{2\varepsilon}[(\Delta r_j)^2 + \tilde{r}_j^2(\Delta\theta_j)^2] - \varepsilon \tilde{V}_j$$

$$- \frac{m}{8\varepsilon}[(\Delta r_j)^2(\Delta\theta_j)^2 + \frac{1}{3}\tilde{r}_j^2(\Delta\theta_j)^4] \ . \tag{2.5}$$

We may therefore write

$$\exp\left[\frac{i}{\hbar}S_j\right] = \exp\left[\frac{i}{\hbar}\left\{\frac{m}{2\varepsilon}[(\Delta r_j)^2 + \tilde{r}_j^2(\Delta\theta_j)^2] - \varepsilon\tilde{V}_j\right\}\right]$$

$$\times \left\{1 - \frac{im}{8\hbar\varepsilon}[(\Delta r_j^2)(\Delta\theta_j^2) + \frac{1}{3}\tilde{r}_j^2(\Delta\theta_j)^4]\right\} \ . \tag{2.6}$$

In this expansion we have retained only the terms $(\Delta r_j)^2(\Delta\theta_j)^2$ and $(\Delta\theta_j)^4$ which are all of order ε^2. The measure is transformed as

$$\prod_{j=1}^{N-1} dx_j\, dy_j = \prod_{j=1}^{N-1} r_j\, dr_j\, d\theta_j \ . \tag{2.7}$$

This expression which describes the mapping of the jth interval has to be symmetrized about the points (j-1, j) so that none of the end points of the interval is preferred when we expand about the mid-point of the interval. We write

$$\prod_{j=1}^{N-1} dx_j dy_j = d\mu \prod_{j=1}^{N}(r_j r_{j-1})^{1/2} = d\mu \prod_{j=1}^{N}\tilde{r}_j(1 - (\Delta r)^2/8\tilde{r}_j^2) \ . \tag{2.8}$$

where for brevity we write $d\mu = (r'r'')^{-1/2} \prod_{j=1}^{N-1} dr_j d\theta_j$. We insert Eqs.(2.6) and (2.8) in the expression (2.1) to arrive at the correct time-sliced form of the propagator in polar coordinates:

$$K(r'',\theta'',r',\theta';T) = \lim_{N\to\infty}(r''r')^{-1/2}A_N \int\cdots\int \prod_{j=1}^{N-1} dr_j\, d\theta_j \prod_{j=1}^{N}\tilde{r}_j$$

$$\times \exp\left[\frac{i}{\hbar}\left\{\left(\frac{m}{2\varepsilon}\right)\left[(\Delta r_j)^2 + \tilde{r}_j^2(\Delta\theta_j)^2\right] - \varepsilon V(\tilde{r}_j, \tilde{\theta}_j)\right\}\right] \tag{2.9}$$

$$\times \prod_{j=1}^{N}\left\{1 - \frac{1}{8}(\Delta r_j)^2\tilde{r}_j^{-2} - \frac{im}{8\hbar\varepsilon}[(\Delta r_j^2)(\Delta\theta_j^2) + \frac{1}{3}\tilde{r}_j^2(\Delta\theta_j)^4]\right\} \ .$$

In order to cast this in a more useful form, we make use of the fact that $(\Delta u_j)^2 \approx \varepsilon$. The way to do this correctly was shown by Schulman and McLaughlin. The idea is to use the identity

$$\int_{-\infty}^{\infty} u^{2n} \exp\left[\frac{-\alpha}{2\beta} u^2\right] du = \frac{(2n-1)!!}{(\alpha/\beta)^n} \int_{-\infty}^{\infty} \exp\left[\frac{-\alpha}{2\beta} u^2\right] du. \qquad (2.10)$$

In our case, $\beta = (m/2\hbar\varepsilon)$ and α is either 1 or $\tilde{r}_j^2$ and the approximate forms of the various $(\Delta u_j)^n$ are

$$(\Delta r_j)^2 \propto \frac{i\hbar\varepsilon}{m}, \quad (\Delta\theta_j)^2 \propto \frac{i\hbar\varepsilon}{m} \tilde{r}_j^{-2}, \quad (\Delta\theta_j)^4 \propto 3\left(\frac{i\hbar\varepsilon}{m}\right)^2 \tilde{r}_j^{-4}. \qquad (2.11)$$

If we insert (2.11) in (2.9) and lift it up into the exponential, we get the polar form of the discretized propagator

$$K(r'',\theta'',r',\theta';T) = \lim_{N\to\infty} (r''r')^{-1/2} A_N \int \cdots \int \prod_{j=1}^{N-1} dr_j d\theta_j \prod_{j=1}^{N} \tilde{r}_j$$

$$\times \exp\left[\frac{i}{\hbar} \left\{\frac{m}{2\varepsilon}\left[(\Delta r_j)^2 + \tilde{r}_j^2 (\Delta\theta_j)^2\right] - \varepsilon(V_j + \Delta V_j)\right\}\right]. \qquad (2.12)$$

The first factor in the exponential is the discretized form of the polar action

$$S = \int_0^T \left[\frac{m}{2}(\dot{r}^2 + r^2 \dot{\theta}^2) - V(r,\theta)\right] dt \qquad (2.13)$$

over the interval $[j-1,j]$. The potential term contains the additional factor

$$\Delta V_j = -\hbar^2/8m\tilde{r}_j^2 \qquad (2.14)$$

which does not depend on the original potential and is of order $\hbar^2$. This term is the required "quantum correction" to ensure that the transformed propagator describes the same quantum problem as the original one. This can be easily verified by the reader by showing that the short time propagator leads to the correct Schrödinger equation and is left as an exercise. However, the origin of quantum correction may also be understood in the following way. In polar coordinates the classical Hamiltonian

$$H = \frac{1}{2m} [p_r^2 + p_\theta^2] + V(r,\theta). \tag{2.15}$$

The quantum Hamiltonian operator reads as

$$\hat{H} = - \frac{\hbar^2}{2m} \Delta + V(r,\theta) , \tag{2.16}$$

where Δ is the Laplace-Beltrami operator which has the form

$$\Delta = \frac{\partial^2}{\partial r^2} + \frac{1}{r} \frac{\partial}{\partial r} + \frac{1}{r^2} \frac{\partial^2}{\partial \theta^2} . \tag{2.17}$$

If we write the Hamiltonian (2.16) in terms of the Hermitian momenta operators

$$p_r = - i\hbar(\partial/\partial r + 1/2r) \; ; \quad p_\theta = - i\hbar \, \partial/\partial\theta , \tag{2.18}$$

we have

$$\hat{H} = \frac{1}{2m} [p_r^2 + p_\theta^2/r^2] + (V + \Delta V) , \tag{2.19}$$

where $\Delta V = - \hbar^2/8mr^2$, precisely the same as in (2.14).

The form (2.12) is too complicated for explicit calculations. For the latter purpose, we show that the path integral can be written in an alternative form. Retracing the steps to Eq.(2.3), we rewrite

$$(\Delta \vec{q}_j)^2 = (\Delta r_j)^2 + r_{j-1} r_j (\Delta \theta_j)^2 - \frac{1}{12} r_{j-1} r_j (\Delta \theta_j)^4 , \tag{2.20}$$

where the arithmetic mean $\tilde{r}_j$ in Eq.(2.3) is now replaced by the geometric mean $(r_j r_{j-1})^{1/2}$. With this, the measure remains as in (2.7) and the path integral (2.12) can be expressed by an equivalent form

$$K = \lim_{N \to \infty} A_N \int \prod_{j=1}^{N-1} r_j dr_j d\theta_j \prod_{j=1}^{N} \exp\left[\frac{i}{\hbar} \left\{\frac{m}{2\varepsilon}\left[(\Delta r_j)^2 + r_j r_{j-1}(\Delta \theta_j)^2\right] - \varepsilon(V_j + \Delta V'_j)\right\} \right]. \tag{2.21}$$

The correction term $\Delta V' = - \hbar^2/8mr_j r_{j-1}$ differs from ΔV_j of (2.14) by an order ε and hence is equivalent to ΔV_j.

Finally, for explicit evaluation, the correction term disappears if we write equivalently

$$K = \lim_{N\to\infty} A_N \int \prod_{j=1}^{N-1} dR_j \prod_{j=1}^{N} \exp\left[\frac{i}{\hbar}\left\{\frac{m}{2\varepsilon}\left(r_j^2 + r_{j-1}^2 - 2r_{j-1}r_j \cos(\Delta\theta_j)\right) - \varepsilon V_j\right\}\right],$$

(2.22)

where dR_j is a short-hand notation for $r_j dr_j d\theta_j$.

This is the starting point for naive evaluation of the propagator in polar coordinates. This is because Euclidean space remains flat irrespective of the coordinates used. In order to bring out the subtle nature of quantum correction, we consider the problem of a particle moving on a circle of radius R. Here the Lagrangian

$$L = \frac{1}{2} m R^2 \dot{\theta}^2 \qquad (2.23)$$

and the propagator can be written in the path integral form

$$K(\theta'',\theta';T) = \lim_{N\to\infty} \tilde{A}_N \int \prod_{j=1}^{N-1} d\theta_j \prod_{j=1}^{N} \exp\left\{\frac{imR^2}{2\hbar\varepsilon}(\Delta\theta_j)^2\right\}, \qquad (2.24)$$

where $\tilde{A}_N = (mR^2/2\pi i\hbar\varepsilon)^{N/2}$. There is no quantum correction term here. For explicit evaluation, we use the reverse of the step (2.21), that is, write

$$(\Delta\theta_j)^2 = 2[(1 - \cos\Delta\theta_j) + (\Delta\theta_j)^4/24] \qquad (2.25)$$

and replace $(\Delta\theta_j)^4$ by $3(i\hbar\varepsilon/mR^2)^2$. The resulting propagator takes the form

$$K(\theta'',\theta';T) = \lim_{N\to\infty} \tilde{A}_N \int \prod_{j=1}^{N-1} d\theta_j \prod_{j=1}^{N} \exp\left\{\frac{im}{\hbar\varepsilon} R^2(1-\cos(\Delta\theta_j)) - \frac{i\varepsilon}{\hbar}\Delta V_j\right\} \qquad (2.26)$$

where $\Delta V_j = \hbar^2/8mR^2$. It may be noted here that a naive replacement $(m/2\pi i\hbar\varepsilon)^{1/2}\exp[-\frac{i\varepsilon}{\hbar}V_j] \to \delta(r_j-R)$ as a constraint in (2.22) does not lead to (2.26). There is no genuine quantum correction but additional term ΔV appears by virtue of the replacement (2.25).

4.2.2. Polar coordinates in three dimensions

Development of a path integral in three-dimensional polar coordinates is similar to that in two-dimensions. The starting point is as before the mid-point expansion of the classical discretized action expressed in polar coordinates

$$(\Delta \vec{q}_j)^2 = (\Delta r_j)^2 + 2r_{j-1} r_j (1 - \cos \psi_{j, j-1}) \qquad (2.27)$$

where $\psi_{j, j-1}$ is the angle between the vectors $\vec{q}_{j-1}$ and $\vec{q}_j$. The well known addition theorem yields

$$\cos \psi_{j, j-1} = \cos \theta_{j-1} \cos \theta_j + \sin \theta_{j-1} \sin \theta_j \cos(\psi_j - \psi_{j-1}). \qquad (2.28)$$

Expanding each of the quantities around the mid-point, we can derive

$$(\Delta \vec{q}_j)^2 = (\Delta r_j)^2 + \tilde{r}_j^2 (\Delta \theta_j)^2 + \tilde{r}_j^2 \sin^2 \tilde{\theta}_j (\Delta \phi_j)^2 - \frac{1}{4}\Big[(\Delta r_j)^2 + (\Delta \theta_j)^2$$

$$+ (\Delta r_j \Delta \phi_j)^2 \sin^2 \tilde{\theta}_j + (\Delta \theta_j \Delta \phi_j)^2 \tilde{r}_j^2 + \frac{1}{3}(\Delta \theta_j)^4 \tilde{r}_j^2 + \frac{1}{3}(\Delta \phi_j)^4 \tilde{r}_j^2 \sin^2 \tilde{\theta}_j \Big].$$

$$(2.29)$$

Inserting this in $\exp [\frac{i}{\hbar} S_j]$, we can write

$$\exp\left[\frac{i}{\hbar} S_j\right] = \left[1 - \frac{im}{8\hbar\epsilon} \Big\{ (\Delta r_j^2)(\Delta \theta_j^2) + (\Delta r_j)^2 (\Delta \phi_j)^2 \sin^2 \tilde{\theta}_j^2 \right.$$

$$\left. + (\Delta \theta_j \Delta \theta_j)^2 \tilde{r}_j^2 + \frac{1}{3}(\Delta \theta_j)^4 \tilde{r}_j^2 + \frac{1}{3}(\Delta \phi_j)^4 \tilde{r}_j^2 \sin^2 \tilde{\theta}_j \Big\} \right]$$

$$\times \exp\left[\frac{i}{\hbar} \Big\{ \frac{m}{2\epsilon}\Big((\Delta r_j)^2 + \tilde{r}_j^2 (\Delta \theta_j)^2 + \tilde{r}_j^2 \sin^2 \tilde{\theta}_j^2 (\Delta \phi_j)\Big) - \epsilon V(\tilde{r}_j, \tilde{\theta}_j, \tilde{\phi}_j) \Big\} \right].$$

$$(2.30)$$

The transformation of measure to symmetric form proceeds as before and we have

$$\prod_{j=1}^{N-1} dx_j dy_j dz_j = \prod_{j=1}^{N-1} r_j^2 \sin \theta_j \, dr_j d\theta_j d\phi_j = (r''r' \sin \theta'' \sin \theta')^{-1/2}$$

$$\times \prod_{j=1}^{N-1} \tilde{r}_j^2 \sin \tilde{\theta}_j \Big(1 - \frac{1}{4}\Big[(\Delta r_j)^2 + (\Delta \theta_j)^2/2 \sin^2 \tilde{\theta}_j \Big] \Big). \qquad (2.31)$$

Inserting the results (2.30) and (2.31) in the propagator and using the Schulman-McLaughlin procedure, we obtain the following expression for the propagator $K(r'',\theta'',\phi'',r',\theta',\phi';T)$:

$$K = \lim_{N \to \infty} A_N (\sin \theta'' \sin \theta')^{-1/2} (r''r')^{-1} \int \prod_{j=1}^{N-1} dr_j d\theta_j d\phi_j \prod_{j=1}^{N} \tilde{r}_j^2 \sin \tilde{\theta}_j$$

$$\times \exp\left[\frac{i}{\hbar}\left\{\frac{m}{2\varepsilon}\left((\Delta r_j)^2 + \tilde{r}_j^2 (\Delta\theta_j)^2 + \tilde{r}_j^2 \sin^2\tilde{\theta}_j (\Delta\phi_j)^2\right) - \varepsilon(V_j + \Delta V_j)\right\}\right], \quad (2.32)$$

where ΔV_j is the quantum correction given by

$$\Delta V_j = -\hbar^2 (1 + 1/\sin^2\theta_j)/8m\tilde{r}_j^2. \quad (2.33)$$

Alternatively we could have used the geometric mean instead of the arithmetic mean, that is, replace $\tilde{r}_j^2$ by $r_j r_{j-1}$ and $\sin^2\tilde{\theta}_j$ by $\sin\theta_j \sin\theta_{j-1}$ in which case the transformed measure need not be symmetrized. The following equivalent form results from this

$$K = \lim_{N \to \infty} A_N \int d\mu \prod_{j=1}^{N} \exp(\frac{i}{\hbar} R_{j, j-1}) \quad (2.34)$$

where

$$d\mu = \prod_{j=1}^{N-1} r_j^2 \sin \theta_j dr_j d\theta_j d\phi_j, \quad (2.35a)$$

$$R_{j,j-1} = \frac{m}{2\varepsilon}\left((\Delta r_j)^2 + r_j r_{j-1} (\Delta\theta_j)^2 + r_j r_{j-1} \sin \theta_{j-1} \sin \theta_j (\Delta\phi_j)^2\right)$$

$$- \varepsilon(V_j + \Delta V_j). \quad (2.35b)$$

Finally the quantum correction cancels out when we use the conventional form of the path integral

$$K = \lim_{N \to \infty} A_N \int d\mu \prod_{j=1}^{N} \exp\left[\frac{i}{\hbar}\left\{\frac{m}{2\varepsilon}\left(r_j^2 + r_{j-1}^2 - 2 r_j r_{j-1} \cos\psi_{j,j-1}\right) - \varepsilon V_j\right\}\right]. \quad (2.36)$$

It is left as an exercise to show that correct Schrödinger equation results when any of the above three forms of the propagator is used. This is another example to convince us that transformation of coordinates in a flat space does not change its character.

The form of the quantum correction ΔV may also derived by taking appeal to the Schrödinger equation in polar coordinates. Here the classical Lagrangian is

$$L = \frac{m}{2}(\dot{r}^2 + r^2\dot{\theta}^2 + r^2\sin^2\theta\,\dot{\phi}^2) - V(r,\theta,\phi) \quad . \tag{2.37}$$

The corresponding Schrödinger Hamiltonian operator has the form

$$\hat{H} = \frac{-\hbar^2}{2m}\Delta + V(r,\theta,\phi). \tag{2.38}$$

The Laplace-Beltrami operator is central to Schrödinger theory. In polar coordinates, this operator can be expressed as

$$\Delta = \frac{\partial^2}{\partial r^2} + \frac{2}{r}\frac{\partial}{\partial r} + \frac{1}{r^2}\hat{L}^2 \tag{2.39}$$

where the operator $\hat{L}^2$ has the standard form

$$\hat{L}^2 \equiv \left(\frac{\partial^2}{\partial\theta^2} + \cot\theta\frac{\partial}{\partial\theta}\right) + \frac{1}{\sin^2\theta}\frac{\partial^2}{\partial\phi^2}. \tag{2.40}$$

The quantum correction can be obtained when we write $\hat{H}$ in terms of the canonical momentum defined by

$$\hat{P}_r = -i\hbar\left(\frac{\partial}{\partial r} + \frac{1}{2r}\right)\ ;\ \hat{P}_\theta = -i\hbar\left(\frac{\partial}{\partial\theta} + \frac{\cot\theta}{2}\right)\ ;\ \hat{P}_\phi = -i\hbar\frac{\partial}{\partial\phi}, \tag{2.41}$$

that is,

$$\hat{H} = \frac{1}{2m}\left[P_r^2 + \frac{1}{r^2}P_\theta^2 + \frac{1}{r^2\sin^2\theta}P_\phi^2\right] + \Delta V(r,\theta,\phi) \tag{2.42}$$

where

$$\Delta V = -\hbar^2[1 + \mathrm{cosec}^2\theta\,]/8mr^2 \tag{2.43}$$

which has the same form as in Eq.(2.33).

Rigid rotor

In order to emphasize the role of the quantum correction, let us consider the case of a rigid rotor, a particle of mass m constrained to move on the surface of a sphere of radius R. The classical Lagrangian in this case reads as

$$L = \frac{mR^2}{2}(\dot{\theta}^2 + \sin^2\theta\,\dot{\phi}^2) \ . \tag{2.44}$$

It is easy to see from (2.43) that the quantum correction here is

$$\Delta V = -\hbar^2 [1 + \cosec^2\theta\,]/8mr^2 \ . \tag{2.45}$$

The propagator can be expressed as the following path integral

$$K = \left(\frac{mR^2}{2\pi i\hbar\varepsilon}\right)^N \int d\mu \prod_{j=1}^{N} \exp\left[\frac{i}{\hbar}\left\{\frac{m}{2\varepsilon}R^2[(\Delta\theta_j)^2 + \sin\theta_j\sin\theta_{j-1}(\Delta\phi_j)^2] - \varepsilon\Delta V_j\right\}\right]$$

$$d\mu = \prod_{j=1}^{N-1} \sin\theta_j\,d\theta_j\,d\phi_j \ . \tag{2.46}$$

We now try to replace the Lagrangian (2.44) by another one, $\bar{L}$, such that

$$L \to \bar{L} = \frac{m}{2}R^2\,\dot{\vec{\Omega}}^2 - (\Delta V)' \ . \tag{2.47}$$

The replacement presumably requires a correction term $(\Delta V)'$ to be determined and $\vec{\Omega}$ is a two-dimensional unit vector on the sphere. We have

$$(\Delta\vec{\Omega}_j)^2 = (\vec{\Omega}_j - \vec{\Omega}_{j-1})^2 = 2(1 - \cos\psi_{j,j-1}) \tag{2.48}$$

where by addition theorem

$$\cos\psi_{j,j-1} = \cos\theta_j\cos\theta_{j-1} + \sin\theta_j\sin\theta_{j-1}\cos(\phi_j - \phi_{j-1}). \tag{2.49}$$

Expanding $\cos(\Delta\theta_j)$ and $\cos(\Delta\phi_j)$ and retaining terms up to fourth order in $(\Delta\theta_j)$ and $(\Delta\phi_j)$, we may write

$$\frac{mR^2}{2\varepsilon}[(\Delta\theta_j)^2 + \sin\theta_{j-1}\sin\theta_j(\Delta\phi_j)^2] = \frac{mR^2}{\varepsilon}(1 - \cos\psi_{j,j-1}) - \varepsilon(\Delta V)' \tag{2.50}$$

where

$$(\Delta V)' = \hbar^2[1 + 1/(\sin\theta_j\sin\theta_{j-1})]/8mR^2 \ . \tag{2.51}$$

We notice here that $(\Delta V)' + (\Delta V) = 0$, that is, the quantum correction is exactly cancelled. It makes us write

$$K = \left(\frac{mR^2}{2\pi i\hbar\varepsilon}\right)^N \int d\mu \prod_{j=1}^{N} \exp\left[\frac{i}{\hbar}\left\{\frac{m}{\varepsilon}R^2(1 - \cos\psi_{j,j-1}) - \varepsilon V(\theta_j, \phi_j)\right\}\right]. \quad (2.52)$$

It is interesting to note here that the naive application of the rotor constraint

$$\left(\frac{m}{2\pi i\hbar}\right)^{1/2} e^{-\varepsilon V(r_j, \theta_j, \phi_j)} = \delta(r_j - R) \quad (2.53)$$

in Eq.(2.36) leads to the propagator (2.52). This is because of the accidental cancellation of the quantum correction when one implements the replacement (2.53). We have already seen that it fails in the two-dimensional case and in the next section we shall show that it fails also in the case of dimensions higher than three.

4.2.3. Generalization to d-dimensional polar coordinates

The treatment presented in the preceding sections can be easily extended to d-dimensional polar coordinates. We define the relation between the cartesian and polar coordinates in a d-dimensional space as follows:

$$x_j = r \prod_{k=1}^{j-1} \sin\theta_k \cos\theta_j \ ; j = 1, 2, \ldots d-1 \ ; \ x_d = r \prod_{k=1}^{d-1} \sin\theta_k \sin\phi$$

$$(2.54)$$

where for notational convenience we have set $\theta_0 = \pi/2$, $\theta_{d-1} = \phi$ and

$$0 \le \theta_k \le \pi \ ; (k = 1, 2\ldots, d-2), \quad 0 \le \phi \le 2\pi,$$

$$r = \left(\sum_{k=1}^{d} x_k^2\right)^{1/2}. \quad (2.55)$$

It is simpler to work out a quantum correction by taking recourse to the Schrödinger equation. First the classical Lagrangian in this case has the form

$$L = \frac{m}{2}\left[\dot{r}^2 + r^2\left(\sum_{k=1}^{d-1} \dot{\theta}_k^2 \prod_{j=0}^{k-1} \sin^2\theta_j\right)\right] - V(r, \{\theta_k\}). \quad (2.56)$$

The quantum Hamiltonian operator reads as

$$\hat{H} = \frac{-\hbar^2}{2m}\Delta_{(d)} + V \qquad (2.57)$$

where the Laplace-Beltrami operator has the form

$$\Delta_{(d)} = \frac{\partial^2}{\partial r^2} + \frac{d-1}{r}\frac{\partial}{\partial r} + \frac{1}{r^2} L^2_{(d)} . \qquad (2.58)$$

Here $L^2_{(d)}$ is the Legendre operator in d-dimensions :

$$L^2_{(d)} = \sum_{k=1}^{d-1} \left(\prod_{j=0}^{k-1} \operatorname{cosec}^2 \theta_j \right) \left[\frac{\partial^2}{\partial \theta_k^2} + (d-k-1) \cot \theta_k \frac{\partial}{\partial \theta_k} \right] . \qquad (2.59)$$

On the other hand, the canonical momenta are defined by the following hermitian operators :

$$p_r = -i\hbar [\partial/\partial r + (d-1)/2r]$$

$$p_{\theta_k} = -i\hbar [\partial/\partial \theta_k + \cot \theta_k (d-1-k)/2]$$

$$p_\phi = -i\hbar \partial/\partial \phi \qquad . \qquad (2.60)$$

If we express the operator $\hat{H}$ in terms of the canonical momenta, we have

$$\hat{H}(r,\{p_\theta\}) = \frac{1}{2m} p_r^2 + \frac{1}{2mr^2} p_\Omega^2 + V(r,\{\theta\}) + \Delta V(r,\{\theta\}) \qquad (2.61)$$

where the quantity p_Ω^2 is defined as

$$p_\Omega^2 = \sum_{k=1}^{d-1} \left(\prod_{j=0}^{k-1} \operatorname{cosec}^2 \theta_j \right) p_{\theta_k}^2 . \qquad (2.62)$$

The quantum correction is given by

$$\Delta V(r,\{\theta\}) = -\frac{\hbar^2}{8mr^2} \sum_{k=1}^{d-1} \left(\prod_{j=0}^{k-1} \operatorname{cosec}^2 \theta_j \right) , \qquad (2.63)$$

The propagator can then be written in the form

$$K^{(d)} = \lim_{N\to\infty} A_N (g'g'')^{-1/4} \int \prod_{j=1}^{N-1} dr_j \prod_{\nu=1}^{d-1} d\theta_\nu^{(j)} \prod_{j=1}^{N} \left(\tilde{g}^{(j)}\right)^{1/2} \exp\left\{\frac{i}{\hbar} S_j\right\}$$

(2.64)

where

$$S_j = \frac{m}{2\varepsilon}\left[(\Delta r_j)^2 + \tilde{r}_j^2 \sum_{k=1}^{d-1}\left(\prod_{l=1}^{k-1}\sin^2\tilde{\theta}_l^{(j)}\right)(\Delta\theta_k^{(j)})^2\right] - \varepsilon\,(V_j + \Delta V_j)$$

(2.65a)

$$\left(\tilde{g}^{(j)}\right)^{1/2} \equiv \tilde{r}_j^{d-1} \prod_{\nu=1}^{d-1}(\sin\tilde{\theta}_\nu^{(j)})^{d-1-\nu} \quad;\quad \Delta V_j = \Delta V(\tilde{r}_j, \{\tilde{\theta}^{(j)}\})$$

(2.65b)

and $A_N = (m/2\pi i\hbar)^{Nd/2}$.

As in three-dimensional case, it is possible to use geometric mean rather than the arithmetic mean, that is, replace $\tilde{r}_j^2$ by $r_j r_{j-1}$ and $\sin^2\tilde{\theta}_\nu^{(j)}$ by $\sin\theta_\nu^{(j)}\sin\theta_\nu^{(j-1)}$ in (2.64). The result is equivalent to (2.64) and may be easily written down.

The form (2.64) may also be easily derived by starting with the path integral in cartesian coordinates $\{x_k\}$ and making use of the transformation equations and the mid-point rule. In fact it is easy to verify that (2.64) is equivalent to (in the short-hand notation)

$$K^{(d)} = \int \mathcal{D}\,[r(t)]\,\mathcal{D}\,[\Omega(t)]\exp\left\{\frac{i}{\hbar}\int_{t'}^{t''}\left[\frac{m}{2}\dot{\vec{x}}^2 - V(r,\{\theta\})\right]dt\right\}$$

(2.66)

where in discretized form

$$\mathcal{D}\,[r(t)]\,\mathcal{D}\,[\Omega(t)] \longrightarrow A_N \prod_{j=1}^{N-1} r_j^{d-1}\,dr_j\,d\Omega^{(j)}\;,$$

(2.67a)

$$d\Omega^{(j)} = \prod_{\nu=1}^{d-1}(\sin\theta_\nu^{(j)})^{d-1-\nu}\,d\theta_\nu^{(j)}\;.$$

(2.67b)

Further the term $\dot{\vec{x}}^2$ can be expressed as

$$\dot{\vec{x}}^2 \longrightarrow (r_j^2 + r_{j-1}^2 - 2\,r_j r_{j-1}\cos\psi_{j,j-1})/\varepsilon^2$$

(2.68a)

$$\cos \psi_{j,j-1} = \sum_{\nu=0}^{d-2} \cos \theta_{\nu+1}^{(j)} \cos \theta_{\nu+1}^{(j-1)} \prod_{\mu=0}^{\nu} \sin \theta_{\mu}^{(j)} \sin \theta_{\mu}^{(j-1)}$$

$$+ \prod_{\nu=1}^{d-1} \sin \theta_{\nu}^{(j)} \sin \theta_{\nu}^{(j-1)} \quad . \tag{2.68b}$$

This is the form which is useful for actual computations. The quantum correction disappears in this form of the path integral. We will see some examples of explicit evaluation in the next section.

Rotor in d-dimensions

We now consider the motion of a particle on a curved manifold of S^{d-1} sphere. It is clear that in this case, the quantum Hamiltonian is simply given by

$$\hat{H} = -\frac{\hbar^2}{2mR^2} L^2_{(d)} \tag{2.69}$$

where R is the fixed radius of the sphere S^{d-1} and $L^2_{(d)}$ is defined in (2.59). The quantum correction now takes the form

$$\Delta V(\{\theta_\nu\}) = -\frac{\hbar^2}{8mR^2} \left[(d-2)^2 + \frac{1}{\sin^2\theta_1} + \ldots + \frac{1}{\sin^2\theta_1 \ldots \sin^2\theta_{d-2}} \right] . \tag{2.70}$$

The classical Lagrangian

$$L \equiv \frac{mR^2}{2} [\dot{\theta}_1^2 + \sin^2\theta_1 \dot{\theta}_2^2 + \ldots + \sin^2\theta_1 \ldots \sin^2\theta_{d-2} \dot{\phi}^2] . \tag{2.71}$$

We can use either the mid-point prescription or the geometric mean prescription to write the appropriate propagator. Using, for instance, the geometric mean prescription and using the short-hand notation, we obtain the following Lagrangian path integral

$$K(\{\theta''\},\{\theta'\};T) = \int \mathcal{D} [\Omega(t)] \exp\left\{\frac{i}{\hbar} \int_{t'}^{t''} \left[L(\{\theta,\dot\theta\}) - \Delta V(\{\theta\}) \right] dt \right\} \tag{2.72}$$

where the path differential measure takes the form

$$\mathcal{D} [\Omega(t)] \longrightarrow \left(\frac{m R^2}{2\pi i \hbar \epsilon} \right)^{N(d-1)/2} \prod_{j=1}^{N-1} d\Omega^{(j)} \tag{2.73}$$

and $d\Omega^{(J)}$ is the (d-1) dimensional surface element on the unit sphere S^{d-1} as defined (2.67).

As in the case of two and three-dimensions, we convert the path integral (2.72) into another one by replacing the original Lagrangian L by another Lagrangian $\bar{L}$ that is simpler to handle. This introduces a correction factor $(\Delta V)'$. Thus

$$L \longrightarrow \bar{L} = \frac{m}{2} R^2 \dot{\vec{\Omega}}^2 - (\Delta V)' \qquad (2.74)$$

where Ω is the d-dimensional unit vector on the S^{d-1} sphere. Now

$$(\Delta \vec{\Omega}^{(J)})^2 = (\vec{\Omega}^{(J)} - \vec{\Omega}^{(J-1)})^2 = 2(1 - \cos \psi_{J,J-1}). \qquad (2.75)$$

Using the addition theorem (2.68), and after some careful but straightforward algebra (which follows the same pattern as in Sec.2.2) we can establish the equivalence

$$\exp[\frac{i}{\hbar} L] = \exp[\frac{i}{\hbar} \{mR^2(1 - \cos \psi_{J,J-1}) - \varepsilon (\Delta V)'\}] \qquad (2.76)$$

where the correction term has the explicit form

$$(\Delta V)' = \frac{\hbar^2}{8mR^2} \left[1 + \sum_{k=1}^{d-2} \prod_{l=1}^{k} \left(\sin \theta_l^{(J)} \sin \theta_l^{(J-1)} \right)^{-1} \right]. \qquad (2.77)$$

The correction term (apart from the sign) is the same as (ΔV) in equation (2.63) and may be obtained from the latter by letting r = R. From (2.70), (2.77) we see that

$$\Delta V + (\Delta V)' = - \frac{\hbar^2}{8mR^2} (d-1)(d-3). \qquad (2.78)$$

Hence the final form of the path integral for the d-dimensional rotor reads as

$$K = \int \mathcal{D}[\Omega(t)] \exp \left\{ \frac{i}{\hbar} \int_{t'}^{t''} \left[\frac{m}{2} R^2 \dot{\vec{\Omega}}^2 + \frac{(d-1)(d-3)\hbar^2}{8mR^2} \right] dt \right\}. \qquad (2.79)$$

Note that for d = 3, the quantum correction vanishes. It is also of the right sign and magnitude for d = 2. Thus, if one wants to obtain the path integral for a d-dimensional rotor from the path integral (2.66) in polar coordinates, the correct replacement is

$$\left(\frac{m}{2\pi i\hbar\varepsilon}\right)^{1/2} \exp\left[\frac{-i\varepsilon}{\hbar} V_j\right] \longrightarrow \delta(r_j - R) \exp\left[\frac{i\varepsilon\hbar^2(d-1)(d-3)}{8mR^2}\right]. \quad (2.80)$$

4.3. Examples of Explicit Evaluation

We list in this section some examples where the path integrals developed in Sec.2 may be explicitly evaluated to yield exact analytical results.

4.3.1. Free particle propagator

Consider first the path integral (2.22) with $V \equiv 0$. This represents a free particle moving in a two-dimensional space. We use the formula

$$\exp(z \cos \psi) = \sum_{k=-\infty}^{\infty} I_k(z) e^{ik\psi} \quad (3.1)$$

where I_k is the modified Bessel function, to write

$$\exp\left[-\frac{im\, r_j r_{j-1}}{\hbar\varepsilon} \cos(\phi_j - \phi_{j-1})\right] = \sum_{k=-\infty}^{\infty} I_k\left(-\frac{im\, r_j r_{j-1}}{\hbar\varepsilon}\right) e^{ik(\phi_j - \phi_{j-1})}$$

in the path integral. The integrations over intermediate angular variables can be performed readily using the orthogonality property of the functions $\{e^{ik\phi}\}$ over the interval $(0, 2\pi)$. The propagator then looks like

$$K(r'', \phi'', r', \phi'; T) = \sum_{l=-\infty}^{\infty} K_l(r'', r'; T) e^{il(\phi'' - \phi')} \quad (3.2)$$

where the radial propagator K_l reads as

$$K_l = \lim_{N \to \infty} A_N (2\pi)^N \int \prod_{j=1}^{N-1} r_j dr_j \prod_{j=1}^{N} \exp\left[\frac{im}{2\hbar\varepsilon}(r_j^2 + r_{j-1}^2)\right] I_l\left(-\frac{im\, r_j r_{j-1}}{\hbar\varepsilon}\right). \quad (3.3)$$

Integrations over r_j in (3.3) may be carried out using another formula

$$\int_0^\infty e^{i\alpha x^2} I_\nu(-ibx) I_\nu(-icx) \, x \, dx = \frac{i}{2\alpha} e^{-i(b^2+c^2)/4\alpha} I_\nu\left(\frac{-ibc}{2\alpha}\right) \quad (3.4)$$

valid for $Re(\alpha) > 0$, $Re(\nu) > -1$. This leads to the final result

$$K_1 = \frac{m}{i\hbar T} \exp\left\{\frac{im}{2\hbar\varepsilon}(r''^2 + r'^2)\right\} I_1\left(\frac{mr''r'}{i\hbar T}\right). \quad (3.5)$$

Indeed, we can check by inserting (3.5) in (3.2) and summing over l (using the reverse of formula (3.1)) that we arrive at the free particle propagator

$$K(r'', r'; T) = \left(\frac{m}{2\pi i\hbar T}\right) \exp\left\{\frac{im}{2\hbar T}(\vec{r}'' - \vec{r}')^2\right\} \quad (3.6)$$

originally derived by Feynman.

In d-dimensional polar coordinates, the derivation of the radial propagator runs similar. We separate the radial part from the angular part by making use of another formula

$$e^{z\cos\psi} = (z/2)^{-\nu} \Gamma(\nu) \sum_{l=0}^\infty (l+\nu) I_{l+\nu}(z) C_l^\nu(\cos\psi) \quad (3.7)$$

where C_l^ν are Gegenbauer polynomials. In our case $\nu = (d-2)/2$. For $d = 2$, this formula leads to (3.1). On the other hand for $d = 3$, $\nu = 1/2$, $C_l^{1/2}(\cos\psi) = P_l(\cos\psi)$ (Legendre polynomial) and (3.7) reduces to another familiar formula

$$e^{z\cos\psi} = \sqrt{(\pi/2z)} \sum_{l=0}^\infty (2l+1) I_{l+1/2}(z) P_l(\cos\psi). \quad (3.8)$$

Now if $\psi_{j,j-1}$ is the angle between two d-dimensional unit vectors $\vec{\Omega}^{(j-1)}$ and $\vec{\Omega}^{(j)}$ the following addition theorem applies

$$\sum_{\mu=1}^M S_l^\mu(\vec{\Omega}^{(j-1)}) S_l^\mu(\vec{\Omega}^{(j)}) = \frac{\Gamma(d/2)}{2\pi^{d/2}} \frac{(2l+d-2)}{(d-2)} C_l^{(d-2)/2}(\cos\psi_{j,j-1}) \quad (3.9)$$

where $S_l^\mu(\vec{\Omega})$ are the real hyperspherical harmonics of degree l associated with unit vector $\vec{\Omega}$. $l = 0, 1, 2, \ldots, \infty$ while $\mu = 1, 2, \ldots, M$, with

$$M = \frac{(2\mathfrak{l} + d - 2)(\mathfrak{l} + d - 3)!}{\mathfrak{l}!(d - 2)!} . \tag{3.10}$$

The functions $S_{\mathfrak{l}}^{\mu}(\Omega)$ satisfy the orthonormality condition

$$\int d\Omega \ S_{\mathfrak{l}}^{\mu}(\Omega) \ S_{\mathfrak{l}'}^{\mu'}(\Omega) = \delta_{\mathfrak{l}\mathfrak{l}'} \delta_{\mu\mu'} . \tag{3.11}$$

For $d = 3$, the formula (3.9) reduces to

$$\sum_{m=-\mathfrak{l}}^{\mathfrak{l}} Y_{\mathfrak{l}m}^{*}(\vec{\Omega}^{(j-1)}) \ Y_{\mathfrak{l}m}(\vec{\Omega}^{(j)}) = \frac{2\mathfrak{l}+1}{4\pi} P_{\mathfrak{l}}(\cos \psi_{j-1,j}) \tag{3.12}$$

where $Y_{\mathfrak{l}m}(\vec{\Omega})$ are the usual spherical harmonics. Using (3.7) and (3.9) in the path integral (2.65) and performing the angular integrations with the help of the relation (3.11), we obtain

$$K^{(d)} = \frac{\Gamma(d/2)}{2\pi^{d/2}} \sum_{\mathfrak{l}=0}^{\infty} \frac{(2\mathfrak{l}+d-2)}{(d-2)} C_{\mathfrak{l}}^{(d-2)/2} (\cos \psi_{0,N}) \ K_{\mathfrak{l}}^{(d)}(r'',r';T) \tag{3.13}$$

where the radial propagator is defined by the path integral

$$K_{\mathfrak{l}} = (r'r'')^{-(d-1)/2} \lim_{N \to \infty} \left(\frac{m}{i\hbar\varepsilon}\right)^{N/2} \int_{0}^{\infty} \prod_{j=1}^{N-1} dr_{j} \prod_{j=1}^{N} \mathcal{R}_{j,j-1} \tag{3.14}$$

where

$$\mathcal{R}_{j,j-1} = \left(\frac{mr_{j}r_{j-1}}{i\hbar\varepsilon}\right)^{1/2} I_{\mathfrak{l}+(d-2)/2}\left(\frac{mr_{j}r_{j-1}}{i\hbar\varepsilon}\right) \exp\left\{\frac{im}{2\hbar\varepsilon}(r_{j}^{2} + r_{j-1}^{2})\right\} . \tag{3.15}$$

The integrations over $\{r_{j}\}$ may be performed as before using the formula (3.4) to arrive at the compact expression

$$K_{\mathfrak{l}} = \left(\frac{m}{i\hbar T}\right)(r''r')^{(2-d)/2} \exp\left\{\frac{im}{2\hbar T}(r''^{2} + r'^{2})\right\} I_{\mathfrak{l}+(d-2)/2}\left(\frac{mr''r'}{i\hbar T}\right) . \tag{3.16}$$

Inserting (3.16) in (3.13) and summing over $\mathfrak{l}$ leads to the usual free particle propagator.

4.3.2. Rotor in d-dimensions

For a particle moving on the surface of sphere S^{d-1} in d-dimensions, one has to use the expression (2.79) for deriving the propagator. Written in discretized form

$$K = \lim_{N \to \infty} \left(\frac{mR^2}{2\pi i\hbar\varepsilon} \right)^{N(d-1)/2} C(R) \int \prod_{j=1}^{N-1} d\Omega^{(j)} \exp\left[\frac{imR^2}{\hbar\varepsilon} \sum_{j=1}^{N} (1 - \cos\psi_{j-1,j}) \right]$$

(3.17)

where the quantity $C(R)$ has the form

$$C(R) = \exp\left\{ \frac{i\hbar T(d-1)(d-3)}{8mR^2} \right\}.$$ (3.18)

The next step is to use (3.7) and (3.9), and carry out the angular integrations successively with the help of the orthogonality relations (3.11). The result may be expressed in the following form

$$K = \frac{\Gamma(d/2)}{2\pi^{d/2}} \sum_{l=0}^{\infty} \frac{(2l+d-2)}{(d-2)} C_l^{(d-2)/2} (\cos\psi_{0,N}) K_l^{(d)}(R;T) \quad (3.19)$$

where

$$K_l^{(d)}(R;T) = \lim_{N \to \infty} \left[\left(\frac{2\pi mR^2}{\iota\varepsilon} \right)^{1/2} e^{imR^2/\hbar\varepsilon} I_{l+(d-2)/2}\left(\frac{mR^2}{i\hbar\varepsilon} \right) \right]^N C(R). \quad (3.20)$$

In order to evaluate the limit, we make use of the asymptotic expansion

$$I_\nu(z) \approx (2\pi z)^{-1/2} \exp\{z - (\nu^2 - \tfrac{1}{4})/2z\} \quad (3.21)$$

for the modified Bessel function (valid for $|z| \to \infty$, $|\arg z| < \pi/2$). Letting $z = -imR^2/\hbar\varepsilon$ along with $N \to \infty$, $\varepsilon \to 0$ such that $\varepsilon N = T$, the limit in question leads to

$$K_l^{(d)}(R;T) = \exp\left\{ -\frac{i\hbar T\, l(l+d-2)}{2mR^2} \right\}. \quad (3.22)$$

This results in the final expression

$$K = \sum_{l=0}^{\infty} \sum_{\mu=1}^{M} S_l^\mu(\vec{\Omega}'') S_l^\mu(\vec{\Omega}') \exp\left\{ -\frac{i\hbar T}{2mR^2} l(l+d-2) \right\}. \quad (3.23)$$

This leads to the correct energy values $E_l = \hbar^2 l(l+d-2)/2mR^2$, and the wave functions $\psi_l^\mu = S_l^\mu(\vec{\Omega})$ of the corresponding Schrödinger equation.

4.3.3. Central potentials

For potentials of the form $V = V(r)$, all the steps used in the case of a free particle propagator go through. The radial propagator is still given by Eq.(3.14) except that the function $\mathcal{R}_{j,j-1}$ gets slightly modified. The exponential factor in the defining relation (3.15) for $\mathcal{R}_{j,j-1}$ gets replaced by

$$\exp\left\{\frac{i}{\hbar}\left[\frac{m}{2\varepsilon}(r_j^2 + r_{j-1}^2) - \varepsilon V(r_j)\right]\right\}. \qquad (3.24)$$

The integrations over r_j can be performed readily for potential of the general form

$$V(r) = \frac{1}{2}\omega^2 r^2 + g/r^2. \qquad (3.25)$$

This corresponds in fact to d-dimensional isotropic harmonic oscillator perturbed by an inverse square potential. In general ω may depend on time t but g is a constant. For simplicity we assume ω to be a constant and introduce a constant "a" such that $g = a^2\hbar^2/2m$. Now, the potential $V(r_j)$ in (3.24) is replaced by

$$V(r_j) = \frac{1}{2}\omega^2 r_j^2 + \frac{a^2\hbar^2}{2mr_j r_{j-1}} \qquad (3.26)$$

which differs from the form given in Eq.(3.25) by an order ε. Writing $z = mr_j r_{j-1}/i\hbar\varepsilon$ and noting that as $\varepsilon \to 0$ we may use the asymptotic formula (3.21) to express the radial propagator in the new form

$$K_1 = \lim_{N\to\infty} \mathcal{A}\,(-i\beta)^N \int_0^\infty \prod_{j=1}^{N-1} r_j dr_j\, e^{i\alpha_j r_j^2} \prod_{j=1}^{N} I_\mu(-i\beta r_j r_{j-1}) \qquad (3.27)$$

where the constants $\mathcal{A} = (r'r'')^{-(d-2)/2} e^{i\beta(r'^2+r''^2)/2}$, $\alpha_j = \beta(1-\omega^2\varepsilon^2/2)$, $\beta = m/\hbar\varepsilon$ and $\mu = \{[1+ (d-2)/2]^2 + 2mg/\hbar^2\}^{1/2}$. The integrations in (3.27) can be performed successively using the formula (3.4) yielding the result

$$K_1 = -i(r'r'')^{-(d-2)/2} \lim_{N\to\infty} a_N\, e^{i(p_N r''^2 + q_N r'^2)}\, I_\mu(-ia_N r''r'). \qquad (3.28)$$

Here the quantities a_N, p_N, q_N are defined by

$$a_N = \beta \prod_{j=1}^{N-1} (\beta/2\gamma_j) \ ; \ p_N = \frac{\beta}{2} - \sum_{j=1}^{N-1} (\beta_j^2/4\gamma_j) \ ; \ q_N = \frac{\beta}{2} - \frac{\beta^2}{4\gamma_{N-1}} \quad (3.29)$$

while for $2 \leq j \leq N-1$,

$$\gamma_1 = \alpha_1 \ , \ \gamma_j = \alpha_j - \beta^2/4\gamma_{j-1} \ , \quad (3.30)$$

$$\beta_1 = \beta \ , \ \beta_j = \beta \prod_{k=1}^{j-1} (\beta/2\gamma_k). \quad (3.31)$$

Thus the task of evaluating the propagator K_1 reduces to the computation of a_N, p_N, and q_N and taking the limit in (3.28). It is convenient for this purpose to introduce a new quantity Q_j such that

$$Q_{j+1} Q_j^{-1} = 2\beta^{-1} \gamma_j = 2\beta^{-1} \alpha_j - \beta Q_{j-1} Q_j^{-1} \quad (3.32)$$

which can be written in more suggestive form after inserting the definitions of α_j and β:

$$(Q_{j+1} - 2Q_j + Q_{j-1})/\varepsilon^2 = -\omega^2 Q_j \ . \quad (3.33)$$

This difference equation reduces in the limit $\varepsilon \to 0$ to the differential equation

$$\ddot{Q}(t) + \omega^2 Q(t) = 0 \quad (3.34)$$

with the initial condition $Q(t') = 0$. The limits of the various quantities may be expressed in terms of the solution $Q(t)$ of Eq. (3.34). It is easy to see that

$$\lim_{N \to \infty} a_N = \frac{m \dot{Q}(t')}{\hbar Q(t'')} \ ; \ \lim_{N \to \infty} p_N = \lim_{N \to \infty} q_N = \frac{m \dot{Q}(t'')}{2 \hbar Q(t'')} \ . \quad (3.35)$$

Inserting these limits in Eq. (3.28), we have→

$$K_1 = \left(\frac{m}{i\hbar}\right) (r'r'')^{(2-d)/2} \left(\frac{\dot{Q}'}{Q''}\right) \exp\left[\frac{im}{2\hbar} \frac{\dot{Q}''}{Q''}(r''^2 + r'^2)\right] I_\mu\left[\frac{m\dot{Q}'}{i\hbar Q''}(r''r')\right]. \quad (3.36)$$

This general expression is valid even when the oscillator frequency varies with time. For constant frequency, we can simplify the result further. In this case, the solution of the oscillator equation with the initial condition $Q(t') = 0$, is given by $Q(t) = C \sin \omega(t-t')$, where C is a constant. This leads to the following expression

$$K_1 = (r'r'')^{(2-d)/2} \left(\frac{m\omega}{i\hbar \sin \omega T}\right)^{d/2} \exp\left[\frac{im\omega}{2\hbar}(r'^2 + r''^2)\cot\omega T\right]$$

$$\times I_\mu\left[\frac{mr'r''\omega}{i\hbar \sin \omega T}\right] \quad (3.37)$$

where $\mu = \{[1 + (d-2)/2]^2 + 2mg/\hbar^2\}^{1/2}$.

Problem 3.1

Starting from Eq.(3.14) (with $R_{j,j-1}$ modified according to Eq.(3.24)) derive the expression for K_1 for the case (a) $V = \omega^2 r^2/2$ and (b) $V = g/r^2$. Further show that in one-dimensional case of (b), the condition $g > -\hbar^2/8m$ is necessary.

4.3.4. Non-spherically symmetric potentials

We present here a few cases of non-spherically symmetric potentials. Consider, for example, the path integral (2.22) in polar coordinates and let

$$V(r,\theta) = g(r^2 \sin^2 \theta)^{-1}, \quad 0 < \theta < \pi. \quad (3.38)$$

Writing for brevity $W_j = m(r_j^2 + r_{j-1}^2)/2\varepsilon$, we express

$$S_j = W_j - \frac{m}{\varepsilon} r_j r_{j-1} \cos(\theta_j - \theta_{j-1}) - \frac{\hbar^2(a^2 - 1/4)\varepsilon}{2mr_j r_{j-1} \sin \theta_j \sin \theta_{j-1}}$$

which implies

$$\exp\left[\frac{i}{\hbar} S_j\right] = \exp\left[\frac{i}{\hbar} W_j\right] \exp\left[\frac{m}{i\hbar\varepsilon} r_j r_{j-1} \cos \theta_j \cos \theta_{j-1}\right]$$

$$\times \exp\left[\frac{mr_j r_{j-1}}{i\hbar\varepsilon} \sin \theta_j \sin \theta_{j-1} - \frac{i\hbar(a^2 - 1/4)\varepsilon}{2mr_j r_{j-1} \sin \theta_j \sin \theta_{j-1}}\right] \quad (3.39)$$

where for convenience we have introduced $a = (1 + 8mg/\hbar^2)/2$. The last exponential can now be replaced by an appropriate Bessel function to $o(\varepsilon)$ by means of the asymptotic formula (3.21). Thus

$$\exp[\tfrac{i}{\hbar} S_j] = \left(\frac{2\pi m}{i\hbar\varepsilon} r_j r_{j-1} \sin\theta_j \sin\theta_{j-1}\right)^{1/2} \exp[\tfrac{i}{\hbar} W_j]$$

$$\times \exp\left(\frac{m}{i\hbar\varepsilon} r_j r_{j-1} \cos\theta_j \cos\theta_{j-1}\right) I_a\left(\frac{m}{i\hbar\varepsilon} r_j r_{j-1} \sin\theta_j \sin\theta_{j-1}\right),$$

(3.40)

In order to carry out the integration over angles θ_j, it is necessary to use the expansion formula

$$(\sin\alpha \sin\beta)^{1/2-\lambda} I_{\lambda-1/2}(z \sin\alpha \sin\beta) \exp(z \cos\alpha \cos\beta)$$

$$= 2^{2\lambda} \frac{[\Gamma(\lambda)]^2}{\sqrt{2\pi z}} \sum_{l=0}^{\infty} \frac{\Gamma(l+1)(\lambda+l)}{\Gamma(2\lambda+l)} I_{l+\lambda}(z) C_l^\lambda(\cos\alpha) C_l^\lambda(\cos\beta) \quad (3.41)$$

where $C_l^\lambda(\cos\theta)$ are Gegenbauer polynomials obeying the orthogonality property

$$\int_0^\pi d\theta (\sin\theta)^{2\lambda+1} C_l^\lambda(\cos\theta) C_{l'}^\lambda(\cos\theta) = \delta_{ll'}/N_{l,\lambda}^2 \quad (3.42a)$$

where the normalization constant $N_{l,\lambda}^2$ is given by

$$N_{l,\lambda}^2 = \frac{2^{2\lambda}}{\pi} \frac{[\Gamma(\lambda)]^2 \Gamma(l+1)(l+\lambda)}{\Gamma(2\lambda+l)}. \quad (3.42b)$$

Using the orthogonality properties of Gegenbauer polynomials the angular integrations are carried out successively. In our case $\lambda = a + 1/2$ and the propagator reads as

$$K = \sum_{l=0}^{\infty} K_l N_{l,(a+1/2)}^2 (\sin\theta'' \sin\theta')^{a+1/2} C_l^{a+1/2}(\cos\theta'') C_l^{a+1/2}(\cos\theta').$$

(3.43)

The radial propagator K_l now reads as

$$K_1(r'',r';T) = \left(\frac{m}{i\hbar T}\right) \exp\left\{\frac{im}{2\hbar T}(r''^2 + r'^2)\right\} I_{1+a+1/2}\left(\frac{mr'r''}{i\hbar T}\right). \qquad (3.44)$$

Problem 3.2

Derive the form of the radial propagator when

$$V = \frac{1}{2}\omega^2 r^2 + g(r^2 \sin^2\theta)^{-1}, \qquad (0 < \theta < \pi). \qquad (3.45)$$

Problem 3.3

Show that the above derivation may be used to obtain also the propagator when

$$V = g(r^2 \sin^2 c\theta)^{-1}, \qquad 0 < \theta < \pi/c \qquad (3.46)$$

Hint : Express $\cos(\Delta\theta_j)$ in terms of $\cos[c(\Delta\theta_j)]$ and follow the derivation above.

Problem 3.4

Show that the propagator for the potential

$$V = g(r^2 \sin^2\theta)^{-1} + f(r^2 \cos^2\theta)^{-1}, \qquad 0 < \theta < \pi \qquad (3.47)$$

has the form

$$K = \sum_{l=0}^{\infty} K_1 N_l^2 (\sin\theta'' \sin\theta')^{a+1/2} (\cos\theta'' \cos\theta')^{b+1/2}$$

$$\times P_l^{a,b}(\cos\theta'') P_l^{a,b}(\cos\theta') \qquad (3.48)$$

where K_1 has the same form as in Eq.(3.44) with the order of the Bessel function given by $(2l + a + b + 1)$ instead of $(l + a + 1/2)$. Generalize this to the case where θ is replaced by $c\theta$ with $0 < \theta < \pi/c$ as in problem (3.3).

Hint : Use the expansion theorem

$$\frac{z}{2} I_\mu(z \cos\alpha \cos\beta) I_\nu(z \sin\alpha \sin\beta) (\cos\alpha \cos\beta)^\mu (\sin\alpha \sin\beta)^\nu$$

$$= \sum_{l=0}^{\infty} (-1)^l (\mu+\nu+2l+1) I_{2l+\mu+\nu+1}(z) P_l^{\mu,\nu}(\cos\alpha) P_l^{\mu,\nu}(\cos\beta) \qquad (3.49)$$

where $P_l^{\mu,\nu}(\cos\alpha)$ are Jacobi polynomials with the orthogonality property

$$\int_0^\pi (\sin\theta/2)^\alpha (\cos\theta/2)^\beta P_l^{\mu,\nu}(\cos\theta) P_{l'}^{\mu,\nu}(\cos\theta) d\theta = \frac{\delta_{ll'}}{N_l^2} \quad (3.50)$$

where $N_l^2 = (-1)^l (2l+\mu+\nu+1)$.

Problem 3.5

Use the path integral in polar coordinates to derive explicitly the propagator for a harmonically bound charged particle of mass m and charge e in a uniform magnetic field $\vec{B}$.

Hint: Resolve the motion into two parts, one along the direction of the magnetic field and the other in a plane perpendicular to the field. It turns out that

$$K = \frac{m\Omega'}{2\pi i\hbar \sin\Omega'T} \exp\left[\frac{im\Omega'}{2\pi i\hbar}\left((r''^2 + r'^2)\cot(\Omega'T) - 2r'r'' \frac{\cos(\Delta\phi)}{\sin(\Omega'T)}\right)\right] K(z)$$

(3.51)

where $K(z)$ is the usual propagator for an oscillator moving along z-axis. Further the quantity $\Delta\phi = (\theta'' - \theta' + \omega T)$ and $\Omega' = (\Omega^2 + \omega^2)^{1/2}$. Here $\omega = eB/2m$ is the Larmor frequency and Ω is the frequency of harmonic motion.

We might mention here that an alternative approach known as the "unconventional dimensional extension" has been used in literature to path integrate some non-central potentials. The reader may consult the references cited at the end of this chapter.

4.4. Path Integration in General Curved Spaces

4.4.1. Quantization of classical Hamiltonian

Consider a dynamical system with the general Lagrangian of the form

$$L = \frac{1}{2} g_{ij} \dot{q}^i \dot{q}^j + b_i \dot{q}^i - V \quad (4.1)$$

where the metric g_{ij}, the coefficients b_i and the potential V are functions of the coordinates q^i ($i = 1, 2, \ldots, n$). These quantities transform like a covariant tensor, a covariant vector and a scalar respectively. For simplicity, the usual summation convention is assumed, that is, repeated index implies a sum over the index. The Lagrangian (4.1) represents the motion of a non-relativistic particle in a curved space of metric g_{ij}.

We first write the Hamiltonian corresponding to (4.1). The generalized canonical momenta have the form

$$p_i = \partial L/\partial \dot{q}^i = g_{ij} \dot{q}^j + b_i \qquad (4.2)$$

and hence the Hamiltonian takes the form

$$H = p_i \dot{q}^i - L = \frac{1}{2} g^{ij}(p_i - b_i)(p_j - b_j) + V. \qquad (4.3)$$

Here (g^{ij}) is the inverse of the matrix (g_{ij}).

Before we proceed further, let us briefly review the procedure of quantizing the classical Hamiltonian (4.3). In quantum mechanics the coordinates and momenta are hermitian operators and are denoted by $\hat{q}_i$ and $\hat{p}_i$ respectively. In Schrödinger representation, we introduce the eigenstates $|q,t\rangle$ and $|p,t\rangle$ of these operators, so that

$$\hat{q}^i(t)|q,t\rangle = q^i|q,t\rangle, \qquad \hat{p}_i(t)|p,t\rangle = p_i|p,t\rangle. \qquad (4.4)$$

Here $q \equiv (q^1, \ldots, q^n)$ and $p \equiv (p_1, \ldots, p_n)$. These operators satisfy the commutation relations

$$[\hat{q}^i(t), \hat{q}^j(t)] = [\hat{p}_i(t), \hat{p}_j(t)] = 0 \quad [\hat{q}^i(t), \hat{p}_j(t)] = i\hbar \delta^i_j \qquad (4.5)$$

while the eigenstates $|q,t\rangle$ are assumed to form a complete orthonormal set, that is,

$$\langle q'', t | q', t \rangle = g^{-1/2}(q') \delta(q'' - q'),$$

$$\int g^{1/2}(q) \, d^n q \, |q,t\rangle\langle q,t| = 1. \qquad (4.6)$$

Here $g(q)$ is the determinant of the metric tensor and $\delta(q)$ is the n-dimensional Dirac delta function. Note that the invariant volume element is $g^{1/2}(q)dq^1 \ldots dq^n$. The eigenvectors of the momentum operator in the coordinate representation have the plane wave structure, that is,

$$\langle q,t|p,t\rangle = \langle q|p\rangle = (2\pi\hbar)^{-n/2}(f(p)g(p))^{-1/4} \exp(\frac{1}{\hbar}(p.q)) \qquad (4.7)$$

where $(p.q) = p_i q^i$ and $f(p)$ is an arbitrary function chosen to define $[f(p)]^{1/2} d^n p$ as the volume element in the momentum space.

Equation (4.7) helps us to obtain the matrix elements of any function of momentum operators. Thus

$$\langle q',t'|(\hat{p}_i)^r|q,t\rangle = \int [f(p)]^{1/2} d^n p \, \langle q'|(\hat{p}_i)^r|p\rangle\langle p|q\rangle$$

$$= [g(q)g(q')]^{-1/2} \int (2\pi\hbar)^{-n} dp (p_i)^r \exp[\frac{i}{\hbar} p.(q'-q)]. \qquad (4.8)$$

The coordinate representation of the momentum operator $\hat{p}_i$ follows from its hermitian character and the commutation relations. Thus

$$\hat{p}_i = -i\hbar(\partial_i + \frac{1}{2}\Gamma_i), \quad \Gamma_i = \Gamma_{ki}^k = \partial_i \ln \sqrt{g} \qquad (4.9)$$

with the notation $\partial_i = \partial/\partial q^i$. Γ_{ij}^k is the affine connection or the Christoffel symbol of the second kind defined by

$$\Gamma_{ij}^k = \frac{1}{2} g^{kl}(\partial_i g_{jl} + \partial_j g_{il} - \partial_l g_{ij}). \qquad (4.10)$$

The quantum dynamics is now described by the time-dependent Schrödinger equation

$$i\hbar \frac{\partial \psi(q,t)}{\partial t} = \hat{H} \psi(q,t) \qquad (4.11)$$

where $\hat{H}$ is the quantum Hamiltonian. Since $\hat{H}$ is hermitian the norm of the wavefunction

$$\langle \psi|\psi\rangle = \int g^{1/2}(q) \, d^n q \, |\psi(q,t)|^2. \qquad (4.12)$$

Since we are considering the non-relativistic case (with no non-classical degrees of freedom like spin), the wavefunction is a scalar. The next question is: what is the form of $\hat{H}$? However, in absence of experimental evidence we have no unambiguous answer. A natural assumption to make is that $\hat{H}$ is invariant under general coordinate transformations. If, in addition we assume that the Hamiltonian operator in a curved space has the same form as in a flat space, then the quantum Hamiltonian of the system takes the form

$$\hat{H} = -\frac{\hbar^2}{2}\Delta + i\hbar b^i \partial_i + \frac{i\hbar}{2} g^{-1/2} \partial_i (g^{1/2} b^i) + \frac{1}{2} b^i b_i + V. \qquad (4.13)$$

Here Δ is the covariant Laplace-Beltrami operator

$$\Delta = g^{-1/2} \partial_i (g^{1/2} g^{ij} \partial_j) \qquad (4.14)$$

and $b^i = g^{ij} b_j$. The expression (4.13) can now be cast in the following operator form with the help of (4.9),

$$\hat{H} = g^{-1/4}(\hat{q})(\hat{p}_i - b_i(\hat{q})) g^{1/2}(\hat{q}) g^{ij}(\hat{q})(\hat{p}_j - b_j(\hat{q})) g^{-1/4}(\hat{q})/2 + V(\hat{q}).$$

$$(4.15)$$

This form of the Hamiltonian corresponds to a particular ordering of $\hat{q}$'s and $\hat{p}$'s.

We now introduce another prescription known as the Weyl ordering rule. This rule is as follows. Consider a classical quantity $q^m p^n$. The Weyl ordered quantal expression $(\hat{q}^m \hat{p}^n)_W$ is just the arithmetic mean of all the possible orderings of the factors $\hat{q}$'s and $\hat{p}$'s, each different ordering being counted once. For example

$$(F(\hat{q})\hat{p})_W = [F(\hat{q})\hat{p} + \hat{p}F(\hat{q})]/2 \qquad (4.16)$$

$$(F(\hat{q}) \hat{p}^2)_W = [F(\hat{q}) \hat{p}^2 + 2\hat{p} F(\hat{q}) \hat{p} + \hat{p}^2 F(\hat{q})]/4 \qquad (4.17)$$

and in particular, it is easy to show that

$$(\hat{q}^m \hat{p}^n)_W = (1/2)^m \sum_{l=0}^{m} {}^m C_l \, \hat{q}^{m-l} \hat{p}^n \hat{q}^l. \qquad (4.18)$$

We now construct the Weyl ordered operator corresponding to (4.3). First note that

$$(b^i(\hat{q}) b_i(\hat{q}))_w = b^i(\hat{q}) b_i(\hat{q}); \qquad (V(\hat{q}))_w = V(\hat{q}) \qquad (4.19)$$

since there are no factors involving momentum operators $\hat{p}_i$. On the other hand, it is easy to verify that

$$[g^{ij}(\hat{q})b_i(\hat{q})\hat{p}_j]_w = [g^{ij}(\hat{q})b_i(\hat{q})\hat{p}_j + \hat{p}_j g^{ij}(\hat{q})b_i(\hat{q})]/2$$

$$= -i\hbar [b^i \partial_i + g^{-1/2}\partial_i(g^{1/2}b^i)/2] \qquad (4.20)$$

and that

$$(g^{ij}(\hat{q})\hat{p}_i\hat{p}_j)_w = \frac{1}{4}\left(\hat{p}_i\hat{p}_j g^{ij}(\hat{q}) + 2\hat{p}_i g^{ij}(\hat{q})\hat{p}_j + g^{ij}(\hat{q})\hat{p}_i\hat{p}_j\right)$$

$$= -\frac{\hbar^2}{4}\left(4g^{ij}[\partial_i\partial_j + \Gamma_i\partial_j + \frac{1}{2}(\partial_i\Gamma_j) + \frac{1}{4}\Gamma_i\Gamma_j] + 4(\partial_i g^{ij})(\partial_j + \frac{1}{2}\Gamma_j) + (\partial_i\partial_j g^{ij})\right)$$

$$= -\hbar^2(\Delta + \frac{1}{4}R + \frac{1}{4} g^{ij}\Gamma^k_{il}\Gamma^l_{jk}). \qquad (4.21)$$

Here, Δ is the Laplace-Beltrami second order differential operator defined in (4.14) and $R = g^{ij}R_{ij}$ is the scalar curvature. The Ricci tensor $R_{ij} = R^k_{ij,k}$ is defined as

$$R_{ij} = \partial_j \Gamma^k_{ik} - \partial_k \Gamma^k_{ij} + \Gamma^l_{ik}\Gamma^k_{jl} - \Gamma^l_{ij}\Gamma^k_{kl}. \qquad (4.22)$$

Combining the results of Eqs. (4.19)-(4.22), we may write the Weyl ordered Hamiltonian operator as

$$\hat{H}_w = \frac{1}{2}(g^{ij}\hat{p}_i\hat{p}_j)_w - (g^{ij}(\hat{q})b_i(\hat{q})\hat{p}_j)_w + \frac{1}{2}(b^i(\hat{q})b_i(\hat{q}))_w + (V(\hat{q}))_w$$

$$= -\frac{\hbar^2}{2}\Delta + i\hbar b^i \partial_i + \frac{i\hbar}{2} g^{-1/2}\partial_i(g^{1/2}b^i) + \frac{1}{2} b^i b_i + V - \frac{\hbar^2}{8}(R + g^{ij}\Gamma^l_{ik}\Gamma^k_{jl}). \qquad (4.23)$$

Comparing this with the Hamiltonian operator $\hat{H}$ of Eq. (4.13), we see that

$$\hat{H} = \hat{H}_W + \Delta V \qquad (4.24)$$

where ΔV represents the quantum correction given by

$$\Delta V = \frac{\hbar^2}{8}(R + g^{ij}\Gamma^k_{il}\Gamma^l_{jk}). \qquad (4.25)$$

This correction is proportional to $\hbar^2$ and vanishes in the classical limit ($\hbar \to 0$). We shall use this result in the next section for constructing a suitable path integral.

4.4.2. Derivation of a path integral

We now use the standard procedure for deriving a path integral based on the Kato-Trotter product formula discussed in chapter 1. The propagator from q' at time t' to q'' at time t'' is the transition amplitude

$$K(q'',t'';q',t') = \langle q''|\exp[\frac{-i}{\hbar}(t''-t')]\hat{H}|q'\rangle \qquad (4.26)$$

where $\hat{H}$ is defined by (4.13). We split the time interval $[t',t'']$ into a large number N of short intervals and use the completeness relation (4.6) to rewrite (4.26) as

$$K(q'',t'';q',t') = \lim_{N\to\infty} \int \prod_{a=1}^{N-1} [g(q_a)]^{1/2} d^n q_a \prod_{b=1}^{N} \langle q_b|\exp(-\frac{i}{\hbar}\varepsilon\hat{H})|q_{b-1}\rangle$$

$$(4.27)$$

where $\varepsilon = (t''-t')/N$ and $q_0 = q'$, $q_N = q''$.

Our task is to evaluate the short time propagator

$$K(q_b, q_{b-1}; \varepsilon) = \langle q_b|\exp(-\frac{i}{\hbar}\varepsilon\hat{H})|q_{b-1}\rangle. \qquad (4.28)$$

For this purpose, it is convenient to write $\hat{H}$ in the form (4.24) and keep terms $O(\varepsilon)$ in (4.28). Thus

$$K(q_b, q_{b-1}; \varepsilon) \approx \langle q_b|\left[1 - \frac{i\varepsilon}{\hbar}\left(\frac{1}{2}(\hat{p}_k g^{kl}(\hat{q})\hat{p}_l)_W - (b_j(\hat{q})g^{jk}(\hat{q})\hat{p}_k)_W\right.\right.$$

$$\left.\left. + \frac{1}{2}b_j(\hat{q})b^j(\hat{q}) + V(\hat{q}) + \Delta V(\hat{q})\right)\right]|q_{b-1}\rangle. \qquad (4.29)$$

Now, we use the relations (4.18) and (4.8) to derive the formula

$$\langle q'|(\hat{q}^m \hat{p}^r)_W|q\rangle = [g(q)g(q')]^{-1/4} \int \frac{d^n p}{(2\pi\hbar)^n} p^r \left(\frac{q+q'}{2}\right)^m e^{ip\cdot(q'-q)/\hbar} .$$

(4.30)

This leads to the derivation of the more general formula:

$$\langle q'|(F(\hat{q})\hat{p}^r)_W|q\rangle = [g(q)g(q')]^{-1/4} F\left(\frac{q'+q}{2}\right) \int \frac{d^n p}{(2\pi\hbar)^n} p^r e^{ip\cdot(q'-q)/\hbar} .$$

(4.31)

In deriving (4.31), we first write $F(\hat{q})$ as a Taylor series in powers of $\hat{q}$ and then evaluate every term by using (4.30). With the help of this result it is now easy to write from (4.29) the expression for the short time propagator as

$$K(q_b, q_{b-1}, \varepsilon) = (g(q_b)g(q_{b-1}))^{-1/4} \int \frac{d^n p}{(2\pi\hbar)^n} \left\{ 1 - \frac{i}{\hbar} \varepsilon \left[\frac{1}{2} p_k g^{kl}(\bar{q}_b) p_l \right.\right.$$

$$\left.\left. - b_k(\bar{q}_b) g^{kl}(\bar{q}_b) p_l + \frac{1}{2} b_k(\bar{q}_b) b^k(\bar{q}_b) + V(\bar{q}_b) + \Delta V(\bar{q}_b) \right] \right\}$$

$$\times e^{ip\cdot(q_b - q_{b-1})/\hbar}$$

$$= (g(q_b)g(q_{b-1}))^{-1/4} \int \frac{d^n p}{(2\pi\hbar)^n} \exp\left\{ \frac{i}{\hbar} \left[p\cdot(q_b - q_{b-1}) - \varepsilon H_{eff}(\bar{q}_b, p) \right] \right\} .$$

(4.32)

Here $\bar{q}_b = (q_b + q_{b-1})/2$ and $H_{eff}(\bar{q}_b, p)$ is the effective Hamiltonian function

$$H_{eff} = H + \Delta V = g^{kl}(p_k - b_k)(p_l - b_l) + V + \Delta V \qquad (4.33)$$

evaluated at the mid-point $\bar{q}_b$. H_{eff} is a function of c-number variables p_k and q_k like the classical Hamiltonian but contains an extra potential depending on the square of the Planck's constant. The form of ΔV is the same as in (4.25).

The complete propagator takes the following form after (4.32) is inserted in (4.27)

$$K(q'',t'';q',t') = \lim_{N\to\infty} [g(q'')g(q')]^{-1/4} \int \prod_{a=1}^{N-1} d^n q_a \prod_{b=1}^{N} \frac{d^n p_b}{(2\pi\hbar)^n}$$

$$\times \exp\left\{\frac{i}{\hbar} \varepsilon \sum_{b=1}^{N} [p_b(q_b - q_{b-1})/\varepsilon - H_{eff}(\bar{q}_b, p_b)]\right\}. \quad (2.34)$$

This is, in fact, the phase space form of the path integral. In the continuum limit, this can be written in the form

$$K = (g(q'')g(q'))^{-1/4} \int \mathcal{D}[q(t)] \, \mathcal{D}[p(t)] \, \exp\left(\frac{i}{\hbar} \int_{t'}^{t''} dt [p\dot{q} - H_{eff}(p,q)]\right)$$

$$(4.35)$$

where

$$\mathcal{D}[q(t)] = \prod_t d^n q(t); \quad \mathcal{D}[p(t)] = \prod_t \left(\frac{d^n p(t)}{(2\pi\hbar)^n}\right), \quad t' \leq t \leq t''.$$

$$(4.36)$$

This integral is over all paths in the phase space with fixed values for the coordinates at the end points, $q^i(t') = q'^i$ and $q^i(t'') = q''^i$. However, the integrations over momenta are unrestricted. Operator ordering through Weyl rule has led to the natural choice of mid-point rule for specifying the point at which the functions are to be evaluated.

In order to obtain the Lagrangian formulation of the path integral, we perform the integrations over momenta in expression (4.32). The result is the short time propagator

$$K(b,b-1,\varepsilon) = g^{1/2}\left(\frac{\hbar}{i\varepsilon}\right)^{n/2} \exp\left\{\frac{i\varepsilon}{\hbar}\left[\frac{1}{2}(q_b^k - q_{b-1}^k)g^{kl}(q_b^l - q_{b-1}^l)\right.\right.$$

$$\left.\left. + b_k(q_b^k - q_{b-1}^k)/\varepsilon - (V + \Delta V)\right]\right\} \quad (4.37)$$

where all functions are evaluated at the mid-point $\bar{q}_b$. The finite time propagator is obtained by inserting (4.37) in (4.27). The result is

$$K = \lim_{N \to \infty} \mathcal{C}_N \int \prod_{a=1}^{N-1} d^n q_a \prod_{b=1}^{N} g^{1/2}(\bar{q}_b) \exp\left\{ \frac{i\varepsilon}{\hbar} L_{eff} [\bar{q}_b, \frac{1}{\varepsilon}(q_b - q_{b-1})] \right\}$$

(4.38)

where the normalization constant $\mathcal{C}_N = [g(q'')g(q')]^{-1/4} (1/2\pi i \hbar \varepsilon)^{nN/2}$ and $L_{eff}(q,\dot{q})$ is the effective Lagrangian,

$$L_{eff} = L - \Delta V = \frac{1}{2} g_{kl} \dot{q}^k \dot{q}^l + b_k \dot{q}^k - (V + \Delta V) .$$

(4.39)

We notice here that the volume element $g^{1/2}(\bar{q}_b) d^n q_b$ occurring in (4.38) is not invariant since $g^{1/2}$ is evaluated at the mid-point $\bar{q}_b$. In order to remedy this, we write

$$[g(q'')g(q')]^{-1/4} \prod_{b=1}^{N} g^{1/2}(\bar{q}_b) = \prod_{a=1}^{N-1} g^{1/2}(q_a) \prod_{b=1}^{N} \mathcal{G}(q_b, q_{b-1})$$

(4.40a)

where the quantity

$$\mathcal{G}(q_b, q_{b-1}) = [g(\bar{q}_b)]^{1/2} [g(q_b)g(q_{b-1})]^{-1/4} .$$

(4.40b)

Finally the expression for $\mathcal{G}(q_b, q_{b-1})$ in (4.40b) can be simplified by expanding $g(q_b)$ and $g(q_{b-1})$ up to second order in the difference $(q_b - q_{b-1})$. The resulting expression can be exponentiated using the standard replacement

$$\Delta^i \Delta^j \longrightarrow i\hbar g^{ij} \varepsilon , \quad (\Delta^i = (q_b^i - q_{b-1}^i))$$

(4.41)

and the final expression for the propagator becomes

$$K(q'',t'';q',t') = \lim_{N \to \infty} (2\pi i \hbar \varepsilon)^{-nN/2} \int \prod_{a=1}^{N-1} (g^{1/2}(q_a) d^n q_a)$$

$$\times \prod_{b=1}^{N} \exp\left\{ \frac{i\varepsilon}{\hbar} \left[L_{eff}[\bar{q}_b, (q_b - q_{b-1})/\varepsilon] - \Delta V'(\bar{q}_b) \right] \right\}$$

(4.42)

where the additional quantum correction

$$\Delta V' = \frac{\hbar^2}{8} g^{ij} \Gamma^l_{li,j} .$$

(4.43)

The continuum limit of this discretized path integral may be formally written as

$$K(q'',t'';q',t') = \int \mathcal{D}[q(t)] \exp\left[\frac{i}{\hbar}\int_{t'}^{t''} \tilde{L}(q,\dot{q})\, dt\right] \quad (4.44)$$

where $\tilde{L} = L - \Delta\tilde{V}$. Here L is the classical Lagrangian (4.1). Further the quantum correction $\Delta\tilde{V} = \Delta V + \Delta V'$ is given explicitly by

$$\Delta\tilde{V} = \frac{\hbar^2}{8}[R + g^{ij}(\Gamma^k_{il}\Gamma^l_{jk} + \Gamma^l_{il,j})]. \quad (4.45)$$

The quantum correction is not invariant under a change of variables. As we have seen in Sec. 4.3, it is present even in a flat space when non-cartesian coordinates such as polar coordinates are used. The reader may easily verify that the expression (4.45) yields the same results as those derived in Sec. 4.3.

Notes and References

Path integrals in polar coordinates were first developed in the paper

S. F. Edwards and Y. V. Gulyaev, Proc. Roy. Soc.(Lond.) **A279**, 229 (1964);
See also, A. M. Arthurs, Proc. Roy. Soc.(Lond.) **A313**, 445 (1969).

The technique of explicit evaluation of the propagator by separation of angular and radial parts was subsequently pointed out in the paper

D. Peak and A. Inomata, J. Math. Phys. **10**, 1422 (1969).
This paper also contains exact results for isotropic oscillator, inverse square well and rigid rotor.

The potential of the form

$$V(r) = \frac{1}{2}\omega^2(t)\, r^2 + g/r^2$$

was treated by

D. C. Khandekar and S. V. Lawande, J. Maths. Phys. **16**, 384 (1975).

Non-spherically symmetric potentials arise naturally in connection of a three body problem with two and three body forces discussed in the paper

F. Calogero, J. Math. Phys. **10**, 2191 (1969).

Exact expressions for the propagators of some solvable cases are obtained in the following papers

D. C. Khandekar and S. V. Lawande, J. Phys. **A5**, 812 (1972);

D. C. Khandekar and S. V. Lawande, J. Phys. **A5**, L57 (1972);

D. C. Khandekar and S. V. Lawande, J. Math. Phys. **18**, 712 (1977).

A non-stationary version of Calogero model was considered by

M. J. Goovaerts, J. Math. Phys. **16**, 720 (1975).

"Unconventional dimensional extension" technique referred to in Sec.4.3 is discussed in

I. Inomata, "Recent developments of techniques for solving non-trivial path integrals", in "Path Summation: Achievements and Goals", Eds. S. Lundqvist, A. Ranfagni, V. Sa-yakamit and L. S. Schulman, (World Scientific, Singapore, 1988).

Generalization to d-dimensional polar coordinates and the role of quantum correction to potential has been discussed extensively by

C. Grösche and F. Steiner, Z. Phys. **C36**, 699 (1987).

This paper contains a large number of references on the subject of path integrals on curved manifolds and also a comparative study of various approaches.

A natural extension of path integral for a rotor in d-dimensions is the group theoretic approach to path integration on spheres. Early references on this subject are covered by

L. S. Schulman, "Techniques and Applications of Path Integration", (Wiley, New York, 1981).

Recent references on the subject are

M. Bohm and G. Junker, "Group theoretical approach to path integration on spheres", in "Path Summation: Achievements and Goals", Eds. S. Lundqvist, A. Ranfagni, V. Sa-yakamit and L. S. Schulman, (World Scientific, Singapore, 1988).

G. Junker, "Path integration on homogeneous spaces", in "Path integrals from meV to MeV", Ed. V. Sa-yakamit et al, (World Scientific, Singapore, 1989).

There are a large number of references on the formulation of path integrals in curved spaces. we mention here only a few of them for a historical perspective and comparative study.

B. S. DeWitt, Rev. Mod. Phys. **29**, 379(1957),

D. C. McLaughlin and L. S. Schulman, J. Math. Phys. **12**, 2520 (1971),

J. S. Dowker and I. W. Mayes, Proc. Roy. Soc. (Lond.) **A327**, 131 (1972),

M. Mizrahi, J. Math. Phys. **16**, 2201 (1975),

J. L. Gervais and A. Jevicki, Nucl. Phys. **B110**, 93 (1976),

M. Omote, Nucl. Phys. **B120**, 325 (1977),

M. S. Marinov, Phys. Rept. **60**, 1 (1980),

T. D. Lee, "Particle Physics and Introduction to Field Theory" (Academic Press, Harwood, 1981),

J. C. D'Olivo and M. Torres, in "Path Summation: Achievements and Goals", Eds. S. Lundqvist, A. Ranfagni, V. Sa-yakamit and L. S. Schulman, (World Scientific, Singapore, 1988).

The main result of all these works is that an appropriate quantum correction $\Delta V \approx \hbar^2$ to the classical potential must be included in the path integral treatment. However, the expressions for ΔV derived by the above authors apparently do not agree. This is due to the different prescriptions used in deriving the discretized version of the path integral. For example, while DeWitt uses a prepoint definition, Mizrahi and Lee use the mid-point prescription.

Various formulae used in the evaluation of exact propagators can be found in the books by

I. S. Gradshteyn and I. M. Ryzhik, "Tables of Integrals, Series and Products" (Academic Press, New York, 1980),

A. Erdelyi, W. Magnus, F. Oberhettinger, and F. G. Tricomi, "Higher Transcendental Functions, Vol II (McGraw Hill, New York, 1985).

CHAPTER 5

COORDINATE TIME TRANSFORMATIONS IN PATH INTEGRALS

5.1. Introduction

The harmonic oscillator and the hydrogen atom are two well-known test problems in quantum mechanics. Feynman's path integral formulation is readily applicable to the former but poses considerable difficulties for the latter. Duru and Kleinert invoked for the first time the Kustaanheimo-Stifel (KS) transformation in path integration of the Coulomb potential. The KS transformation, well-known in celestial mechanics, consists of a space transformation (not necessarily a change of coordinates) followed by a time transformation. This technique falls under a general class of coordinate-time transformation methods discussed in the present chapter. It can be used to path integrate a number of potentials besides the Coulomb potential.

A change of variables is a standard technique in ordinary calculus but creates problems in the path integral calculus. As we shall see subsequently a path integral assumes a more complicated form after a change of coordinates. However, if in addition, we make a transformation of time, it helps to simplify the path integral. In other words, a change of variables of integration in a path integral consists, in general, of a change of coordinates together with a time transformation and brings path integral calculus on par with the ordinary calculus.

While these coordinate-time transformations have deeper roots in classical mechanics, the precise nature of them in the context of usual formulation of quantum mechanics is somewhat obscure. In a sense, Feynman's path integral provides a bridge between classical and quantum mechanics; it is perhaps natural that the transformations find a place in this formulation of quantum mechanics.

There are two types of time transformations, viz., global transformations and local transformations. A global time transformation takes the time parameter t into a new "time" parameter τ by

$$t = \rho(\tau); \quad \tau = \rho^{-1}(t) \quad . \quad (1.1)$$

A local time transformation, on the other hand, is a position dependent time transformation

$$d\tau = f[\vec{q}(t)] dt \quad (1.2)$$

where $f[\vec{q}(t)]$ is a well behaved function of $\vec{q}$. Here the value of the parameter τ depends on the path along which the integration is performed. In quantum mechanics, paths of motion are unspecified. For example, "all paths" contribute to the Feynman propagator. Hence the path-dependent parameter τ cannot simply replace the time parameter t. There is no "a priori" meaning of the local time transformation in quantum mechanics. For instance, if $\vec{q}(t)$ is taken as an operator in the Hilbert space, then the new parameter τ must also be regarded as an operator. This is hardly desirable and it is clear that local time transformation must be regarded as a c-number transformation for the parameter τ to be "time-like". This can be so only if coordinate representation is used. Indeed, Feynman's path integral expresses the propagator in c-number representation and the use of local-time transformation might be legitimate. Both global and local time transformations are most useful only when combined with an appropriate coordinate transformation.

5.2. Local Time Transformations in Classical Mechanics

Local time transformation (1.1) occurs very naturally in classical mechanics. In the standard central force problem, the classical action is

$$S(t'',t') = \int_{t'}^{t''} \left[\frac{m}{2} \dot{\vec{q}}^2 - V(r) \right] dt \quad , \quad r = |\vec{q}|. \quad (2.1)$$

The variational principle $\delta S = 0$ in plane polar coordinates (r,θ), yields two equations, viz.,

$$m\ddot{r} - \frac{\ell^2}{mr^3} + \frac{dV}{dr} = 0 \quad ; \quad mr^2\dot{\theta} = \ell = \text{const}. \qquad (2.2)$$

The conservation of angular momentum is implied in the second equation of the pair (2.2) and is used explicitly in the first. The first integral of the latter expresses the conservation of energy, E, that is

$$\frac{1}{2}m\dot{r}^2 + \frac{\ell^2}{2mr^2} + V(r) = E . \qquad (2.3)$$

The integration of (2.3) yields the classical trajectory r(t) as a function of time t. However, one is often interested in the equation for the orbit as a function of θ, that is, $r = r(\theta)$. This can be achieved by introducing the time transformation $d\theta = \ell \, dt/mr^2$. It is clear that this time transformation is local, that is, it depends on the radial position of the particle r(t) at time t. We also know that this transformation is most useful when we also make the coordinate transformation $u = 1/r$. It is standard to make use of both of these transformations to arrive at the orbit equation

$$\frac{d^2u}{d\theta^2} + u = -\frac{m}{\ell^2}\frac{d}{du}[V(1/u)]. \qquad (2.4)$$

In particular, for the Kepler problem, $V(r) = -K/r$, K being a constant, the equation of the orbit takes a simple form

$$\frac{d^2u}{d\theta^2} + u = \left(\frac{mK}{\ell^2}\right). \qquad (2.5)$$

This is just the equation of a harmonic oscillator of unit frequency acted on by a constant force of magnitude mK/ℓ^2. Thus the original Kepler problem is mapped onto a harmonic oscillator by means of the local-time transformation and the coordinate transformation. However, there is no unique way of doing this. For example, another transformation $d\psi = (-2E/m)^{1/2} r^{-1} dt$ may be applied to reduce the radial equation for the Kepler problem to the form

$$\frac{d^2 r}{d\psi^2} + r = -\frac{K}{2E} . \qquad (2.6)$$

In deriving (2.6) we have made use of the energy conservation equation (2.3).

Another transformation that has been used in celestial mechanics is the so-called Kustaanheimo-Stiefel (KS) transformation. The local time transformation in this case may be written as $ds = (4r)^{-1} dt$ (the factor 1/4 is used for convenience). For planar Kepler problem, this time transformation is used with the Levi-Civita coordinate transformation while for the three-dimensional Kepler problem, a simple change of variable from r to $\rho = r^2$ is sufficient.

The Levi-Civita transformation is a conformal mapping of $z = \omega^2$ ($z = x + iy$ and $\omega = u_1 + iu_2$) and may be written as

$$\begin{pmatrix} x \\ y \end{pmatrix} = A \begin{pmatrix} u_1 \\ u_2 \end{pmatrix} \; ; \; A = \begin{pmatrix} u_1 & -u_2 \\ u_2 & u_1 \end{pmatrix} . \qquad (2.7)$$

The matrix A has the property that, $A^T A = A A^T = u^2 I = rI$. We can also show that $d\vec{q} = 2 A\, d\vec{u}$ which implies that $(d\vec{q})^2 = 4 u^2 (d\vec{u})^2$.

We now consider the modified action $W(t'',t')$ in two dimensions

$$W(t'',t') = E(t'' - t') + \int_{t'}^{t''} \left(\frac{m}{2} \dot{q}^2 + \frac{K}{r} \right) dt \qquad (2.8)$$

and apply the KS time transformation along with the Levi-Civita mapping. We arrive at the following form

$$W(t'',t') = 4\alpha(s'' - s') + \int_{s'}^{s''} \left(\frac{1}{2} m \left(\frac{d\vec{u}}{ds} \right)^2 + 4 E u^2 \right) ds \qquad (2.9)$$

where $s' = s(t')$ and $s'' = s(t'')$. When we extremize the modified action W of (2.8), we obtain the usual equations of motion for the Kepler problem. What is important to note here is that the transformed quantity in (2.9) corresponds to the modified action for a harmonic oscillator of frequency $\omega^2 = -8E/m$. Thus the local time and coordinate transformations have converted the modified action for the planar Kepler problem into that for a two-dimensional harmonic oscillator. In contrast to the previous transformations which converted the orbit equation into an oscillator equation, (cf. Eqs. (2.5) and (2.6)) the new transformation directly simplifies the action.

The generalization of this two-dimensional problem to the three-dimensional case can be done only by a special coordinate mapping from R^3 to R^4. We shall postpone the discussion of this mapping to Sec. 5.7 where the mapping is used to solve the quantal Kepler problem. For the present, we might make the rather simple choice of coordinates, viz., $r = \rho^2$ and consider the following modified radial action (for a fixed angular momentum ℓ)

$$W_1(t''-t') = E(t''-t') + \int_{t'}^{t''} \left(\frac{m\dot{r}^2}{2} - \frac{\ell^2}{2mr^2} + \frac{K}{r} \right) dt \qquad (2.10)$$

and apply the time transformation. The result is

$$W_1(t''-t') = 4\alpha(s''-s') + \int \left[\frac{m}{2} \left(\frac{d\rho}{ds}\right)^2 - \frac{\nu^2}{2m\rho^2} + 4 E\rho^2 \right] ds \qquad (2.11)$$

where $\nu = 2\ell$. Once again this action yields the radial equation for a harmonic oscillator with frequency $\omega^2 = -8E/m$. The combination of KS time transformation and the simple change of variable from r to ρ^2 has reduced the original Kepler problem to the harmonic oscillator problem.

There is another important point here, viz., the use of the Hamilton's characteristic function W rather than the action. The function is defined as

$$W = \int_{t'}^{t''} (L + E) \, dt = S + E(t''-t'). \qquad (2.12)$$

It is clear from the above discussion that simplification results only when time transformation and change of variables are applied to this function.

These ideas can be carried over to quantum mechanics with the help of the concept of an entity called promoter introduced in the subsequent section.

5.3. Concept of the Promotor

The idea of introducing local time transformation and the change of coordinates in a path integral is motivated by the following considerations. We have already introduced the energy dependent Green's function in Chapter 1. It is just the Fourier transform of the propagator. It contains information about the energy eigenvalues and the corresponding eigenfunctions. In fact, the poles of Green's function yield the energy spectrum while the residues at these poles give the energy eigenfunctions. The Green's function has the usual definition

$$G(\vec{q}'',\vec{q}';E) = (i\hbar)^{-1} \int e^{(iE\tau/\hbar)} K(\vec{q}'',\vec{q}';\tau) \, d\tau \ . \tag{3.1}$$

We may rewrite this in the alternate form

$$G(\vec{q}'',\vec{q}';E) = (i\hbar)^{-1} \int P(\vec{q}'',\vec{q}';\tau) \, d\tau \tag{3.2}$$

where $P(\vec{q}'',\vec{q}';\tau)$ is an entity known as the promotor. It is defined as

$$P(\vec{q}'',\vec{q}';\tau) = \int \exp[\frac{iW}{\hbar}] \, \mathcal{D}[\vec{q}(t)] \tag{3.3}$$

where W is Hamilton's characteristic function introduced in Sec.2. The promotor can clearly be expressed as a path integral. Moreover, the expression (3.3) is invariant under a time transformation, that is,

$$G(\vec{q}'',\vec{q}';E) = (i\hbar)^{-1} \int \tilde{P}(\vec{q}'',\vec{q}';\sigma) \, d\sigma \ . \tag{3.4}$$

Thus we can manipulate the path integral for the promotor conveniently without changing the Green's function for the problem. This is the essential rationale for using the local time transformation in the context of path integrals. The end product is the energy Green's function and in rare instances, where the inverse Fourier transform can be performed analytically, we obtain the analytical form of the propagator as well.

Apart from this, there are several other ways of looking at the problem. For example, in Sec.4, we employ the strategy of first introducing the coordinate transformation in the path integral. This procedure is straightforward if midpoint rule is used. The required local time transformation suggests itself when one finds that the mass parameter has become local as a result of the transformation of coordinates. The idea of a promotor appears very naturally here.

As we shall see in subsequent sections the local time transformation merely helps us to recover the eigenvalues and eigenfunctions of the Schrödinger equation for hydrogen atom and other cases via the Greeen's function. The equivalence between Feynman formulation and the usual Schrödinger formulation is thereby established, albeit indirectly.

5.4. Coordinate-Time Transformations in Path Integral

We start with the propagator expressed as the conventional path integral

$$K(x'',t'';x',t') = \int \exp\left[\frac{i}{\hbar}\int_{t'}^{t''} L\, dt\right] \mathcal{D}\,[x(t)] \qquad (4.1)$$

for the Lagrangian

$$L = \frac{m}{2}\dot{x}^2 - V(x)\,. \qquad (4.2)$$

For simplicity, we illustrate the method for a one-dimensional problem. The propagator should be considered as the limit of the discretized form K_N as $N \rightarrow \infty$, where

$$K_N = A_N \int \exp\left\{\frac{i}{\hbar}\sum_{j=1}^{N} S_j\right\} \prod_{j=1}^{N-1} dx_j \qquad (4.3)$$

with the normalization factor

$$A_N = \left(\frac{m}{2\pi i\hbar\varepsilon}\right)^{N/2} \qquad (4.4)$$

and the discretized action in interval $[t_{j-1}, t_j]$ taking the form

$$S_j = \frac{m}{2\varepsilon}(x_j - x_{j-1})^2 - \varepsilon V(x_j) \ . \tag{4.5}$$

Let us apply a coordinate transformation

$$x = f(q) \ . \tag{4.6}$$

In the discretized version, we express the increments $\Delta x_j = (x_j - x_{j-1})$ in terms of the increments $\Delta q_j = (q_j - q_{j-1})$. Here the midpoint rule is a safe bet. Also contributions up to the order ε are to be retained in the action and we have to keep in mind that $(\Delta q_j)^2 \approx \varepsilon$. Expanding $f(q_j)$ and $f(q_{j-1})$ about the midpoint $\bar{q}_j = (q_{j-1} + q_j)/2$, and retaining terms up to third order in Δq_j we have

$$\Delta x_j = f'_j \Delta q_j \left[1 + \frac{1}{24} \frac{f'''_j}{f'_j} (\Delta q_j)^2 \right] \tag{4.7}$$

where primes denote derivatives with respect to q and $f_j = f(\bar{q}_j)$. This yields the following expression for the contribution from kinetic energy term in the action

$$\frac{m}{2\varepsilon}(\Delta x_j)^2 = \frac{m}{2\varepsilon}(f'_j)^2 (\Delta q_j)^2 [\, 1 + f'''_j (\Delta q_j)^2 / 12 f'_j \,] \ . \tag{4.8}$$

The potential energy term takes the simple form

$$\varepsilon V(x_j) = \varepsilon V[f(\bar{q}_j)] + O(\varepsilon^2) = \varepsilon V(f_j) \ . \tag{4.9}$$

Now consider the term

$$\prod_{j=1}^{N-1} dx_j = \prod_{j=1}^{N-1} f'(q_j) \, dq_j \tag{4.10}$$

which needs to be symmetrized about the points q_j, q_{j-1} such that none of the end-points of the interval are preferred when we expand about the mid-point $\bar{q}_j$ of the interval. This may be done by rewriting

$$\prod_{j=1}^{N-1} dx_j = [f'(q_N) f'(q_0)]^{-1/2} \prod_{j=1}^{N} (f'(q_j) f'(q_{j-1}))^{1/2} \prod_{j=1}^{N-1} dq_j \ . \tag{4.11}$$

We expand $f(q_j)$ and $f(q_{j-1})$ up to second order in Δq_j, so that

$$(f'(q_j) f'(q_{j-1}))^{1/2} = f'_j \left[1 + \frac{1}{4} (\Delta q_j)^2 \left\{ \frac{f'''_j}{f'_j} - \left(\frac{f''_j}{f'_j}\right)^2 \right\} \right]$$

and consequently

$$\prod_{j=1}^{N-1} dx_j = [f'(q_N) f'(q_0)]^{-1/2} \prod_{j=1}^{N} f'_j \left[1 + \frac{1}{4} (\Delta q_j)^2 \left\{ \frac{f'''_j}{f'_j} - \left(\frac{f''_j}{f'_j}\right)^2 \right\} \right].$$

(4.12)

The transformation (4.6) has made the discretized form of the path integral sufficiently complicated. Moreover, the mass parameter has turned into a local one, mf'^2_j. We apply the following local time transformation to overcome this difficulty.

$$\frac{dt}{ds} = [f'(q(s))]^2; \qquad t(s_N) = t'', \qquad t(s_0) = t' \qquad (4.13)$$

where s stands for the new "time". For consistency, we have to first symmetrize (4.13) over the interval (j-1, j) in order to avoid any preference of one end-point over the other. This means that

$$\varepsilon = \sigma_j f'(q_j) f'(q_{j-1}) \qquad (4.14)$$

where $\sigma_j = s_j - s_{j-1}$. Expanding $f'(q_j)$ and $f'(q_{j-1})$ around the mid-point $\bar{q}_j$, we have

$$\varepsilon = \sigma_j f'^2 \left[1 + \frac{1}{4} (\Delta q_j)^2 \left(\frac{f'''_j}{f'_j} - \left(\frac{f''_j}{f'_j}\right)^2 \right) \right]. \qquad (4.15)$$

Note that σ_j are no longer of equal length. An immediate consequence of (4.12) and (4.15) is that the path differential measure takes the form

$$A_N \prod_{j=1}^{N-1} dx_j = [f'(q') f'(q'')]^{-1/2} \prod_{j=1}^{N} \left(\frac{m}{2\pi i \hbar \sigma_j}\right)^{1/2} \prod_{j=1}^{N-1} dq_j . \qquad (4.16)$$

Inserting the expression (4.15) for ε in (4.8) and retaining terms up to $(\Delta q_j)^4$, we obtain

$$\frac{m(\Delta x_j)^2}{2\varepsilon} = \frac{m(\Delta q_j)^2}{2\sigma_j} + \frac{m(\Delta q_j)^4 \lambda_j}{8\sigma_j} \qquad (4.17)$$

where

$$\lambda_j = \left(\frac{f''_j}{f'_j}\right)^2 - \frac{2}{3}\frac{f'''_j}{f'_j} . \qquad (4.18)$$

Incidentally, the Schwarzian derivative of $f(x)$ is $-2\lambda/3$, a quantity which remains invariant under all fractional transformations. Also, the potential energy term takes the form

$$\varepsilon V(x_j) = \sigma_j (f'_j)^2 V(f_j) = \sigma_j (f'_j)^2 V_j . \qquad (4.19)$$

Combining all these results, we write

$$\exp\left[\frac{i}{\hbar} S_j\right] = \exp\left[\frac{i}{\hbar}\left\{\frac{m}{2\sigma_j}(\Delta q_j)^2 + \frac{m}{8\sigma_j}\lambda_j(\Delta q_j)^4 - \sigma_j f'^2_j V_j\right\}\right]. \qquad (4.20)$$

We can remove the term in $(\Delta q_j)^4$ by making use of the following formula valid for large a

$$\int_{-\infty}^{\infty} \exp[-ax^2 - bx^4]\, dx = \int_{-\infty}^{\infty} dx\, \exp\left[-ax^2 - \frac{3b}{4a^2}\right] + O(1/a^3). \qquad (4.21)$$

In our case, $a = m/(2i\sigma_j \hbar)$, $b = m\lambda_j/(8i\sigma_j \hbar)$ and hence

$$\exp\left[\frac{i}{\hbar} S_j\right] = \exp\left[\frac{i}{\hbar}\left\{\frac{m}{2\sigma_j}(\Delta q_j)^2 - \sigma_j\left((f'_j)^2 V_j + \frac{3\hbar^2}{8m}\lambda_j\right)\right\}\right]. \qquad (4.22)$$

The last important point is that the new time difference $(s''-s')$ is a path dependent quantity. This dependence must be incorporated by means of the constraint

$$T = t''-t' = \int_{s'}^{s''} ds\, [f'(q(s))]^2 \qquad (4.23)$$

into the path integral. For this, the following identity is used

$$[f'(q'')f'(q')] \int_0^\infty ds \, \delta\left(T - \int_{s'}^{s''} ds [f'(q(s))]^2\right) = 1 \,. \quad (4.24)$$

We can therefore write the propagator

$$K(f(q''),f(q');T) = \lim_{N\to\infty} \int_{s'}^{s''} \delta\left(T - \int d\tau \, [f'(q(\tau))]^2\right) K_N \, ds \quad (4.25)$$

where K_N is the transformed discretized form

$$K_N = \sqrt{f'(q')f'(q'')} \int \prod_{j=1}^{N} \left(\frac{m}{2\pi i\hbar\sigma_j}\right)^{1/2} \prod_{j=1}^{N-1} dq_j \, \exp\left[\frac{i}{\hbar} S_j\right] \,. \quad (4.26)$$

The Fourier representation of the δ-function yields

$$K = \frac{1}{2\pi\hbar} \int_{-\infty}^{\infty} e^{-iTE/\hbar} \, G(x'',x';E) \, dE \quad (4.27)$$

where

$$G(x'',x';E) = [f'(q')f'(q'')]^{1/2} \int_0^\infty ds \, P(q'',q';s) \,. \quad (4.28)$$

The quantity $P(q'',q',s)$ is the promotor which like the propagator is defined as the limit of the discretized form P_N

$$P(q'',q';s) = \lim_{N\to\infty} \int \prod_{j=1}^{N} \left(\frac{m}{2\pi i\hbar\sigma_j}\right)^{1/2} \prod_{j=1}^{N-1} dq_j \, \exp\left[\frac{i}{\hbar} \tilde{S}\right] \quad (4.29a)$$

where the new action $\tilde{S}$ reads as

$$\tilde{S} = \sum_{j=1}^{N} \left\{ \frac{m}{2\sigma_j} (\Delta q_j)^2 - \sigma_j \left(f_j'^2(V-E) + \frac{3\hbar^2}{8m} \lambda_j\right) \right\} \,. \quad (4.29b)$$

The success of the coordinate and local time transformation depends on how far one is able to evaluate the promotor in a closed form. We can also write the expression for the promotor in Feynman form as

$$P(q'',q';s) = \int \exp\left\{\frac{i}{\hbar}\tilde{S}[q(s)]\right\} \mathcal{D}[q(s)] \qquad (4.30)$$

where $\tilde{S}[q(s)]$ is the continuum limit of Eq.(4.29b)

$$\tilde{S}[q(s)] = \int_0^s \left[\frac{m}{2}\left(\frac{dq}{d\sigma}\right)^2 - \tilde{V}(q)\right] d\sigma . \qquad (4.31)$$

The new potential $\tilde{V}(q)$ has the form

$$\tilde{V}(q) = [f'(q)]^2 [V(f(q)) - E] + \frac{3\hbar^2}{8m}\lambda . \qquad (4.32)$$

The original promotor involved the path integration of the classical Hamilton's characteristic function $W = S + Et$ rather than the classical action. The transformed promotor is like the propagator in new coordinates and new time with the action $\tilde{S}[q(s)]$. The formulation discussed here has the merit that the coordinate transformation suggests the local time transformation. One can then ask the question: For what kinds of coordinate transformation can the promotor be explicitly evaluated? The most obvious case is that the new potential $\tilde{V}(q)$ must look like

$$\tilde{V}(q) = aq^2 + bq^{-2} + c \qquad 0 < q < \infty \qquad (4.33)$$

that is, $\tilde{V}(q)$ must be a sum of a quadratic and an inverse quadratic term apart from a constant.

We can illustrate two important cases by examining the following coordinate transformations: I) $x = f(q) = q^2$ and II) $x = f(q) = A \ln q$. In case I), $\lambda = q^{-2}$ and the requirement (4.33) implies that

$$V(f) = (\frac{a}{4} + E) + \frac{\hbar^2}{2mq^4}\left(\frac{mb}{2\hbar^2} - \frac{3}{16}\right) + \frac{c}{4q^2} \qquad 0 < q < \infty . \qquad (4.34)$$

Choosing $a = -4E$, and $b = \frac{2\hbar^2}{m}(\gamma^2 - \frac{1}{16})$ and $c = -4Ze^2$ and inserting the value $q^2 = x$, we arrive at the effective potential for the hydrogen atom

$$V(x) = -\frac{Ze^2}{x} + \frac{\hbar^2}{2mx^2}(\gamma^2 - \frac{1}{4}) . \qquad (4.35)$$

Thus a coordinate transformation $x = q^2$ followed by a local time transformation $ds = dt/4q^2$ maps the Coulomb problem onto that of the harmonic oscillator. We use this to obtain the energy Green's function in the subsequent section.

In case II), $\lambda = -\hbar^2/3q^2$ and the requirement (4.33) leads to

$$V(f) = \frac{1}{A^2}\left(b + E\,A^2 + \frac{\hbar^2}{8m}\right) + \frac{aq^4}{A^2} + \frac{cq^2}{A^2}. \qquad (4.36)$$

Choosing $a/A^2 = V_0$, $c = -2a$, $(b + E\,A^2 + \frac{\hbar^2}{8m}) = a$, $A = -2/\alpha$ and inserting $q = \exp[-x/a]$, we obtain the Morse potential

$$V(x) = V_0(1 - e^{-\alpha x})^2. \qquad (4.37)$$

Thus Morse potential is mapped into the harmonic oscillator by the coordinate transformation $x = -(2/\alpha)\ln q$, together with the local time transformation $ds = \frac{1}{4}\alpha^2 q^2\, dt$. Note that these coordinates and local time transformations are exactly as in the classical mechanical treatment of the Kepler and Morse potentials.

5.5. Illustrative Examples

We illustrate the use of the technique of coordinate-time transformation for solving some non-trivial problems. We can usually get the promotor and the energy Green's function. It is sometimes possible to take the Fourier inverse of the Green's function to obtain the Feynman propagator.

5.5.1. Coulomb potential

The propagator for all spherically symmetric potentials can be separated in radial and angular parts. The angular part of the propagator can be explicitly written and it is necessary to evaluate only the radial propagator corresponding to the effective one dimensional Lagrangian

$$L = \frac{1}{2}m\dot{r}^2 - \left[\frac{(\gamma^2 - 1/4)\hbar^2}{2mr^2} - \frac{Ze^2}{r}\right]. \qquad (5.1)$$

The formal expression for the radial propagator is

$$K_\nu(r'',r',T) = (r''r')^{-(d-1)/2} \int \mathcal{D}[r(t)] \exp\left[\frac{i}{\hbar} S[r(t)]\right] \quad (5.2)$$

where d is the dimensionality of the system (d = 2, 3). Also $\gamma = \ell + \frac{1}{2}$, $\nu = \ell$ if d = 3 and $\gamma = \nu = \ell$ if d = 2.

By means of the transformation $r = q^2$ and $dt = ds/4q^2$, the problem is reduced to the evaluation of the promotor $P(q'',q',s)$ (4.30) with the potential

$$\tilde{V}(q) = -4Eq^2 + \hbar^2 \frac{(4\gamma^2 - 1/4)}{2mq^2} - 4Ze^2 \quad (5.3)$$

where $0 < q < \infty$. The promotor is given by

$$P_\nu(q'',q',s) = \exp\left[\frac{4iZe^2 s}{\hbar}\right] K_\nu(q'',q',s) \quad (5.4)$$

where K_ν is the radial propagator for a harmonic oscillator and has been evaluated earlier in chapter 4. It can be obtained from Eq. (3.37) of this chapter by identifying $\omega^2 = -8E/m$ and replacing $(r,T) = (q,s)$ and letting $\mu = 2\gamma$. The factor in front of (5.2) is included in (5.4).

The Green's function is obtained by inserting (5.4) in (4.28). It is convenient to introduce the following notations

$$\tau = 4q'q''s, \quad 2k = \frac{m\omega}{i\hbar}, \quad \Omega = \frac{\hbar k}{2m\sqrt{r''r'}}, \quad p = -\frac{imZe^2}{\hbar^2 k} \quad (5.5)$$

to write the energy Green's function as

$$G_\nu(r'',r';E) = \frac{-ik}{(r'r'')^{(\alpha-1)/2}} \int_0^\infty d\tau \exp[-2p\Omega\tau] \operatorname{cosech}(\Omega\tau)$$

$$\times \exp[ik(r'+r'')\coth(\Omega\tau)] I_{2\gamma}[-2ik(r'r'')^{1/2}\operatorname{cosech}(\Omega\tau)]. \quad (5.6)$$

To perform the integration over τ we take recourse to the formula

$$\int_0^\infty dq \ e^{-2pq} \ \text{cosech} \ q \ \exp\left[-\frac{x+y}{2} \coth q\right] I_{2\nu}[(xy)^{1/2} \text{cosech} \ q]$$

$$= \left[\frac{\Gamma(p+\nu+1/2)}{(xy)^{1/2}\Gamma(2\nu+1)}\right] M_{-p,\nu}(x) \ W_{-p,\nu}(y) \quad (5.7)$$

where $M_{p,\nu}(x)$ and $W_{p,\nu}(x)$ are the Whittaker functions. The radial Green's function for $r'' > r'$, takes the closed form

$$G_\nu(r'',r';E) = \left[\frac{m}{\hbar k} \frac{\Gamma(p+\gamma+1/2)}{(r'r'')^{(d-1)/2} \ \Gamma(2\gamma+1)}\right] M_{p,\gamma}(-2ikr') \ W_{-p,\gamma}(-2ikr'')$$

$$(5.8)$$

where $p = -i(mZ^2e^4/2\hbar^2 E)^{1/2}$ and $k = (2mE/\hbar^2)^{1/2}$. The expression (5.8) holds over the entire E space by analytic continuation and we can recover both the discrete and continuous eigenfunctions. The discrete energy spectrum is deducible from the poles of the spectral function $G_\nu(E) = \int G_\nu(r,r;E) \ r^2 dr$. In fact the poles arise only from the Γ-function appearing in (5.8) and are given by $(\gamma + p + 1/2) = -n_r$ where ($n_r = 0, 1, 2,...$). This results in the well-known expression for energies for Coulomb potential,

$$E_n = -\frac{mZ^2e^4}{2\hbar^2 n^2} \quad (5.9)$$

where $n = n_r + m + \frac{1}{2}$ in the two-dimensional case and $n = n_r + \ell + 1$ in the usual three-dimensional case. The residues at the poles of the Green's function yield the corresponding eigenfunctions. We show this for three-dimensional case, viz. the hydrogen atom. Denoting the radial eigenfunction by $R_{n\ell}(r)$, we have

$$|R_{n\ell}(r)|^2 = \text{Res} \ [\ G_1(r,r,E)/i\hbar \ ]\big|_{p=-n} \ , \quad (5.10)$$

Now, for integral values of p

$$(\ell + 1 + p) \ \Gamma(\ell + 1 + p) \xrightarrow[p=-n]{} [\ \Gamma(n - \ell) \ ]^{-1} \quad (5.11)$$

$$W_{-p,\,l+1/2}(z) \xrightarrow[p=-n]{} \frac{\Gamma(n+\ell+1)}{\Gamma(2\ell+2)} M_{n,\,l+1/2}(z) \ . \qquad (5.12)$$

Thus the eigenfunction corresponding to energy E_n is

$$R_{nl}(r) = \left(\frac{mZe^2}{\hbar^2 r^2}\right)^{1/2} \left(\frac{(n+\ell)!}{(n-\ell-1)!}\right)^{1/2} \frac{1}{n(2\ell+1)!} M_{n,\,l+1/2}\left(\frac{2mZe^2}{\hbar^2 n} r\right) \ . \qquad (5.13)$$

In order to compare this with the usual solution, we have to use the relation between the Whittaker function and the confluent hypergeometric function. In terms of the latter, the radial wave-function for the hydrogen atom reads as

$$R_{nl}(r) = 2a_0^{-3/2} \frac{n^{-1-2}}{(2\ell+1)!} \left(\frac{(n+\ell)!}{(n-\ell-1)!}\right)^{1/2} \left(\frac{2r}{a_0}\right)^l \exp\left[\frac{-r}{na_0}\right]$$

$$\times F(-n+\ell+1,\ 2\ell+1;\ 2r/a_0) \qquad (5.14)$$

where $a_0 = (\hbar^2/mZe^2)$ is the Bohr radius. $R_{nl}(r)$ is the correctly normalized wave function. The derivation of the energy eigenfunction for the two-dimensional case is left as an exercise.

5.5.2. Morse potential

For the one-dimensional Morse potential (4.37), the transformation $x = -(2/\alpha) \ln q$ leads to the new potential

$$\tilde{V}(q) = \frac{m}{2} \omega^2 q^2 + \frac{\hbar^2(4\nu^2-1/4)}{2mq^2} - m\omega^2 \qquad (5.15)$$

where $\omega^2 = \dfrac{8V_0}{m\alpha^2}$ and $\nu^2 = 2m \dfrac{V_0 - E}{\hbar^2 \alpha^2}$. The promotor has the expression

$$P_\nu(q'',q';s) = \exp\left[\frac{im\omega^2 s}{\hbar}\right] K_\nu(q'',q';s) \qquad (5.16)$$

where the quantity $K_\nu(q'',q';s)$ now represents the radial propagator for a two-dimensional harmonic oscillator and can be obtained from Eq.(3.37) of chapter 4 by letting $(r,T) = (q,s)$ and $\gamma = 2\nu$. Writing $2k = -m\omega/\hbar$

and $\Omega = i\omega$, and inserting (5.16) in (4.8), we obtain the Green's function

$$G(x'',x';E) = -\frac{4k}{\alpha} \exp[\alpha(x'+x'')/4] \int_0^\infty ds \exp[-2k\Omega s] \operatorname{cosech}(\Omega s)$$

$$\times \exp\left[k\left(e^{-\alpha x'} + e^{-\alpha x''}\right) \coth(\Omega s)\right] I_{2\nu}\left[2k\, e^{-\alpha(x'+x'')/2} \operatorname{cosech}(\Omega s)\right].$$

(5.17)

The formula (5.7) reduces the integral to a closed form

$$G_\nu(x'',x',E) = \left(\frac{-2}{\alpha\Omega}\right) \frac{\Gamma(k+\nu+1/2)}{\Gamma(2\nu+1)} \exp[\tfrac{3}{4}(x''+x')\alpha]$$

$$M_{-k,\nu}(-2k e^{-\alpha x'}) \, W_{-k,\nu}(-2k e^{-\alpha x''}) \qquad (x'' > x'). \quad (5.18)$$

The expression holds for the entire E space as before by analytic continuation. The poles of the Green's function yield the energy eigenvalues. These poles are the singularities of the Γ function in the numerator and are given by $k + \nu + \frac{1}{2} = -n$, where $n = 0, 1, \ldots < n_{max}$. This yields

$$E_n = V_0 - V_0\left(1 - \frac{\alpha\hbar}{\sqrt{2mV_0}}(n+\tfrac{1}{2})\right)^2 \qquad (5.19)$$

with $n_{max} = \left(\frac{2mV_0}{\alpha^2\hbar^2}\right)^{1/2} - \frac{1}{2}$.

The corresponding eigenfunctions can also be readily calculated. We have

$$\psi_n^2(x) = \operatorname{Res} \frac{G(x,x;E)}{i\hbar}\bigg|_{k=-(n+\nu+1/2)}. \qquad (5.20)$$

In order to compute this, let us note that

$$(k+\nu+1/2)\,\Gamma(k+\nu+1/2) \xrightarrow[k=-(n+\nu+1/2)]{} \frac{1}{\Gamma(n+1)}, \qquad (5.21)$$

$$W_{-k,\nu}(z) \xrightarrow[k=-(n+\nu+1/2)]{} \frac{\Gamma(n+2\nu+1)}{\Gamma(2\nu+1)} M_{n+\nu+1/2,\nu}(z) \quad , \quad (5.22)$$

$$M_{n+\nu+1/2,\nu}(z) = e^{-z/2} z^{\nu} F(-n, 2\nu+1; z) \quad . \quad (5.23)$$

Thus the wave functions are given by

$$\psi^2(x) = \left(\frac{2}{\alpha\hbar\omega}\right)\left(\frac{\Gamma(n+2\nu+1)}{[\Gamma(2\nu+1)]^2}\frac{1}{n!}\right)\left(M_{n+\nu+1/2,\nu}(z)\right)^2 \quad . \quad (5.24)$$

It is not, however, possible to take the Fourier inverse of (5.18) analytically to obtain closed form of the Feynman propagator.

5.6. Propagators Related to a Rigid Rotor

In the preceding section, we have related Coulomb and Morse potential problems to harmonic plus inverse square potential problem by means of coordinate time transformation. This technique is also useful to treat a large number of the problems. As examples, we consider the three potentials, viz., the Scarf (S), the symmetric Poschl-Teller (PT) and the Rosen-Morse (RM) potential. They are defined as

$$S : \quad V(x) = V_0 \operatorname{cosec}^2 ax \,, \quad V_0 = \varepsilon_0(s^2 - \tfrac{1}{4}), \quad \varepsilon_0 = \frac{\hbar^2 a^2}{2m} \quad (6.1a)$$

$$PT : \quad V(x) = V_0 \tan^2 ax, \quad V_0 = \varepsilon_0 \lambda(\lambda-1) \,, \quad (6.1b)$$

$$RM : \quad V(x) = V_0 \tanh^2(ax), \quad V_0 = \varepsilon_0 \ell(\ell+1). \quad (6.1c)$$

Under suitable transformation these problems can be reduced to an exactly solvable generic auxiliary problem involving the potential

$$\tilde{V}(\theta) = \frac{\hbar^2}{2m}(\mu^2 - \tfrac{1}{4}) \operatorname{cosec}^2 \theta \quad (6.2)$$

where μ is a constant which depends on the type of the potential. The required coordinate transformations are

$$S : \quad x = a^{-1}\theta, \qquad \frac{dt}{ds} = a^{-2} \qquad (6.3a)$$

$$PT : \quad x = a^{-1}\theta - \frac{\pi}{2a}, \qquad \frac{dt}{ds} = a^{-2} \qquad (6.3b)$$

$$RM : \quad x = a^{-1}\tanh^{-1}(\cos\theta), \qquad \frac{dt}{ds} = a^{-2}\csc^2\theta . \qquad (6.3c)$$

The correction term $3\hbar^2\lambda/8m$ is absent in the S and PT cases because of the linearity of the transformations.

The Green's function $G(x'',x';E)$ for the three potentials take the following forms

$$G_S = a^{-1} \int ds \, \exp\left(\frac{i\hbar E s}{2m\varepsilon_0} \right) K_S(\theta'',\theta';s) \qquad (6.4a)$$

$$G_{PT} = a^{-1} \int ds \, \exp\left\{ \frac{i\hbar s}{2m} [\lambda(\lambda-1) + \frac{E}{\varepsilon_0}] \right\} K_{PT}(\theta'',\theta';s) \qquad (6.4b)$$

$$G_{RM} = a^{-1}(\sin\theta'' \sin\theta')^{-1/2} \int ds \, \exp\left[\frac{i\hbar}{2m} (\ell+\tfrac{1}{2})^2 s \right] K_{RM}(\theta'',\theta';s). \qquad (6.4c)$$

Here K_S, K_{PT}, and K_{RM} are the propagators for potentials (6.2) with different values of μ. These values are

$$\mu_S^2 = s^2, \quad \mu_{PT}^2 = (\lambda - \tfrac{1}{2})^2, \quad \mu_{RM}^2 = \ell(\ell+1) - \frac{E}{\varepsilon_0} . \qquad (6.5)$$

Next consider the problem of obtaining the propagator for the potential in (6.2). The potential $\tilde{V}(\theta)$ restricts the motion of the particle within the sector $0 < \theta < \pi$. It is convenient to consider an auxiliary two-dimensional problem in polar coordinates described by the Lagrangian

$$\tilde{L} = \tfrac{1}{2} m (\dot{r}^2 + r^2 \dot{\theta}^2) - \tilde{V}(\theta) . \qquad (6.6)$$

The idea is to obtain the propagator for (6.6) and insert the constraint $r = 1$. The resulting expression can be inserted in (6.4a-c) with the corresponding values of μ. The propagator for (6.6) has already been obtained in chapter. 4 (See Eq.(3.43)). It is of the form

$$K = \sum_{n=0}^{\infty} K_n(r'',r';s) \, N_n^2 \, (\sin\theta''\sin\theta')^{\mu+1/2} \, C_n^{\mu+1/2}(\cos\theta'') \, C_n^{\mu+1/2}(\cos\theta') \tag{6.7}$$

where $C_n^{\mu+1/2}(\cos\theta)$ are Gegenbauer polynomials. Now, the propagator, for the constrained problem is to be obtained from the path integral representation of $K_n(r'',r',T)$. The result is derived in Sec. 4.3.2 of chapter 4. The expression for K_n after inserting the constraint $r = 1$ reads as

$$K_n = \exp\left[-\frac{i\hbar s}{2m}[n + \mu + \frac{1}{2}]^2\right]. \tag{6.8}$$

The normalization factor N_n is given by

$$N_n^{-2} = \left\{\frac{\pi}{2^{2\mu}} \, \frac{\Gamma(n + 2\mu + 1)}{n!(n + \mu + \frac{1}{2})\,[\Gamma(\mu + \frac{1}{2})]^2}\right\}. \tag{6.9}$$

Inserting the appropriate expression for K in (6.4) and integrating over "new" time s, we have the various Green's functions :

$$G_S(x'',x';E) = \mathcal{A}_S \sum_{n=0}^{\infty} \frac{(N_n^s)^2 \, C_n^{s+1/2}(\cos ax'') \, C_n^{s+1/2}(\cos ax')}{E - \varepsilon_0[(n+s+\frac{1}{2})^2]} \tag{6.10}$$

$$G_{PT}(x'',x';E) = \mathcal{A}_{PT} \sum_{n=0}^{\infty} \frac{(N_n^{PT})^2 \, C_n^{\lambda}(-\sin ax'') \, C_n^{\lambda}(-\sin ax')}{E - \varepsilon_0[(n+\lambda)^2 - \lambda(\lambda - 1)]} \tag{6.11}$$

where the constants $\mathcal{A}_S$ and $\mathcal{A}_{PT}$ have the following expressions

$$\mathcal{A}_S = i\hbar a \, [\sin(ax'')\sin(ax')]^{s+1/2}, \tag{6.12a}$$

$$\mathcal{A}_{PT} = i\hbar a \, [\cos(ax'')\cos(ax')]^{\lambda}. \tag{6.12b}$$

Further the expression for G_{RM} reads as

$$G_{RM}(x'',x';E) = \mathcal{A}_{RM} \sum_{n=0}^{\infty} \frac{(N_n^{RM})^2 \, C_n^{\mu+1/2}(\tanh ax'') \, C_n^{\mu+1/2}(\tanh ax')}{\varepsilon_0 \, (n+\ell+\mu+1) \, (\ell-n-\mu)} \tag{6.13}$$

where

$$\mathcal{A}_{RM} = i\hbar a \, [\text{sech}\,(ax'') \, \text{sech}\,(ax')]^\mu \,. \tag{6.14}$$

The energy eigenvalues are obtained from the poles of the Green's function G in the complex energy plane. These poles are directly displayed in G_S and G_{PT} as the transformations are linear in these two cases. Therefore

$$E_n^S = \varepsilon_0 (n + s + \tfrac{1}{2})^2 \tag{6.15}$$

$$E_n^{PT} = \varepsilon_0 [n^2 + (2n + 1)\lambda]. \tag{6.16}$$

On the other hand, the bound state spectrum of the RM potential is not displayed explicitly. The poles of G_{RM} are given by $(\ell - n - \mu) = 0$, where μ is a function of energy $\mu^2 = [\ell(\ell+1)] - E/\varepsilon_0$. Solving for energy E, we have

$$E_n^{RM} = -\varepsilon_0 \, (n^2 - 2n\ell - \ell^2) \,. \tag{6.17}$$

The corresponding normalized eigenfunctions are obtained from the residue of $G/i\hbar$ at the poles. One can easily read them as

$$\psi_n^s (x) = a \, N_n^{\tilde{s}} \, [\sin(ax)]^{s+1/2} \, C_n^{s+1/2}(\cos ax) \tag{6.18}$$

$$\psi_n^{PT}(x) = a \, N_n^{PT} \, [\cos(ax)]^\lambda \, C_n^\lambda \, (-\sin ax) \tag{6.19}$$

$$\psi_n^{RM}(x) = (a/\varepsilon_0) \, N_n^{RM} \, [\text{cosech}\,(ax)]^{1-n} \, C_n^{1-n+1/2}(\text{cosech}\,ax). \tag{6.20}$$

An alternative way of writing these wave functions is in terms of the associated Legendre polynomials. The required formula for accomplishing this is

$$C_n^\lambda(z) = \frac{\Gamma(2\lambda+n)}{\Gamma(n+1)} \frac{\Gamma(\lambda+1/2)}{\Gamma(2\lambda)} \left\{\frac{z^2-1}{4}\right\}^{\frac{1}{4} - \frac{\lambda}{2}} P_{\lambda+n-1/2}^{-\lambda+1/2}(z) \,. \tag{6.21}$$

It is left as an exercise to show that the eigenfunctions $\psi_n(x)$ may also be written as

$$\psi_n^s(x) = \left[a(n+s+\tfrac{1}{2}) \, \frac{\Gamma(n+2s+1)}{\Gamma(n+1)} \right]^{1/2} (\sin ax)^{1/2} \, P_{n+s}^{-s}(\cos ax) \qquad (6.22)$$

$$\psi_n^{PT}(x) = \left[a(n+\lambda) \, \frac{\Gamma(n+2\lambda)}{\Gamma(n+1)} \right]^{1/2} (\cos ax)^{1/2} \, P_{n+\lambda-1/2}^{-\lambda+1/2}(\sin ax) \qquad (6.23)$$

$$\psi_n^{RM}(x) = \left[a(\ell-n) \, \frac{\Gamma(-n+2\ell+1)}{\Gamma(n+1)} \right]^{1/2} P_\ell^{1-n}(\tanh ax) \, . \qquad (6.24)$$

5.6.1. Infinite square well

We now show one application of these results. The infinite square well potential (ISW) problem is perhaps the simplest bound-state problem in quantum mechanics. Its path integral treatment is, however, not easy. One way is to use the free particle propagator and the method of images. Another way is to use coordinate and time transformation to relate it to a solvable problem. We use the second approach here. The ISW potential is defined as

$$V(x) = 0 \qquad 0 < x < L$$

$$= \infty \qquad x \le 0 \text{ and } x \ge L \, . \qquad (6.25)$$

Now introduce the transformation

$$x = f(q) = (2L/\pi) \tan^{-1} [\exp(-aq)] \quad (-\infty < q < \infty) \qquad (6.26a)$$

$$\frac{dt}{ds} = \left(\frac{aL}{\pi}\right)^2 \text{sech}^2(aq) \, . \qquad (6.26b)$$

The new potential can be written as

$$\tilde{V}(q) = \varepsilon_0(\tfrac{1}{2} - k^2) + \varepsilon_0(k^2 - \tfrac{1}{4}) \tanh^2(aq) \qquad (6.27)$$

where $\varepsilon_0 = \hbar^2 a^2/2m$ and $\hbar^2 k^2/2m = (L/\pi)^2 E$. The transformed Green's function for the infinite square well may now be written in terms of the propagator K_{RM} of the Rosen-Morse potential

$$V_{RM} = \varepsilon_0(k^2 - \tfrac{1}{4}) \tanh^2(aq) \qquad (6.28)$$

that is,

$$G_{ISW} = \frac{aL}{\pi} (\text{sech}(aq'')\text{sech}(aq'))^{1/2} \int_0^\infty ds \exp\left[-\frac{i}{\hbar}\varepsilon_0(\tfrac{1}{2} - k^2)s\right] K_{RM} . \qquad (6.29)$$

The kernel K_{RM} can be written in terms of the Green's function $G_{RM}(q'',q';E_{RM})$ as

$$K_{RM} = \int_{-\infty}^{\infty} \frac{dE_{RM}}{2\pi\hbar} \exp\left[-\frac{i}{\hbar} E_{RM} s\right] G_{RM}(q'',q';E_{RM}) . \qquad (6.30)$$

Green's function G_{RM} can be written down from its expression (6.13) by the replacement $x \to q$ and $\ell \to k - \tfrac{1}{2}$. Inserting the resulting expression in (6.29) and carrying out the integration over s we obtain $E_{RM} = \varepsilon_0(k^2 - \tfrac{1}{2})$. This implies $\mu = \tfrac{1}{2}$ and the expression for G_{ISW} takes the simple form ($(N_n^{RM})^2 = 2/\pi$),

$$G_{ISW} = \left(\frac{2aL}{\pi^2 \varepsilon_0}\right) \mathcal{A}_{ISW} \sum_{n=0}^{\infty} \frac{1}{k^2 - (n+1)^2} C_n^1(\tanh(aq'')) C_n^1(\tanh(aq')) \qquad (6.31)$$

where the constant $\mathcal{A}_{ISW}$ can be obtained from Eq.(6.14) by the replacement $\mu = 1/2$. Going over to the x-coordinate with the help of the relations $\tanh(aq) = \cos(\pi x/L)$ and the formula

$$C_n^1(\cos\theta) = \frac{\sin(n+1)\theta}{\sin\theta} , \qquad (6.32)$$

we obtain the Green's function for the infinite square well

$$G_{ISW} = \frac{2i\hbar}{L} \sum_{n=1}^{\infty} (E - E_n)^{-1} \sin\left(\frac{n\pi x''}{L}\right) \sin\left(\frac{n\pi x'}{L}\right) . \qquad (6.33)$$

Thus the energy eigenvalues E_n and the wave functions $\psi_n(x)$ are

$$E_n = \frac{\hbar^2}{2m}\left(\frac{n\pi}{L}\right)^2 ; \quad \psi_n(x) = \sqrt{(2/L)} \sin(n\pi x/L) . \qquad (6.34)$$

Alternatively we may take the Fourier inverse of the Green's function (6.33) to write the propagator as

$$K(x'',x';T) = \sum_{n=1}^{\infty} \exp\left[-\frac{i}{\hbar} E_n T\right] \sin\left(\frac{n\pi x''}{L}\right) \sin\left(\frac{n\pi x'}{L}\right). \quad (6.35)$$

Problem

Show that the propagator for infinite square well may be written as

$$K(x'',x';T) = \sum_{\nu=-\infty}^{\infty} [K_f(x'', x'+ 2\nu L;T) - K_f(x'', -x'+ 2\nu L;T)] \quad (6.36)$$

where K_f is the free particle propagator.

5.7. Coulomb Problem Based on KS Transformation

In this section we outline the derivation of Hydrogen atom problem using the KS time and coordinate transformations. The essential idea involved here is the mapping of the path integral for the Coulomb problem onto the exactly known path integral for the harmonic oscillator. Instead of using the coordinate-time transformation discussed in Sec. 3, we employ a related but somewhat different approach. We first re-parameterize the paths according to a new "time" and subsequently make a change of variables by means of KS transformation. We start with the Lagrangian path integral

$$K(\vec{x}'',\vec{x}';T) = \int \exp\left[\frac{i}{\hbar} \int_{t'}^{t''} L(\vec{x},\dot{\vec{x}}) \, dt\right] \mathcal{D}[\vec{x}(t)] \quad (7.1)$$

where the attractive Coulomb Lagrangian

$$L(\vec{x},\dot{\vec{x}}) = \frac{1}{2} m \dot{\vec{x}}^2 + \frac{Ze^2}{r} \quad (r=|\vec{x}|) \quad (7.2)$$

and $T = t''-t'$. First we parameterize paths by means of a new "time"

$$s(t) = \int \frac{dt}{r(t)} \quad ; \quad T = \int_0^{\tau} r(s) \, ds \, . \quad (7.3)$$

The constraint (7.3) is incorporated by means of Dirac δ-function so that the propagator

$$K(\vec{x}'',\vec{x}';T) = \int_0^\infty \delta\left(T - \int_0^\tau r(s)ds\right) K(\vec{x}'',\vec{x}';\tau)\, r''d\tau \ . \quad (7.4)$$

Inserting the Fourier representation of the δ-function, we may write (7.4) in the form

$$K(\vec{x}'',\vec{x}';T) = \frac{1}{2\pi\hbar} \int_{-\infty}^\infty dE\, e^{-iET/\hbar}\, G(\vec{x}'',\vec{x}';E) \quad (7.5)$$

where $G(x'',x';E)$ is the energy Green's function

$$G(\vec{x}'',\vec{x}';E) = \int_0^\infty \exp\left(\frac{iE}{\hbar} \int_0^\tau r(s)ds\right) K(\vec{x}'',\vec{x}',\tau)\, r''\, d\tau \ . \quad (7.6)$$

Inserting the path integral form (7.1) in (7.6) and rearranging the terms, the notion of promotor arises. That is

$$G(\vec{x}'',\vec{x}';E) = \int_0^\infty \exp\left(\frac{i\alpha\tau}{\hbar}\right) P(\vec{x}'',\vec{x}',\tau)\, d\tau \quad (7.7)$$

where $\alpha = Ze^2$ and the promotor is defined by the path integral

$$P(\vec{x}'',\vec{x}',\tau) = \lim_{\varepsilon \to 0} \left(\frac{m}{2\pi i\hbar\varepsilon}\right)^{3N/2} r_N(\bar{r}_N)^{-3/2}$$

$$\times \int \exp\left(\frac{i}{\hbar} \sum_{j=1}^N \tilde{S}(\vec{x}_j)\right) \prod_{j=1}^{N-1} (\bar{r}_j)^{-3/2}\, d\vec{x}_j \ . \quad (7.8)$$

Here the discretized action is

$$\tilde{S}(\vec{x}_j) = (m/2\varepsilon\bar{r}_j)(\Delta\vec{x}_j)^2 + \varepsilon\, E\, \bar{r}_j \quad (7.9)$$

where the time step $\varepsilon = (t_j - t_{j-1})/\bar{r}_j$ and $\bar{r}_j$ is related to r_j and shall be defined later.

Our objective is to reduce each short time integral in (7.8) to a Gaussian form and it is advisable to take recourse to a change of variables. A suitable transformation which maps $x^a = (x,y,z)$ in $\mathbb{R}^3$ to the Cartesian coordinates $q^\mu = (q^1, q^2, q^3, q^4)$ in $\mathbb{R}^4$. This transformation may be written as

$$x^a = \sum_{\mu=1}^{4} A^{a\mu}(q) \, q^\mu \qquad (a = 1, 2, 3) \qquad (7.10)$$

where the matrix $A(q)$ is given by

$$A(q) = \begin{pmatrix} q^3 & q^4 & q^1 & q^2 \\ q^4 & -q^3 & -q^2 & q^1 \\ q^1 & q^2 & -q^3 & -q^4 \\ q^2 & -q^1 & q^4 & -q^3 \end{pmatrix} . \qquad (7.11)$$

The matrix has the property that

$$\tilde{A} A = \left(\sum_{\mu=1}^{4} (q^\mu)^2 \right) I = r \, I . \qquad (7.12)$$

The matrix (7.11) maps only a subspace of $\mathbb{R}^4[q]$ onto $\mathbb{R}^3[x]$. A constraint such as

$$\sum_{\mu=1}^{4} A^{a\mu} \, dq^\mu = 0 \qquad (7.13)$$

is necessary to specify the subspace. For path integration we are interested in the change of intervals $\Delta x_j^a = x_j^a - x_{j-1}^a$. It is easy to see by explicit calculation that

$$\Delta x_j^a = 2 \sum_{\mu=1}^{4} A^{a\mu}(\bar{q}_j) \, \Delta q_j^\mu \qquad (7.14)$$

where

$$\bar{q}_j^\mu = \frac{1}{2} \left(q_j^\mu + q_{j-1}^\mu \right) \quad ; \quad \Delta q_j^\mu = q_j^\mu - q_{j-1}^\mu . \qquad (7.15)$$

We notice, however, that

$$\sum_{\mu=1}^{4} A^{4\mu}(q) q^\mu = 0 \qquad (7.16)$$

but a quantity

$$\eta_j = 2 \sum_{\mu=1}^{4} A^{4\mu}(\bar{q}) \Delta q_j^{\mu} \tag{7.17}$$

need not vanish. When q_j is constrained, η_j is a function of Δx_j^a, otherwise η_j is an independent variable. The matrix $A(\bar{q})$ therefore maps $(\Delta q^1, \Delta q^2, \Delta q^3, \Delta q^4)$ onto $(\Delta x, \Delta y, \Delta z, \eta)$, even though $A(q)$ maps the vector, (q^1, q^2, q^3, q^4) onto $(x,y,z,0)$. We use this fact to convert the three-dimensional path integral (7.8) into an equivalent four-dimensional form by inserting the representation of unity, viz.,

$$\prod_{j=1}^{N} \left[\left(\frac{m}{2\pi i \hbar \varepsilon}\right)^{1/2} \int_{-\infty}^{\infty} \exp\left(\frac{i m \eta_j^2}{2 \hbar \varepsilon \bar{r}_j}\right) (\bar{r}_j)^{-1/2} d\eta_j \right] = 1. \tag{7.18}$$

This insertion simply converts the path integral for the promotor in the four-dimensional form

$$P(\vec{x}'',\vec{x}',\tau) = \lim_{\varepsilon \to 0} \mathcal{P}_N \int \exp\left(\frac{i}{\hbar} \sum_{j=1}^{N} \tilde{S}(\vec{x}_j)\right) \left[\prod_{j=1}^{N-1} (\bar{r}_j)^{-2} d^3 x_j d\eta_j \right] d\eta_N \tag{7.19}$$

where

$$\tilde{S}(\vec{x}_j) = (m/2\varepsilon \bar{r}_j)[(\Delta \vec{x}_j)^2 + \eta_j^2] + \varepsilon E \bar{r}_j \tag{7.20a}$$

and

$$\mathcal{P}_N = \left(\frac{m}{2\pi i \hbar \varepsilon}\right)^{2N} r_N (\bar{r}_N)^{-2}. \tag{7.20b}$$

The next step is to specify the midpoint $\bar{r}_j$. A consistent choice is to set

$$\bar{r}_j = \sum_{\mu=1}^{4} (\bar{q}_j^{\mu})^2 = \bar{q}_j^2 \tag{7.21}$$

so that

$$\tilde{A}(\bar{q}_j) A(\bar{q}_j) = \bar{r}_j I \quad ; \quad (\Delta \vec{x}_j)^2 + \eta_j^2 = 4 \bar{r}_j (\Delta q_j)^2. \tag{7.22}$$

The Jacobian of the transformation is

$$\partial(\vec{x},\eta)/\partial(q)_j = 2^4 \bar{r}_j^2. \tag{7.23}$$

The expression (7.8) now takes the simple form

$$P(\vec{x}'',\vec{x}';\tau) = 2^{-4}\int \hat{K}(\vec{q}'',\vec{q}',\tau)(r'')^{-1}d\eta'' \qquad (7.24)$$

where

$$\hat{K} = \lim_{\varepsilon\to 0}\left(\frac{2m}{\pi i\hbar\varepsilon}\right)^{2N}\int \exp\left[\frac{i}{\hbar}\sum_{j=1}^{N}\left(\frac{2m}{\varepsilon}(\Delta q_j)^2 + \varepsilon E \bar{q}_j^2\right)\right]\prod_{j=1}^{N-1}d^4(q_j). \qquad (7.25)$$

The path integral (7.25) is just the propagator for an isotropic oscillator of mass $M = 4m$ and frequency $\omega = (-E/2m)^{1/2}$ in $\mathbb{R}^4$ and has the standard expression

$$\hat{K} = N^4(\tau)\exp\left\{-\pi N^2(\tau)[(\vec{q}''^2 + \vec{q}'^2)\cos\tau - 2\vec{q}''\cdot\vec{q}']\right\} \qquad (7.26)$$

where

$$N(\tau) = \left(\frac{M\omega}{2\pi i\hbar \sin\tau}\right)^{1/2}. \qquad (7.27)$$

We have to insert (7.26) in (7.24) and perform the integration over η''. This may be carried out by making use of the transformation to polar coordinates

$$q^1 = \sqrt{r}\sin\left(\frac{\theta}{2}\right)\cos\left(\frac{\beta+\phi}{2}\right), \quad q^2 = \sqrt{r}\sin\left(\frac{\theta}{2}\right)\sin\left(\frac{\beta+\phi}{2}\right) \qquad (7.28a)$$

$$q^3 = \sqrt{r}\cos\left(\frac{\theta}{2}\right)\cos\left(\frac{\beta+\phi}{2}\right), \quad q^4 = \sqrt{r}\cos\left(\frac{\theta}{2}\right)\sin\left(\frac{\beta+\phi}{2}\right) \qquad (7.28b)$$

where β is an additional angular variable ($0 \leq \beta \leq 4\pi$). We may check that $q^2 = r$ and that $\partial\eta/\partial\beta = r$. Also

$$\vec{q}'\cdot\vec{q}'' = (r'r'')^{1/2}\left\{\sin\left(\frac{\theta'}{2}\right)\sin\left(\frac{\theta''}{2}\right)\cos\left(\frac{\beta''-\beta'+\phi''-\phi'}{2}\right)\right.$$

$$\left.+\cos\left(\frac{\theta'}{2}\right)\cos\left(\frac{\theta''}{2}\right)\cos\left(\frac{\beta'-\beta''+\phi''-\phi'}{2}\right)\right\}. \qquad (7.29)$$

The integration over η'' can now be converted as the integration over β''. We may therefore rewrite (7.24) in the form

$$P(\vec{x}'',\vec{x}',\tau) = \left(\frac{N(\tau)}{2}\right)^4 \exp[-\pi N^2(\tau)(r''+r')\cos\omega\tau\,]$$

$$\times \int_0^{4\pi} \exp\left[u\cos\left(\frac{\beta''-\beta'+\phi''-\phi'}{2}\right) + v\cos\left(\frac{\beta'-\beta''+\phi''-\phi'}{2}\right)\right] d\beta'' \quad (7.30)$$

where, for brevity we use the definitions

$$u = 2\pi(r''r')^{1/2} N^2(\tau) \sin\left(\frac{\theta'}{2}\right) \sin\left(\frac{\theta''}{2}\right) \quad (7.31a)$$

$$v = 2\pi(r''r')^{1/2} N^2(\tau) \cos\left(\frac{\theta'}{2}\right) \cos\left(\frac{\theta''}{2}\right) \,. \quad (7.31b)$$

Next use the expansion formula given in Eq.(3.1) of chapter 4 to write the integral over β'' (denoted by J) in the form

$$J = \sum_{m,m'=-\infty}^{\infty} e^{[i(m+m')(\phi''-\phi')/2]} I_m(u) I_{m'}(v) \int_0^{4\pi} e^{[i(m-m')(\beta''-\beta')]} d\beta''$$

$$= 4\pi I_0([u^2+v^2+2uv\cos(\phi''-\phi')]^{1/2}) \,. \quad (7.32)$$

The last step follows from the addition theorem of the modified Bessel function. Inserting the values of u and v, we can evaluate J as

$$J = 4\pi I_0(2\pi N^2(\tau)(r''r')^{1/2}\cos(\gamma/2)) \,, \quad (7.33)$$

where

$$\cos\gamma = \cos\theta'\cos\theta'' + \sin\theta'\sin\theta''\cos(\phi''-\phi') \,. \quad (7.34)$$

Inserting this value of J in (7.30), we have

$$P = \frac{\pi N^4(\tau)}{4} e^{-\pi N^2(\tau)(r''+r')\cos\omega\tau} I_0[2\pi N^2(\tau)\sqrt{r''r'}\cos(\gamma/2)] \,. \quad (7.35)$$

Now we go back to Eq.(7.7) and write

$$G = \frac{k^2}{4\pi} \int_0^\infty d\tau\, e^{[i\alpha\tau/\hbar]} \operatorname{cosech}^2\left(\frac{\hbar k\tau}{2m}\right) \exp\left[ik(r'+r'')\coth\left(\frac{\hbar k\tau}{2m}\right)\right]$$

$$\times I_0\left[-2ik(r'r'')^{1/2}\cos(\gamma/2)\operatorname{cosech}\left(\frac{\hbar k\tau}{2m}\right)\right] \quad (7.36)$$

where $\omega = i\hbar k/2m = i(E/2m)^{1/2}$. This is the complete expression for the energy Green's function for the hydrogen atom. We had previously obtained the radial Green's function in Eq. (5.8). In order to make contact with this, we use the formula

$$zJ_0[z\cos(\gamma/2)] = \sum_{l=0}^\infty (2l+1) J_{2l+1}(z) P_l(\cos\gamma) \quad (7.37)$$

and the expression

$$G = \sum_{l=0}^\infty \frac{2l+1}{4\pi} G_l(r'',r';E) P_l(\cos\gamma) . \quad (7.38)$$

This yields immediately the radial Green's function

$$G_l = \frac{k}{2}(r'r'')^{-1/2} \int_0^\infty d\tau\, e^{i\alpha\tau/\hbar} \operatorname{cosech}\left(\frac{\hbar k\tau}{2m}\right) \exp\left[ik(r'+r'')\coth\left(\frac{\hbar k\tau}{2m}\right)\right]$$

$$\times J_{2l+1}\left[2k(r'r'')^{1/2}\operatorname{cosech}\left(\frac{\hbar k\tau}{2m}\right)\right] . \quad (7.39)$$

The use of the formula (5.7) then yields the result (5.8).

The merit of employing the full KS coordinate transformation is that we obtain the entire Green's function. On the other hand, if we first expand the Feynman propagator in terms of the spherical harmonics and consider the path integral for the radial propagator $K_l(r'',r';T)$ we are dealing with an object that does not strictly exist mathematically because of the singular nature of the attractive Coulomb potential around $r = 0$. The latter leads to collapse of paths in the $1/r$ hole. Local time parameterization using the KS time and coordinate transformation alleviates this difficulty.

Problems

7.1 Use KS time transformation to derive an exact energy Green's function for the potentials

(i) $V(\vec{r}) = (-a/r) + (b/r^2\sin^2\theta)$,

(ii) $V(\vec{r}) = (-a/r) + (b + \cos\theta)/r^2\sin^2\theta)$.

7.2 Use Levi-Civita transformation (2.7) to derive the energy Green's function for attractive Coulomb potential in two-dimensions.

Notes and References

KS transformations as applied to classical Kepler problem are beautifully discussed in the paper

P. Kustaanheimo and E. Stifel, J. Reine. Angew. Math. **218**, 204 (1965).

Use of this transformation in classical and quantum Kepler problem is discussed in the following papers

M. M. Boiteux and M. Rene Lucas, C. R. Acad. Sci. **274**, 867 (1972),

A. O. Barut, C. K. E. Schreider and R. Wilson, J. Math. Phys. **20**, 2244 (1979).

KS transformation was first used to path integrate hydrogen atom problem in the paper by

I. H. Duru and H. Kleinert, Phys. Lett. **84B**, 185 (1979).

This paper uses the Hamiltonian path integral formulation. Subsequent calculations using Lagrangian formulations have been reported in the following papers

R. Ho and A. Inomata, Phys. Rev. Lett. **48**, 231 (1982),

A. Inomata, Phys. Lett. **87A**, 387 (1983).

Coordinate-time transformations in path integrals were originally examined as point canonical transformation (PCT) in the path integral by

N. Pak and I. Sokman, Phys. Rev. **A30**, 1629 (1984).

Applications of this approach to Coulomb and Morse potentials and a variety of other cases is contained in the following articles

I. H. Duru, Phys. Rev. **D28**, 2689 (1983),

N. Pak and I. Sokman, Phys. Lett. **100A**, 327 (1984),

N. Pak and I. Sokman, Phys. Lett. **103A**, 298 (1984) ,

A. Inomata, Phys. Lett. **101A**, 253 (1984),

P. Y. Cai, A. Inomata and R. Wilson, Phys. Lett **96A**, 117 (1983),

F. Steiner, Phys. Lett. **106A**, 363 (1984) ,

A. Inomata and M. A. Kayed, Phys. Lett. **108A**, 9 (1985).

Path integral solution to infinite square well is treated in the article

I. Sokman, Phys. Lett. **106A**, 212 (1984).

Method of images to treat the square-well problem is reported by

M. Goodman, Am. J. Phys. **49**, 9 (1981).

A path integral solution for the Coulomb plus Aharonov-Bohm potential is found in

L. Chetouani, L. Guechi and T. F. Hammann, J. Math. Phys. **30**, 655 (1989).

Exact path integral solution for a charged particle moving in the field of a dyon or a heavy magnetic Dirac monopole carrying a double charge is obtained in

L. Chetouani, L. Guechi, M. Letlout and T. F. Hammann, Il. Nuovo Cimento **105B**, 387 (1990).

Radial Green's function for the Coulomb problem has been generalized to n-dimensions in the paper

L. Chetouani and T. F. Hammann, J. Math. Phys. **27**, 2944 (1986).

Other references can be found in the following articles

A. Inomata, "Recent developments of techniques of solving non-trivial path integrals", in "Path Summation:Achievements and Goals", Eds. S. Lundquist, A. Ranfagni, V. Sa-yakanit, and L. S. Schulman (World Scientific, Singapore, 1988), pp.~114-146.

A. Inomata, "Time transformation techniques in path integration", in "Path integrals from meV to Mev", (Eds. V. Sa-yakanit et al), (World Scientific, Singapore, 1989) pp.~112-146.

CHAPTER 6

CONSTRAINED PATH INTEGRALS

In the last two decades, theoretical physics has been confronted with several new problems where path integrals have played a major role in their proper understanding. These problems involve evaluation of path integrals obeying certain constraints.

6.1. Examples of Constrained Path Integrals

6.1.1. Problems in Polymer Physics

Let us consider a problem concerning the physics of polymers. The basic concepts have been already discussed in chapter 3. Let $\{A\}$ denote a large ensemble of planar molecules of type A. Assume that the monomers constituting A do not interact. In the chain model of polymers a polymer configuration was simulated by regarding it as a random walk of N steps each step being of size ℓ along the length of the polymer. Further it was shown that in the limit as the distance ℓ between successive monomers is small compared to total length of the polymer ($N\ell$), the fraction of the molecules of type A (also called as the configuration sum) $P(\vec{q},\vec{q}_0;N)$ with their end-points fixed at $\vec{q}_0$ and $\vec{q}$ respectively is described by the functional integral

$$P(\vec{q},\vec{q}_0;N) = \int \exp\left[-\frac{3}{2\ell^2}\int_0^N \left(\frac{d\vec{q}}{d\nu}\right)^2\right] \mathcal{D}\,[q(\nu)]$$

$$\equiv \left(\frac{3}{2\pi\ell^2 N}\right)^{3/2} \exp\left(\frac{-3(q-q_0)^2}{2N\ell^2}\right). \qquad (1.1)$$

Here ν is a parameter running along the length of the polymer $0 < \nu \leq N$. The relation (1.1) implies that a free polymeric system with fixed end-points is distributed in a Gaussian manner. Consider now a slightly different situation. Let another molecule B modelled by an infinite straight line perpendicular to the plane and intersecting the plane at some point (say the origin) is added to the system. (see Fig. 6.1). The presence of B leads to a reduction in the number of configurations available to the molecules of type A because A cannot pass through B. Consequently the entropy of the molecules of type A is lowered while their free energy is increased.

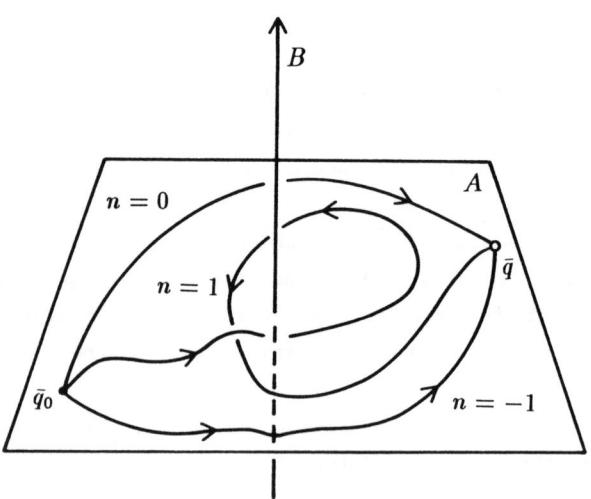

Fig. 6.1. Simple entanglement problem

The different configurations of type A molecules can be classified by the number 'm' of windings around the molecule B (around the origin).

Hence in order to compute the physical properties of type A molecules one has to compute the fraction of type A molecules not only with fixed end-points but also corresponding to a distinct winding index.

6.1.2. Aharonov-Bohm effect

Another example in which the evaluation of a constrained path integral becomes necessary is related to the explanation of the so called Aharonov-Bohm effect (A-B).

Aharonov and Bohm created a great deal of surprise and controversy three decades ago when they showed that electromagnetic potentials (EMP) can lead to observable effects on an electron passing through a field-free region. Hence they concluded that EMP have more fundamental consequence in quantum mechanics than in classical mechanics. Recall that in classical physics EMP are introduced merely to simplify solution of Maxwell's equations. More explicitly, it asserts that the dynamics of a charged particle moving in a region in which both the electric field and the magnetic field vanish is influenced through the presence of the vector potential. This assertion is at variance with classical physics because in such a region the total electromagnetic force being zero the particle travels freely. However, according to quantum mechanics the evolution of the particle is described through the Feynman propagator $K(\vec{q}'',\vec{q}',T)$ associated with the action functional

$$S = \frac{m}{2} \int_0^T \left(\frac{d\vec{q}}{dt}\right)^2 dt + \frac{e}{c} \int \vec{A}.d\vec{q} \ . \qquad (1.2)$$

Here $\vec{A}$ is the vector potential associated with the electromagnetic field and the integral in the second term is evaluated along the trajectory $\vec{q}(t)$ of the particle. Now consider a situation in which a beam of electrons travels in x-y plane in presence of an infinitely long and impenetrable solenoid parallel to z direction confining a constant magnetic flux φ.

The vector potential $\vec{A}(\vec{q})$ corresponding to this set-up can be found in any text book of electrodynamics and reads as

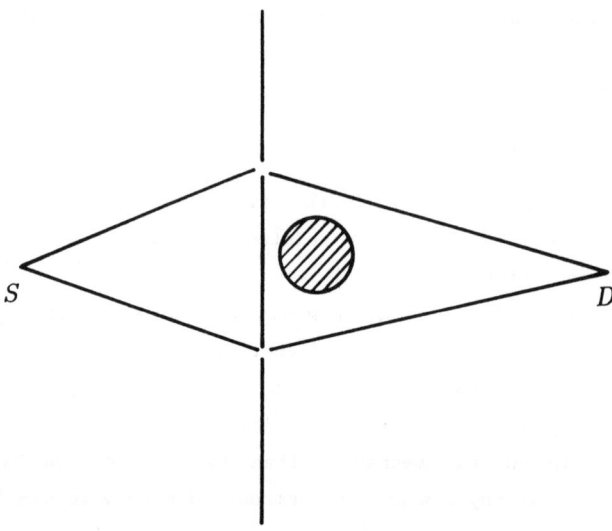

Fig. 6.2. A typical A-B set up: S is a source of charged particles travelling in space towards a detector D. The magnetic flux is confined to a small region shown by the shaded portion.

$$A(\vec{q}) = \frac{\varphi}{4\pi} \int_{\mathcal{R}} |(\vec{q}-\vec{s})|^{-3} (\vec{q}-\vec{s}) \times d\vec{s} \qquad (1.3)$$

where $d\vec{s}$ denotes a line element along the solenoid represented by the closed curve $\mathcal{R}$. The function $\vec{A}(\vec{q})$ has the property

$$\int_{\mathcal{R}} A(\vec{q}) d\vec{q} = \int_{\Gamma_0} \vec{A} \cdot d\vec{q} + n\varphi \qquad (1.4)$$

where Γ_0 denotes some fixed contour and the integer n denotes the number of times the contour $\mathcal{R} - \Gamma_0$ winds through $\mathcal{R}$. Thus, the total amplitude $K(\vec{q}'',\vec{q}';T)$ for a charged particle to go from $\vec{q}'$ to $\vec{q}''$ can be written as

$$K(\vec{q}'',\vec{q}';T) = \exp\left(\frac{ie}{\hbar c} \int_{\Gamma_0} \vec{A} \cdot d\vec{q}\right) \sum_{n=-\infty}^{\infty} \exp\left[\frac{ie\varphi}{\hbar c} n\right] K_n(\vec{q}'',\vec{q}';t) , \qquad (1.5)$$

where K_n denotes the amplitude for the charged particle to go from q' to q'' encircling the solenoid n times.

The above example shows that in order to explain the A-B effect we must evaluate a path integral over only those set of trajectories which encircle a given point a fixed number times. This implies that we must evaluate the path integral

$$K_n(\vec{q},\vec{q}_0;T) = \int \exp(iS/\hbar) \, \mathcal{D}[\vec{q}(t)] \qquad (1.6)$$

over the set of trajectories which apart from assuming fixed values at two points also encircle another fixed point n times. This poses two problems. First we must be able to find a functional of trajectories $\mathcal{L}[\vec{q}(t)]$ such that it is non-zero only when the trajectories obey the constraint. Secondly having found the functional one has to evaluate the unconstrained path integral

$$K_n(\vec{q},\vec{q}_0;T) = \int \mathcal{L}[\vec{q}(t)] \exp(iS/\hbar) \, \mathcal{D}[\vec{q}(t)]. \qquad (1.7)$$

6.2. The Constraint As a Functional

For the problem when the paths entangle with a closed curve Γ, the functional $\mathcal{L}$ should have the following property.

$\mathcal{L}[\vec{q}(t)]$ = n : If $\vec{q}(t)$ winds n times around the curve Γ

= 0 otherwise, . $\qquad (2.1)$

which will be fulfilled if we write $\mathcal{L}$ as :

$$\mathcal{L}[\vec{q}(t)] = \int_c \vec{A}(\vec{q}).d\vec{q} \qquad (2.2)$$

where c is a path along the trajectory $\vec{q}$. The vector field $\vec{A}$ must satisfy the following properties

$$\vec{\nabla}.\vec{A}(\vec{q}) = 0, \qquad (2.3a)$$

$$\vec{\nabla} \times \vec{A}(\vec{q}) = \vec{0} \text{ if } \vec{q} \text{ does not belong to } \Gamma, \quad (2.3b)$$

$$\vec{\nabla} \times \vec{A}(\vec{q}) = \infty, \text{ if } \vec{q} \text{ belongs to } \Gamma. \quad (2.3c)$$

Note that the three properties quoted in Eqs. (2.3) are precisely the properties obeyed by vector potential $\vec{A}$ associated with a current carrying coil along Γ. Therefore finding the functional $\mathcal{L}$ is equivalent to finding the vector potential when a current flows through Γ. The expression for the vector potential for the case when a current flows through a closed curve Γ is given in Eq. (1.3) with Γ coinciding with $\mathcal{R}$. Moreover we normalize $\vec{A}$ such that $\varphi = 1$. One can immediately verify that

$$\oint \vec{A}[\vec{q}] \cdot d\vec{q} = m \quad (2.4)$$

around any closed curve which winds m times around the line element of Γ and zero otherwise. The integral in Eq. (2.4) is known as a Gauss looping integral in electrodynamics. Since the integral is an invariant for all paths that link Γ through m windings it is also called a linking number. Having found $\vec{A}$ and hence $\mathcal{L}$, the propagator associated with n windings can be expressed as

$$K_n = \int \delta(\mathcal{L} - n) \exp(iS/\hbar) \, \mathcal{D}[q(t)]. \quad (2.5)$$

6.3. Evaluation of The Path Integral

To evaluate the path integral in Eq. (2.5), we represent the Dirac δ-function as a Fourier integral and write

$$K_n = \int e^{-i\lambda n} K_\lambda(\vec{q}, \vec{q}_0; T) \, d\lambda \quad (3.1)$$

where K_λ is the propagator

$$K_\lambda = \int \exp\left[\frac{i}{\hbar} S_\lambda\right] \mathcal{D}[\vec{q}(t)] \quad (3.2)$$

with

$$S_\lambda = S + \lambda \hbar \mathcal{L}$$

$$= S + \lambda \hbar \int \vec{A}(\vec{q}) \cdot d\vec{q}$$

$$= S + \lambda \hbar \int_0^T \vec{A}(\vec{q}) \cdot \dot{\vec{q}} \, dt. \qquad (3.3)$$

The explicit evaluation of K_λ will depend both on S and A. For the systems where S is given by

$$S = \int_0^T \left(\frac{m}{2} \dot{\vec{q}}^2 - V(q) \right) dt, \qquad (3.4)$$

K_λ is Green's function of the time-dependent Schrödinger equation described by the Hamiltonian operator

$$\hat{H} = \frac{1}{2m} (\hat{p} - \lambda \hat{A})^2 + V(\vec{q}) \qquad (3.5)$$

where $\hat{p} = -i\hbar \partial/\partial \vec{q}$. If $\hat{H}$ admits a complete set of eigenfunctions φ_n corresponding to eigenvalues λ_n, the Green's function K_λ has the form

$$K_\lambda = \sum_{\{n\}} \varphi_n(\vec{q}) \, \varphi_n^*(\vec{q}_0) \, e^{-i\lambda_n T}. \qquad (3.6)$$

6.3.1. *A simple entanglement problem*

For the specific problem mentioned in 6.1, where the planar electron trajectories wind around an infinitely long straight impenetrable solenoid placed at the origin, the vector potential $\vec{A}$ is given by

$$A_1 = -\frac{y}{2\pi(x^2+y^2)} \quad ; \quad A_2 = \frac{x}{2\pi(x^2+y^2)} \quad ; \quad A_3 = 0 \qquad (3.7)$$

and the associated Schrödinger operator may be written as

$$\frac{i}{\hbar}\frac{\partial}{\partial t} - \frac{1}{2m}\left\{\frac{\partial}{\partial \vec{q}} - \frac{i\lambda}{\hbar}\vec{A}(\vec{q})\right\}^2 . \tag{3.8}$$

Choosing the plane polar coordinates $\vec{q} = (r,\theta)$, the equation for K_λ can be written as

$$-\frac{i}{\hbar}\frac{\partial}{\partial t} - \frac{1}{2m}\left\{\frac{\partial^2}{\partial r^2} + \frac{1}{r}\frac{\partial}{\partial r} + \frac{1}{r^2}\left(\frac{\partial}{\partial \theta} + \lambda\right)^2\right\} K_\lambda = \frac{\delta(r-r')}{\sqrt{rr'}}\delta(\theta-\theta')\delta(t) \tag{3.9}$$

which can be separated into radial and angular parts by using the usual ansatz

$$K_\lambda = \sum_p K_{\lambda p}\, e^{ip(\theta - \theta')} . \tag{3.10}$$

The resulting radial equation can be solved for $K_{\lambda p}$ to arrive at

$$K_{\lambda p} = \frac{1}{2\pi i\hbar t}\exp\left[\frac{im}{2\hbar t}(r^2+r'^2)\right] I_{|\lambda+p|}\left(-\frac{imrr'}{\hbar t}\right) . \tag{3.11}$$

Having evaluated $K_{\lambda p}$, and hence K_λ, K_n can be evaluated as a quadrature by using the defining relation (3.1) with K_λ given by (3.10).

The constraint and topology

The forgoing evaluation of the propagator provides an explicit example of a constrained propagator. In this case the trajectories entering into the path integral were winding around a fixed point a given number of times. This was possible because the trajectories could be classified through their winding numbers. For more complicated situations the classification of trajectories is done by dividing them into distinct 'homotopy classes'. Homotopy is a concept in algebraic topology.

Two objects are said to topologically equivalent if they can be transformed into one another by continuous transformation. At intuitive level a continuous transformation can be thought of as bending, stretching, compressing or twisting or any combination of these without cutting the object and then gluing it. As an example, a sphere and a cube are topologically equivalent, whereas a sphere and a torus are not. A circle is equivalent to a point.

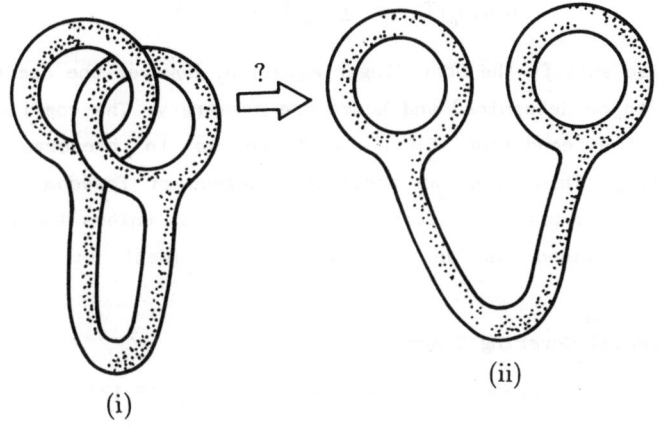

Fig. 6.3.   *The ring puzzle*

However, the classification is not always as transparent as above. Consider for example, the famous ring puzzle as shown in Fig. 6.3. Imagine the object in (i) to be made from a perfect elastic material. Can one deform it so as to unlink the two rings as in (ii) by a continuous transformation? Apparently this might seem to be impossible at first sight but we assure the reader that such a thing is indeed possible and leave the actual manipulations as an exercise.

Thus topologically (i) and (ii) are equivalent whereas the simple surfaces of sphere and torus are not.

Now we return to the relevant example of trajectories. It is easy to verify that two closed trajectories winding around the origin different number of times are topologically inequivalent. This is due to the fact that without opening the loop or removing the solenoid (implying thereby that the origin is no more a singular point) it is impossible to change the linking number. Let K_n be the propagator associated with trajectories characterized by the linking number n. Since K_n is a propagator, it must satisfy the Schrödinger (Diffusion) equation and hence

$$K(\vec{q},\vec{q}_0;T) = \sum_n C_n K_n(\vec{q},\vec{q}_0;T) \qquad (3.12)$$

would also satisfy the Schrödinger equation. However the coefficients C_n are yet to be determined and hence are arbitrary. The coefficients must satisfy the condition $\sum_n |C_n|^2 = 1$ so as to preserve the total probability. This gives an additional degree of freedom to the total propagator and brings in certain non-uniqueness unless the coefficients C_n are provided as an input from experiment or certain other considerations.

The Universal Covering Space

The non-uniqueness in the propagator in (3.12) arises from the basic fact that the underlying dynamics is being studied on a multiply-connected space. In the simple entanglement problem, the presence of the molecule B (the solenoid at origin) makes the origin inaccessible to polymer configurations (Electron trajectories). We call this point (origin) a singular point. Each configuration winding around this singular point different number of times belongs to a different homotopy class. Thus two closed circles of equal radius having different winding number (though in no way different on our physical space) must be treated differently. This can be done by studying the problem not on the physical space but on the so-called "covering space".

To illustrate the construction of covering space consider the simple entanglement problem. Let a plane polar coordinate system be fixed at the origin. Imagine a trajectory originating from r',θ' and terminating at r',θ. On the physical space the angle θ varies between $(0, 2\pi)$. Select a sector between $(0, 2\pi)$ for θ' and define a trajectory going to r',θ after n windings by the map P_n: θ'

$$P_n : \theta' \longrightarrow 2n\pi + (\theta-\theta'), \quad \theta < 2\pi . \qquad (3.13)$$

where $n \geq 0$ if the trajectories wind anticlockwise and $n < 0$ when they wind clockwise.

Now consider two scenarios. In one case the trajectories go to θ after n windings whereas in the other case the trajectories wind m number of times around the origin. In the first case the map P_n takes the point θ' to $2n\pi + (\theta - \theta')$ while the second case corresponds to the point $2m\pi + (\theta - \theta')$. Thus though after every complete winding we arrive at the same point on our physical space, the mapping P_n associates a distinctly different value with each point (r,θ) after n windings on the real line. Note, however, P_n^{-1} is not defined which is the root cause of non-uniqueness. Thus for the simple entanglement problem the covering space would be specified by the parameters $r,\theta : 0 < r < \infty, -\infty < \theta < \infty$. To evaluate the propagator $K_n(r,\theta,r',\theta';T)$ we must evaluate $K_n(r,\theta+2\pi n,r',\theta';T)$ on the covering space. The covering space and the trajectories are shown in the Fig. 6.4.

6.4. Evaluation of The Propagator

The Lagrangian for the free particle moving in a plane is

$$L = \frac{m}{2} (\dot{r}^2 + r^2 \dot{\theta}^2) \qquad (4.1)$$

and therefore the propagator associated with n windings is given by

$$K_n(r,\theta,r',\theta'; T) = \int_{r',\theta'}^{r,\theta+2n\pi} \exp\left[\frac{i}{\hbar}\left(\int_0^T L\, dt + \frac{\hbar^2}{8m}\int_0^T \frac{dt}{r^2}\right)\right] \mathcal{D}\, [\vec{q}(t)] , \qquad (4.2)$$

with $0 < r < \infty, -\infty < \theta < \infty$.

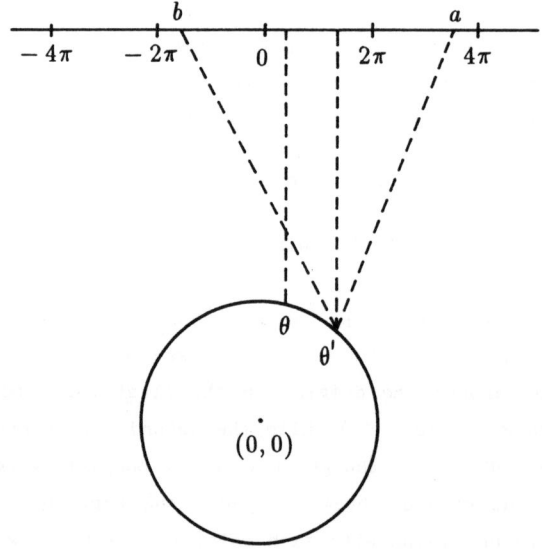

Fig. 6.4 *The covering space*

The second term in the exponential occurring in the integration the right-hand side of Eq.(4.2) arises due to the Jacobian of the transformation from Cartesian to polar coordinates. The Lagrangian L of Eq.(4.1) being quadratic in θ the path integration over θ is straightforward and leads to

$$K_n(r,\theta,r',\theta';T) = \int \mathcal{D}\,[r(t)]\,\exp\left\{\frac{i}{\hbar}\int_0^T\left[\frac{m\dot{r}^2}{2} + \frac{\hbar^2}{8mr^2}\right]dt\right\}$$

$$\times \left(\frac{M}{2\pi i r r'}\right)^{1/2} \exp\left[\frac{iM}{2\hbar}(\theta - \theta' + 2n\pi)^2\right] \quad (4.3)$$

where M is a functional of the radial path and is given by

$$M = m\left(\int_0^T \frac{dt}{r^2}\right)^{-1}. \tag{4.4}$$

In order to perform the radial path integration we use the identity

$$\left(\frac{M}{i\hbar}\right)^{1/2} \exp\left(\frac{iM}{2\hbar}\varphi^2\right) = \frac{1}{\sqrt{2\pi}} \int_{-\infty}^{\infty} \exp\left(-\frac{i\hbar}{2M}\xi^2 + i\xi\varphi\right) d\xi \tag{4.5}$$

and write the path integral in (4.3) as

$$K_n = \int_{-\infty}^{\infty} \exp\left[i\xi(\theta - \theta' + 2\pi n)\right] K_\xi \, d\xi,$$

The path integral K_ξ now reads as

$$K_\xi = \frac{1}{2\pi\sqrt{rr'}} \int \exp\left[\frac{i}{\hbar} S_\xi\right] \mathcal{D}[r(t)]. \tag{4.6}$$

The corresponding radial action functional S_ξ is given by

$$S_\xi = \int_0^T \exp\left\{\frac{m}{2}\dot{r}^2 - \frac{\hbar^2}{2m}\frac{(\xi^2 - 1/4)}{r^2}\right\} dt. \tag{4.7}$$

The path integral in Eq.(4.7) corresponds to the propagator of an action of a fictitious one-dimensional particle of mass m which moves on the half line $0 < r < \infty$ under the influence of the inverse square potential of strength $(\hbar^2/2m)(\xi^2-1/4)$. We have already evaluated the propagator corresponding to this system while studying the path integral in polar coordinates. Therefore we write

$$K_n = \frac{m}{2\pi i\hbar T} \exp\left(\frac{im}{2\hbar T}(r^2 + r'^2)\right) \int_{-\infty}^{\infty} e^{i\xi(\theta - \theta' + 2\pi n)} I_{|\xi|}\left(\frac{mrr'}{i\hbar T}\right) d\xi \tag{4.8}$$

which is the desired result.

In more complicated geometries, the structure of the covering space would, of course, be very complex. However, even then looking at the things in several alternative ways is often rewarding and at least in the case of the simple entanglement problem the approach is very direct as compared to the approach outlined in Sec. 6.1. Moreover, consider a case where the solenoid or the molecule B has finite thickness a. In this case the functional $\mathcal{L}$ may have slightly different structure. However, the covering space is still characterized by the parameters (r,θ); $a < r < \infty$, $-\infty < \theta < \infty$ and hence the angular path integration in (4.3) can still be performed leading to radial path integration as in (4.4). At this stage we can take the advantage of the fact that the radial propagator is the Green's function of the appropriate Schrödinger equation, whose eigenfunctions vanish both at $r = a$ and $r = \infty$. We leave the tedious algebra as an exercise for the reader. The main motivation for drawing attention to this problem was to point out that considerable progress could be made towards the solution of a constrained problem by an appropriate choice of approach.

The mathematicians have studied the related problems in the context of Brownian motion. In particular Spitzer has considered the distribution of winding number of a Brownian particle. However, these problems remained exclusively within the purview of mathematicians until Edwards showed their connection to the problems of theoretical physics.

6.5. Propagators corresponding to More Than One Constraints

Next, let us consider a situation where the problem involves more than one constraints. Let the planar configurations wind with more than one infinitely long straight lines labelled by i (be it the molecular species B or the impenetrable solenoids) intersecting the plane at the point (x_i, y_i), $i = 1, 2, \ldots$ and pose the following question:

What are the fractions of polymer configurations of type A, winding about the singularity (x_i, y_i) n_i times with their ends fixed at two points ? In the context of quantum dynamics the relevant question would be "What is the amplitude $K_{n_1,n_2,\ldots,n_N}(\vec{q},\vec{q}';T)$ associated with trajectories that wind around the point (x_1,y_1) n_1 times"?

In order to answer this question we must find a functional $\mathcal{L}(r(\nu))$ which is non-zero only if the configuration $r(\nu)$ wind around (x_1,y_1) n_1 times. This poses no problem since we already know that the relevant functional can be generated by evaluating the vector potential $\vec{A}$ associated with currents along the respective wires, and can be immediately written with the help of Eq. (3.7). The expression for the vector potential $\vec{A}_1 = (A_{11},A_{21},A_{3i})$ reads as

$$A_{11} = -\frac{(y-y_1)}{2\pi[(x-x_1)^2 + (y-y_1)^2]} \quad (5.1)$$

$$A_{21} = \frac{(x-x_1)}{2\pi[(x-x_1)^2 + (y-y_1)^2]} \quad (5.2)$$

$$A_{31} = 0 \quad (5.3)$$

and the necessary probability amplitude is given by

$$K = \int \prod_{i=1}^{n} \delta(\mathcal{L}_i - n_i) \exp(iS/\hbar) \, \mathcal{D}\,[\vec{q}(t)] \quad (5.4)$$

where the functional $\mathcal{L}_1$ is given by Eq.(2.5) with $\vec{A}$ replaced by $\vec{A}_1$. At this stage, we would like to remind the reader that we have been successful in discovering only one topological invariant whereas for complete specification of the distinct 'homotopy classes' we need more. Therefore, the configurations specified by the windings numbers $(n_1,n_2,\ldots,n_N)$ do not form a distinct homotopy class.

The reason lies in the fact that we have ignored the relative order amongst different windings. For example, consider two configurations winding around two singularities one at (x_1,y_1) and the other at (x_2,y_2) as shown in Fig. 6.5. The configuration 121 winds first around (x_1,y_1) once, then once around (x_2,y_2) and then again once around (x_1,y_1) before

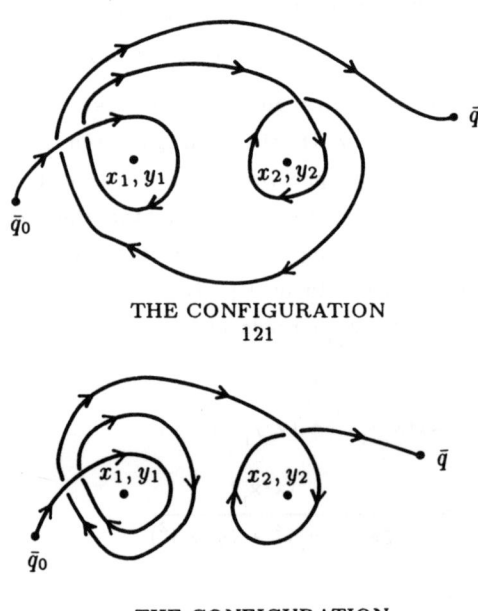

Fig. 6.5. *The configurations 121 and 112 are homotopically distinct but they are assigned the same winding sequence (2,1).*

connecting $\vec{q}$. The configuration 112 winds twice around (x_1, y_1) in succession and then once around (x_2, y_2). Hence 112 and 121 are topologically distinct. However, both have been assigned the sequence (2,1). In many physical situations, such a difference may not be manifest in terms of observable effects. For example if the straight lines were impenetrable solenoids and configurations represent the electron trajectories, the total amount of phase which an electron acquires while moving along a configuration is proportional to total winding number and magnetic flux in the solenoid. However, in case of statistical mechanical studies, the entropy of the configuration 121 is distinctly different than 112 and therefore it leads to observable effects. Even if the complications posed in the above paragraph are ignored, we still have to find the propagator associated with the winding sequence $(n_1, n_2, \ldots, n_N)$. Even for two singularities, this task is difficult and we shall not discuss it here any further.

Problem 6.1

What is the universal covering space for a problem involving motion in a plane where two points have been removed?

6.6. Total Winding Index and Stochastic Area

Instead of asking for the amplitude $K[r,\theta,r'\theta';T]$ associated with the sequence $(n_1, n_2, \ldots, n_N)$ suppose we ask for the amplitude associated with trajectories where the total winding number (algebraic) is fixed, i.e., $n_1 + n_2 + \ldots n_N = N$. The desired amplitude will then be described by the propagator $K_N(\vec{q}, \vec{q}'; T)$

$$K_N(\vec{q}, \vec{q}'; T) = \int \delta(\mathcal{L}_T - N) \exp(iS/\hbar) \, \mathcal{D}\,[\vec{q}(t)] \qquad (6.1)$$

where the functional $\mathcal{L}_T$ now modifies to

$$\mathcal{L}_T = \oint A_T(\vec{q}) \, d\vec{q} \qquad (6.2)$$

where $d\vec{q}$ is the line element along a closed curve winding total N times around various singularities and A_T is the vector field $A_T = (A_{T1}, A_{T2}, A_{T3})$

$$A_{T1} = \sum_i \frac{-(y-y_i)}{2\pi[(x-x_i)^2 + (y-y_i)^2]} \qquad (6.3a)$$

$$A_{T2} = \sum_i \frac{(x-x_i)}{2\pi[(x-x_i)^2 + (y-y_i)^2]} \qquad (6.3b)$$

$$A_{T3} = 0 \qquad (6.3c)$$

Even with this simplification, the analytical evaluation of the K_N is not possible.

However, the situation simplifies considerably if one introduces a further assumption that all the singularities are uniformly distributed in space with a finite surface density σ. In this case we can define the average winding index ϕ as the average of total winding number N over the singularities. First note that

$$\oint_\Gamma \vec{A}_T(\vec{q}) \, d\vec{q} = N \tag{6.4}$$

where Γ is a closed contour enclosing all the singularities. By Stoke's theorem we can write (6.4) as

$$\int_S \vec{\nabla} \times \vec{A}_T(\vec{q}) \, d\vec{S} = N \tag{6.5}$$

where S is the surface enclosed by Γ. Taking the averages over the singularities it is easy to see that

$$2\pi\sigma\mathcal{A} = \langle N \rangle \tag{6.6}$$

where $\mathcal{A}$ is algebraic area enclosed by the curve Γ. Thus we see that when the singularities are uniformly distributed, a configuration can be characterized by an average winding index which is proportional to the area enclosed by the configuration.

The concept of area is really meaningful only for closed curves. However this concept can be extended to open curves by interpreting it as the area enclosed between the curve and the chord joining the end-points.

We must emphasize here that $\mathcal{A}$ is algebraic area. As has been pointed out earlier, this is a direct consequence of the fact that the total winding index does not specify a distinct homotopy class completely. Hence several topologically distinct configurations are not distinguished from one another. This loss is reflected in the area acquiring an algebraic character. For example, though the configurations represented by a point and a figure of 8 belong to topologically

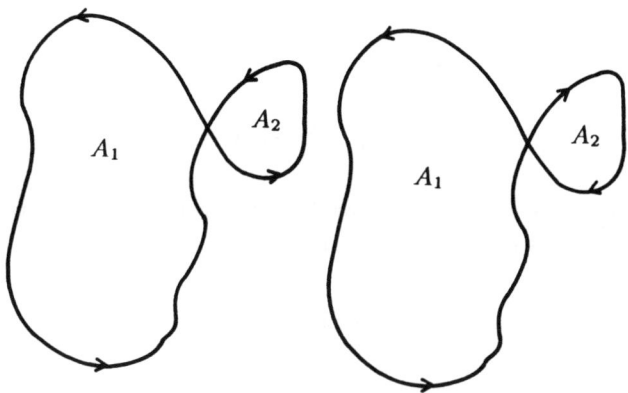

Fig. 6.6. Two configuration having same geometric area but different algebraic area

distinct classes both of them are assigned algebraic area equal to zero. Moreover, the configurations having same geometric area are counted differently (See Fig. 6.6).

With all these drawbacks, the biggest asset of the model is its analytical simplicity because the propagator associated with the trajectories enclosing a fixed algebraic area $\mathcal{A}$ can be expressed as

$$K(\mathcal{A}) = \int \delta\left(\int_0^T \frac{1}{2}(x\dot{y}-y\dot{x})\,dt - \mathcal{A}\right) \exp\left(\frac{i}{\hbar}S\right) \mathcal{D}[\vec{q}(t)]\,, \qquad (6.7)$$

which can be rewritten as

$$K(\mathcal{A}) = \int_{-\infty}^{\infty} e^{i\lambda\mathcal{A}} K_\lambda(\vec{q},\vec{q}';T)\,d\lambda\,. \qquad (6.8)$$

In Eq. (6.8) K_λ represents the propagator

$$K_\lambda = \int \exp\left(\frac{i}{\hbar} S_\lambda\right) \mathcal{D}[\vec{q}(t)] \qquad (6.9)$$

associated with the action S_λ

$$S_\lambda = S + \frac{\lambda \hbar}{2} \int_0^T (x\dot{y} - y\dot{x})\, dt \ . \qquad (6.10)$$

Notice, that S_λ corresponds to the action of a particle characterized by the action S and moving under an uniform magnetic field of strength λ. If S happens to be corresponding to a free particle then we already know the associated propagator. The propagator K_λ is given by

$$K_\lambda = \frac{\lambda}{2\pi i T \sin(\lambda T/2)} \exp\left[\frac{i\lambda m}{T\hbar}(xy' - x'y) - \frac{im\lambda}{2\hbar T}\cot(\lambda T/2)\,(\vec{q}-\vec{q}')^2\right]$$

and therefore, the propagator associated with the configurations of fixed area can be evaluated by using Eq. (6.8).

It might be worthwhile to point out here that all the constrained propagators discussed so far were evaluated by relating them to the appropriate propagators in magnetic fields. This leads to a natural query whether there is an underlying relation between dynamics on multiply connected spaces with dynamics in magnetic fields. It is well known that quantum Hamiltonian on a singly connected space S_c is equivalent to that on a multiply connected space S_m along with an appropriate vector potential. The actual form of the vector potential is decided by the projective mapping $S_c - S_m$ used to describe the motion on S_m.

For the case when the singularities are uniformly distributed, the corresponding propagator is related to the propagator in a uniform magnetic field. This is very natural since each point-singularity is equivalent to a flux line which when smeared out gives rise to uniform magnetic field and hence the propagator for average winding index is related to the propagator in uniform magnetic field. In summary we might surmise that the topology of the space is somehow related to the appearance of magnetic fields.

6.7. Statistical Mechanics of Entangled Polymers

6.7.1. *The properties of entangled polymers*

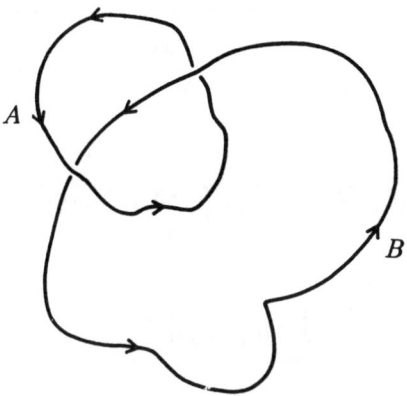

Fig. 6.7. *A Topological bond*

The type of constraints discussed in the previous sections appear very naturally in treatment of the physical properties of entangled polymers. Polymers being long molecules tend to get entangled easily when in solution because of thermal fluctuations.

In Fig. 6.7 we show a typical "topological" bond between two macromolecular rings. Two closed molecular loops A and B are not joined by any chemical bond, yet it is necessary to break a covalent bond to separate them. This in essence is entanglement.

The problem of mutual entanglement of two polymeric components is of direct relevance to the viscous and elastic properties of polymers. When such polymers are subjected to stress, the polymers which are entangled have to be pulled apart and this extra force is purely topological in nature.

To study the statistical mechanics of mutually entangled polymers, one represents one of the molecular species by a fixed closed curve. The other molecular species entangle around this closed curve. As a very special case the fixed curve could be thought of as an infinite straight line. In such a case, we are led directly to the problem mentioned in the introduction.

The propagator for a particle moving in a plane winding around a singularity at the origin has already been found in Sec. 6.4. When translated to polymers by the correspondence mentioned in the introduction, we get the fraction of planar configurations winding around origin as

$$p_n(r,\theta;r',\theta') = \left(\frac{2}{3}\pi Nl^2\right)^{-1} \exp\left(\frac{-3}{2N\ell^2}(r^2+r'^2)\right)$$

$$\times \int_{-\infty}^{\infty} I_{|\xi|}\left(\frac{3rr'}{N\ell^2}\right) e^{i(2\pi n+\theta-\theta')\xi} d\xi . \qquad (7.1)$$

If the polymeric species A wind in three dimensions, p_n gets multiplied by a one-dimensional free propagator $p_n(z, z')$

$$p_n(z, z') = \left(\frac{2\pi N\ell^2}{3}\right)^{-1/2} \exp\left[-3\frac{(z-z')^2}{2N\ell^2}\right] . \qquad (7.2)$$

This is due to the fact that the constraint described by the functional (3.7) when expressed in cylindrical coordinates (r,θ,z) involves r and θ alone. Thus without loss of generality we shall consider only the case of planar molecules.

Let us confine ourselves to the discussion of the statistical mechanical properties of a ring shaped polymer whose one end is fixed at the point r. The fraction of configurations of this polymer which are not entangled with the line from the top is given by $p_0(r,\theta,r,\theta)$ which has the expression

$$p_0 = \frac{3}{2\pi Nl^2} \exp\left[-\frac{3r^2}{Nl^2}\right] \int_{-\infty}^{\infty} I_{|\xi|}\left[\frac{3r^2}{Nl^2}\right] d\xi. \qquad (7.3)$$

The normalized probability P_0 that the polymer is not entangled with the line is given by

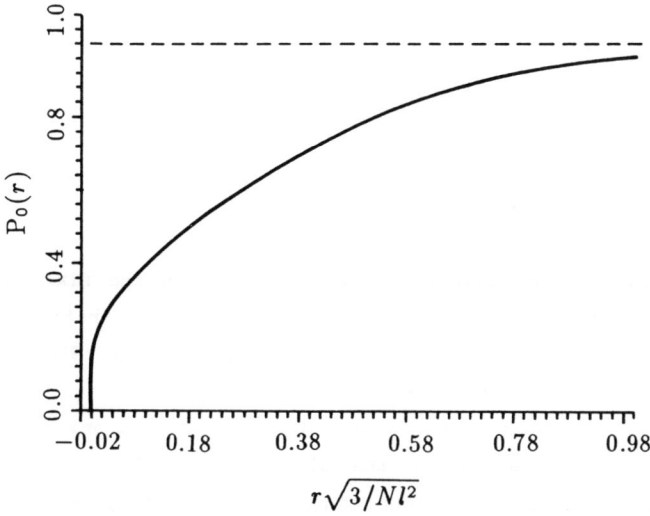

Fig. 6.8. The behaviour of the normalized probability P_0

$$P_0 = P_0 \left(\frac{3}{2\pi Nl^2} \right)^{-1} = \exp\left[-\frac{3r^2}{Nl^2}\right] \int_{-\infty}^{\infty} I_{|\xi|}\left[\frac{3r^2}{Nl^2}\right] d\xi. \qquad (7.4)$$

The probability P_0 as a function of r has been plotted in the Fig. 6.8. It increases from zero with a cusp at the origin. Similarly the probability of entanglement $P_e = 1 - P_0$ with this straight line decreases from unity with a cusp at the origin.

A quantity of physical interest is the elastic force on the polymer. For the configurations which are not entangled, the Helmholtz free energy F_0 of the non-entangled configurations is given by

$$F_0(r) = -kT \ln P_0. \qquad (7.5)$$

Noting that the modified Bessel function I_k for small argument behaves as

$$I_{|\xi|}\left[\frac{3r^2}{Nl^2}\right] \approx \left[\frac{3r^2}{2Nl^2}\right]^{|\xi|} [\Gamma(|\xi| + 1)]^{-1}, \qquad (7.6)$$

we can write F_0 as

$$F_0(r) \propto kT \ln\left[\ln\left(\frac{2N\ell^2}{3r^2}\right)\right] , \quad r \to 0 \qquad (7.7)$$

and hence the force $f(r)$ on the polymer is given by

$$f(r) = -\frac{\partial F_0}{\partial r} = 2kT\left[r \ln\left(\frac{2N\ell^2}{3r^2}\right)\right]^{-1} . \qquad (7.8)$$

Equation (7.8) implies that as $r \to 0$, i.e., when the fixed point is brought near the molecule B (modelled as an infinite straight line), a repulsive force develops which becomes infinitely strong. Since there is no interaction between molecular species A and B this force is purely a consequence of the constraint. Also note that the entropic force is a highly non-linear function of the distance r. This is in complete contrast to the linear behaviour exhibited by free systems. In other words whereas the free polymeric systems obey Hooke's law the entangled polymers do not. This kind of behaviour is indeed shown by rubber like substances.

6.7.2. *The properties of entangled polymers: Entanglement with clusters*

In actual systems like rubber the mutual entanglement is usually with a cluster of molecules. We assume that the molecules of B are again represented by infinite lines perpendicular to the plane of A and the molecular species B are uniformly distributed throughout the space. The properties of the system A entangled with B will again be described by a propagator evaluated along trajectories enclosing a fixed area. This can be obtained from Eq.(6.8) if we identify the propagator $K(A)$ with the fraction of polymer configurations enclosing a fixed area. For the case of polymers the quanatity $K_\lambda(\vec{q},\vec{q}';T)$ appearing under the integral is now denoted by $p(\vec{q},N;\vec{q}',A)$ and is obtained by replacing λ by $-i\omega$. It has the form

$$p(\vec{q},N;\vec{q}',A) = \frac{\omega}{2\pi\ell^2 \sinh\left(\frac{N\omega}{2}\right)} \exp\left[\frac{i\omega}{\ell^2}(xy'-yx') - \frac{\omega}{2\ell^2}\coth\left(\frac{N\omega}{2}\right)(q-q')^2\right].$$

$$(7.9)$$

To evaluate the configuration sum for a ring shaped polymer, we substitute $x = y$ and $x' = y'$. Carrying out the integration over λ in Eq. (6.8) the fraction, $p_c(\mathcal{A})$, of closed molecular loops of area $\mathcal{A}$ is given by

$$p_c(\mathcal{A}) = [2N\ell^2 \cosh^2(2\pi\mathcal{A}/N\ell^2)]^{-1}. \qquad (7.10)$$

The fraction $p_c(\mathcal{A})$ continuoulsy decreases from $(2N\ell^2)^{-1}$ at $\mathcal{A} = 0$. For large values of $\mathcal{A}$ it behaves as $\exp[(-4\pi\mathcal{A}/n\ell^2)/2n\ell^2]$. This is physically reasonable since the configurations occupying large areas for a given length are less probable.

To describe the complete statistical behaviour of these systems one however needs to evaluate a more general functional integral known as generating functional $Z(\vec{q},\vec{q}',N,\vec{\eta})$. It is defined as

$$Z = \int_{-\infty}^{\infty} e^{ig\mathcal{A}} \tilde{Z}(g) \, dg \qquad (7.11)$$

where $\tilde{Z}(g)$ is the functional integral

$$\tilde{Z}(g) = \int \mathcal{D}\,[\{\vec{q}(\nu)\}] \, \exp[\vec{\eta}\cdot\vec{q}_1 - S\,] \qquad (7.12)$$

where $\vec{q}_1$ is the position of a fixed monomer at ν_1 and as explained earlier the action S corresponds to a free particle under the influence of a uniform magnetic field of strength proportional to g (cf. Eq. (6.11). Further $\vec{\eta}$ is a parameter introduced for convenience. Using the Markovian property of path integrals it is easy to see that the desired functional integral of Eq. (7.12) can be expressed as

$$\tilde{Z}(g) = \int e^{\vec{\eta}\cdot\vec{q}_1} \, p(\vec{q},\vec{q}_1,N-N_1,\mathcal{A}) \, p(\vec{q}_1,\vec{q}',N_1,\mathcal{A}) \, d\vec{q}_1 \qquad (7.13)$$

where p is as defined in Eq. (7.9). Moreover since we are interested in the properties of ring shaped polymers we shall evaluate the path integral in Eq. (7.13) under the conditions $\vec{q}' = q = 0$. Amongst the physical properties of polymeric substances, most significant is the average $<q_1^2>$ of the position of the repeating unit squared when both ends are fixed. If the polymer configurations are characterized by the action functional, the desired average is defined as

$$< q_1^2 > \; = \; \int q_1^2 \; \exp \; (-S) \; \mathcal{D} \; [\vec{q}(\nu)] / \; P_c \qquad (7.14)$$

where P_c is the value of $\tilde{Z}(g)$ at $\vec{\eta} = 0$. This can be expressed in terms of Z by the relation

$$< q_1^2 > \; = \; \frac{1}{Z} \left(\frac{\partial^2 Z}{\partial \eta^2} \right)_{\vec{q} = \vec{q}', \vec{\eta} = 0} . \qquad (7.15)$$

A substitution of p from Eq.(7.9) in Eq.(7.13) enables us to write

$$Z(\mathcal{A},N,N_1,\eta) \; = \; \frac{2}{(\pi N \ell^2)^2} \int_{-\infty}^{\infty} du \; \frac{u}{\sinh u} \; \exp\left\{ iu\xi + \frac{\varphi}{u} \; \frac{\sinh \alpha u \; \sinh(1-\alpha)u}{\sinh u} \right\}$$

with

$$\xi \; = \; 4\mathcal{A}/N\ell^2 \quad , \quad \alpha \; = \; N_1/N \quad , \quad \varphi \; = \; N\eta^2 \ell^2/4 \qquad (7.16)$$

and finally using (7.15) we obtain the value of $<q_1^2>$ which reads as

$$<q_1^2> \; = \; 4N\ell^2 \; \cosh^2\!\left(\frac{2\pi\mathcal{A}}{N\ell^2}\right) \left\{ \frac{\xi \; e^{-x}}{1 + e^{-x}} - \frac{\xi \; e^{-x}(e^{-x} + \cos y) - (1-2\alpha)e^{-x}\sin y}{1 + 2e^{-x} \cos y + e^{-x}} \right\}$$

(7.17)

where $x = \pi\xi = 4\pi\mathcal{A}/N\ell^2$, $y = 2\pi\alpha = 2\pi N_1/N$.

The result in Eq. (7.17) already contains a very typical behaviour of rubber like substances in bulk, namely, their rigidity. Notice that for configurations associated with large values of $\mathcal{A}$, $<\vec{q}_1^2>$ is proportional to $\mathcal{A}$. This is in complete contrast to the behaviour of the free polymeric systems where $<\vec{q}_1^2>$ is proportional to N_1. If we again recall that our basic model of polymeric statistical mechanics was essentially based on free random walks along the length of the polymer, the message is obvious. The constraint introduces the rigidity $<q_1^2> \propto \mathcal{A}$

$\propto N_1^2$ in the bulk which is simply missing in free systems. One can derive an analogous expression for the entropic force and convince oneself that the expression for the force is a highly non-linear function of r.

Levy's stochastic area

In the last section we discussed a model where the statistical mechanical properties of the entangled polymer clusters can be related to the area enclosed by closed polymer configurations. The polymer configurations themselves were visualized as random walks. In other words we discussed the distribution of the area enclosed by Brownian paths.

Almost half a century back Paul Levy had posed the following question. A free Brownian particle starts from some point (say origin) at some initial time (t = 0). What is the area enclosed on an average by this particle after time T? The area here is defined as the area enclosed by the trajectory of the particle and the chord connecting initial and final points [see Fig. 6.9].

To evaluate the distribution of the area, we should evaluate the generating functional

$$F(\lambda) = \int \int e^{i\lambda \mathcal{A}} d\mu_w[\vec{q}(t)] d\vec{q} \qquad (7.18)$$

where area $\mathcal{A}$ enclosed by a trajectory can be expressed as

$$\mathcal{A} = \frac{1}{2} \int (\dot{x}y - \dot{y}x) dt \qquad . \qquad (7.19)$$

Further, $d\mu_w[\vec{q}(t)]$ denotes the probability density or Wiener measure associated with the free Brownian motion.

We have already obtained an expression for the configuration sum $p(\vec{q},\vec{q}';N)$ for the polymeric systems enclosing a fixed area $\mathcal{A}$ in Eq.(7.9). This can be integrated over $\vec{q}$ to arrive at the characteristic functional associated with "Levy's stochastic area". The desired functional reads as

$$F(\lambda) = \int p(o,\vec{q},0,T) d\vec{q} = [\cosh(T\lambda/2)]^{-1} \qquad (7.20)$$

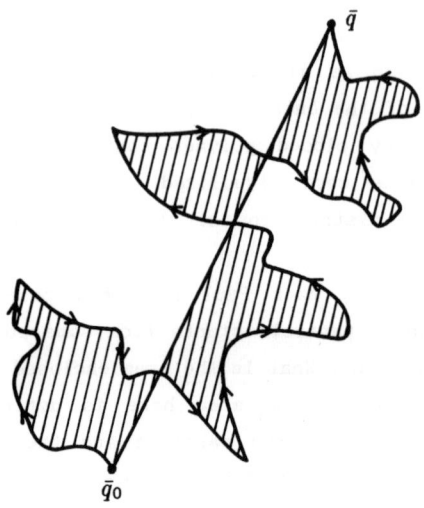

Fig. 6.9. The area enclosed by a Brownian trajectory

which can be used to derive

$$< \mathcal{A} > = \partial F/\partial \lambda \big|_{\lambda=0} = 0 \quad , \tag{7.21}$$

$$< \mathcal{A}^2 > = \partial^2 F/\partial \lambda^2 \big|_{\lambda=0} = (T/2)^2 \quad . \tag{7.22}$$

The expression $F(\lambda)$, is often called the generating functional of Levy's Stochastic area. As can be seen from Eqs. (7.21) and (7.22), the area is itself a stochastic Gaussian random variable. Further, note that the expression for $F(\lambda)$ has a completely different analytical structure as compared to $p(\lambda)$ of Eq. (7.9). This is because Levy has considered paths which assume a fixed value at only one time (t = 0), whereas the polymer configurations are constrained to assume fixed values at two points (t = 0 and t = T). Thus the mathematical structure of the sample paths in two cases is different. Just as a remark we might add that the Brownian motion constrained at two points is called a Brownian bridge. Thus in the case of polymers we are essentially generalizing Levy's results to the case of Brownian bridge.

6.8. Aharonov-Bohm Effect

The basic experimental set-up for an A-B experiment consists of a source of charged particles S passing through a region containing an impenetrable long solenoid confining the magnetic field in a very narrow region and a detector D. The presence of the solenoid makes the region multiply connected. The detector D is used to detect the interference pattern. A shift in the interference pattern is observed as compared with the normal pattern in absence of the magnetic field. The shift is proportional to the magnetic flux ϕ enclosed in the solenoid and vanishes if ϕ is an integral multiple of $2\pi\hbar c/e$.

For a typical A-B set up the classical Lagrangian associated with the electron can be written as

$$L = \frac{m}{2} \vec{q}^2 - \frac{e}{c} \vec{A} \cdot \vec{q} \qquad (8.1)$$

where $\vec{A}$ is vector potential due to the solenoid. If the solenoid is assumed to be very thin and placed perpendicular to x-y plane intersecting it at the origin O the vector potential $\vec{A}$ is given by Eq.(1.3). Assuming a cylindrical coordinate system located at the point O the Lagrangian L takes the form

$$L = \frac{m}{2} (\dot{r}^2 + r^2 \dot{\theta}^2 + \dot{z}^2) - \frac{e\varphi}{2\pi c} \dot{\theta} \qquad (8.2)$$

Since the motion along z direction is not influenced by the magnetic field, the electron evolves freely in this direction. Hence it does not change the interference pattern and we need not consider the motion along z-direction. Effectively the motion is restricted to the motion in two-dimensional (multiply connected) space which is characterized by the Lagrangian L_e

$$L_e = \frac{m}{2}(\dot{r}^2 + r^2\dot{\theta}^2) - \frac{e\varphi}{2\pi c}\dot{\theta} \quad . \tag{8.3}$$

Since the space is multiply connected we must perform path integration on the universal covering space associated with the problem. On this space the last term in Eq.(8.3) only yields an additive constant in action and hence a multiplicative constant to propagator. The propagator K_n associated with n windings for the free Lagrangian on the covering space has already been found out in Eq.(4.8). Since quantum mechanically there is no way to select any path of a particular winding number the total amplitude $K(r'',\theta'',r',\theta';T)$ to reach the detector D placed at r'', θ'' will be given by

$$K(r'',\theta'',r',\theta';T) = \sum_n K_{ne}(r'',\theta'',r',\theta';T) \tag{8.4}$$

where the propagator associated with the trajectories winding n times around the solenoid is given by

$$K_{ne}(r'',\theta'',r',\theta';T) = e^{i\xi(\theta''-\theta'+2\pi n)} K_n(r'',\theta'',r',\theta';T), \quad n > 0 , \tag{8.5}$$

$$K_{ne}(r'',\theta'',r',\theta';T) = e^{i\xi[\theta''-\theta'+(2n-1)\pi]} K_n(r'',\theta'',r',\theta';T), \quad n < 0 \tag{8.6}$$

where the propagator $K_n(r'',\theta'',r',\theta';T)$ is given by Eq.(4.8) and the parameter ξ is a shorthand notation for the quantity $e\varphi/2\pi\hbar c$.

The term $\exp[i\xi(\theta''-\theta')]$ yields only a constant phase factor for the propagator. Now let us consider a case when the parameter ξ assumes an integral value. The use of Poisson's summation formula

$$\sum_{n=-\infty}^{\infty} e^{2\pi i n\alpha} = \sum_{n=-\infty}^{\infty} \delta(n+\alpha) \tag{8.7}$$

enables us to write the expression for the total propagator of (8.4) as

$$K = \frac{m}{2\pi i\hbar T} \sum_{n=-\infty}^{\infty} \exp\left[\frac{im}{2\hbar T}(r'^2 + r''^2)\right] I_n\left(\frac{r'r''m}{i\hbar T}\right) e^{in(\theta''-\theta')} \quad (8.8)$$

apart from a constant phase factor. The propagator in Eq.(8.8) is merely a representation of the free particle propagator in two-dimensional polar coordinates. Hence we can immediately conclude that the total propagator reduces to that corresponding to a free particle whenever ξ assumes integral values. In other words, these values of the parameter ξ which are related to the vector potential do not lead to any observable A-B effect.

Next let us consider the effect when ξ assumes non-integral values. For this case it is not possible to evaluate the integral appearing in the expression of $K_n(r'',\theta'',r',\theta';T)$ in closed analytical form. Therefore we restrict our attention to the case $r'r'' \gg \hbar T$ and approximate the Bessel function I by its asymptotic form

$$I_{|\lambda|}(z) \simeq \frac{1}{\sqrt{2\pi z}} \exp\left[z - (\lambda^2 - \frac{1}{4})\frac{1}{2z}\right]. \quad (8.9)$$

This leads to an approximate expression for $K_n(r'',\theta'';r',\theta';T)$ which reads as

$$K_n = \frac{m}{2\pi i\hbar T} \exp\left[\frac{im}{2\hbar T}(r'-r'')^2\right] e^{i\xi(\theta''-\theta'+2\pi n)}$$

$$\times \exp\left[\frac{imr'r''}{2\hbar T}(\theta''-\theta'+2\pi n)^2\right]. \quad (8.10)$$

The intensity $P(r',\theta',r'',\theta'',T)$ of the interferrence pattern will be given by the square of the modulus of the propagator K in (8.4) and can be written as

$$P(r',\theta',r'',\theta'',T) = \sum_{n,l} \frac{1}{2}(K_n^* K_l + K_l^* K_n). \quad (8.11)$$

Using the approximate value of K_n from (8.10), we can write

$$K_n^* K_1 + K_1^* K_n = 2\left(\frac{m}{2\pi\hbar T}\right)^2 \cos\psi \qquad (8.12)$$

where the quantity ψ is defined as

$$\psi = 2\pi|1-n|\left\{\xi + \frac{mr'r''\bar{\theta}}{\hbar T}\right\} + 2\pi^2\left(\frac{r'r''m}{\hbar T}\right)|1-n|(1+n+1) \qquad (8.13)$$

and $\bar{\theta} = \theta'' - \theta' - \pi$.

One can see that for integral values of ξ, Eq.(8.11) yields a value of P which is independent of ξ. For non-integral values the interference pattern can be generated by using the relation (8.11).

However, in practice the experimental set-up departs considerably from the above mentioned idealized situation. First, the solenoid has a finite thickness. This causes no problem because the vector potential A is still given by the expression (1.3) outside the solenoid. By keeping the distance d between the solenoid and source large compared to the thickness of solenoid r and an electron source of short wave-length one can be very close to ideal situation.

The question of impenetrability of solenoid is another issue. This is achieved by setting up a two slit experiment and keeping the solenoid between the two slits. This geometry also has another feature. Since r << d, the paths with winding number 0 or -1 have a larger amplitude and hence the intensity P of Eq.(8.11) is almost entirely built up of terms n = 0, n = -1. Thus

$$P = |K_0|^2 + |K_{-1}|^2 + 2\,\text{Re}\,K_0^* K_{-1} \qquad (8.14)$$

$$P = 2\left(\frac{m}{2\pi\hbar T}\right)^2 \cos^2\left[\pi\left(\xi + \frac{r'r''\bar{\theta}}{\hbar T}\right)\right] \qquad (8.15)$$

which is the standard A-B result. Again in an actual experiment the values of $\theta'' - \theta'$ are closer to π and hence

$$P = 2\left(\frac{m}{2\pi\hbar T}\right)^2 \cos^2\pi\xi \quad, \qquad (8.16)$$

Notes and References

The best place to get a path integral formulation of the chain model of polymers is a book by

F. W. Wiegel, "Introduction to Path Integral Methods in Physics and Polymer Science" (World Scientific, Singapore, 1986).

A physical problem namely the statistical mechanical properties of entangled polymers was first formulated in terms of topologically constrained path integrals by

S. F. Edwards, Proc. Phys. Soc. **91**, 513 (1967).

The model considered the polymer configurations as random walks on a plane entangling with an infinite straight line placed perpendicular to the plane. The properties of this model were studied by

F. W. Wiegel, Phase Transitions, **7**, 102 (1983).

Later Wiegel also generalizes the model to include attractive interactions on the polymeric system and studies the entanglement probabilities. They are contained in

F. W. Wiegel, J. Chem. Phys., **67**, 469 (1977).

The Feynman propagator associated with a point singularity in a plane was obtained by

A. Inomata and V. A. Singh, J. Math. Phys., **19**, 2318 (1978).

Inomata and Singh used the time-slicing method to arrive at the Feynman propagator. Khandekar et al studied the same problem by considering the dynamics on the covering space to derive the expression for the Feynman propagator. This is contained in

D. C. Khandekar, K. V. Bhagwat and F. W. Wiegel, Phys. Lett. **127A**, 379 (1988).

The point constraint in a plane has also been considered by

S. Prager and H. L. Frisch, J. Chem. Phys. **46**, 1475 (1967),

N. Saitoh and Y. Chen, J. Chem. Phys. **59**, 3701 (1973).

The case involving uniformly distributed singularities as constraints has been considered by Brereton and Butler in the polymer context. It is contained in

M. G. Brereton and Clare Butler, J. Phys. **A20**, 3955 (1987).

However, these authors consider discrete chains and hence their analysis requires numerical methods.

Khandekar and Wiegel use, right from the beginning, continuous random walk model to study the entanglement of a Brownian particle with uniformly distributed singularities in a plane. Their results are contained in

D. C. Khandekar and F. W. Wiegel, J. Phys. **A21**, L573 (1988).

Later Khandekar and Wiegel used these results to study the statistical mechanical properties of polymeric system. They can be found in

D. C. Khandekar and F. W. Wiegel, J. Stat. Mech. **53**, 1073 (1988).

These authors also consider cases where the properties of entangled systems are studied by allowing harmonic interaction between monomers. They form the subject matter of

D. C. Khandekar and F. W. Wiegel, J. de Physique, **50**, 263 (1989).

and in the review article

D. C. Khandekar and F. W. Wiegel, "Path-Integral Treatment of Entangled Polymers" in "Path-Integrals from meV to MeV", Eds. Sa-yakanit *et al* (World Scientific, 1990).

At first glance it is not obvious that the A-B effect is caused by the topology of paths. This was made clear by Gerry and Singh by formulating the problem in the language of path integrals. It can be found in

C. C. Gerry and V. A. Singh, Phys. Rev. **D20**, 2550 (1979),

F. W. Wiegel, Physica, **109A**, 609 (1981).

Later Inomata and Singh used the path integral formulation to obtain explicit expression for Feynman propagator and demonstrate A-B effect. The explicit interference pattern for an idealized A-B set up was examined by

C. C. Bernido and A. Inomata, Phys. Lett. **77A**, 394 (1980),

C. C. Bernido and A. Inomata, J. Math. Phys. **22**, 715 (1981).

The mathematical studies of the winding number distribution were carried out by Spitzer almost half a century before. His results are contained in

F. Spitzer, Trans. Am. Math. Soc. **87**, 187 (1958).

Paul Levy has considered the distribution of area enclosed by a Brownian particle. Paul Levy has studied Brownian motion very extensively and under varying conditions. The famous Levy's Stochastic Area is in fact the distribution of area of a trajectory. One can find its mention in books written by P. Levy.

P. Levy, Am. J. Math. **62**, 487 (1940).

Additional References

A. Inomata, The Aharonov-Bohm Effect with higher winding numbers in "Proc. Second International Symp. on Foundation of Quantum Mechanics", Tokyo 1986.

D. H. Kobe, Ann. Phys. **123**, 381 (1979),

Kazuo Kitihara, "Path integral in a deformed space and application to Aharonov-Bohm Effect in a Dislocated Crystal", in "Path integral from meV to MeV", Eds. Sa-yakanit *et al.* (World Scientific, 1990) p.~97.

The ring puzzle has been discussed in the book

K. Devlin, "Mathematics: The New Golden Age" (Penguin, England, 1988).

CHAPTER 7

TIME DEPENDENT INVARIANTS AND FEYNMAN PROPAGATOR

7.1. Introduction

It is well-known that when Hamiltonian $\hat{H}$ is constant in time the quantal dynamics is treated in terms of the energy eigenfunctions $\phi_n(x)$ of the stationary Schrödinger equation. This method of expanding a time-dependent wave function $\phi(x,t)$ in terms of the energy eigenfunctions and determining the expansion coefficients from the known initial wavefunction $\phi(x,0)$ is no longer applicable when $\hat{H} = \hat{H}(t)$, that is, when the Hamiltonian is time-dependent. The energy is no longer constant and one has to take recourse to other methods.

The time dependence of $\hat{H}(t)$ implies that the system is not closed but is under some external influence (which may not be specified). This changes the parameters of the problem like the total energy or the angular momentum, etc. For example, an atom or molecule may undergo radiative transitions and the explicitly time-dependent term in $\hat{H}(t)$ is the interaction with the radiation field. Another example is that of electromagnetic field in cavity as in the case of a laser, which varies with time under the action of some reservoir. In this case, the field energy $\mathcal{E}$ has the usual expression in terms of the electric ($\vec{E}$) and magnetic ($\vec{H}$) fields given by

$$\mathcal{E} = \frac{1}{2} \int dv \, (\varepsilon_0 E^2 + \mu_0 H^2) \qquad (1.1)$$

where the integration extends over the volume of the cavity and ε_0, μ_0 refer to the dielectric constant and permeability in free space. This problem is usually studied by modelling $\mathcal{E}$ as the Hamiltonian of a harmonic oscillator with a variable mass:

$$\mathcal{E} = H(x,p,t) = \frac{p^2}{2M(t)} + \frac{1}{2} M(t) \omega_0^2 x^2 \qquad (1.2)$$

where the variable mass

$$M(t) = M \exp(-2\gamma t), \qquad \gamma > 0. \qquad (1.3)$$

Another standard example is that of a harmonic oscillator with variable frequency $\omega(t)$. Classically, this example served as a model to describe the lengthening pendulum. Also it served as a model of plasma describing the motion of charged particles in time varying electric and magnetic fields.

The standard methods of solving the quantal time-dependent problems are:

(a) Time-dependent perturbation theory
(b) Adiabatic approximation method
(c) The method of "sudden" approximation.

The perturbation theory is applicable when the time dependent term is small. Hence a few terms of the perturbation series are expected to give a reasonable answer to a physical problem. Adiabatic approximation is valid only when the time scale of the variation of the time dependent term in the Hamiltonian is long compared to all characteristic periods of the system, that is, the external disturbances are acting very slowly. Exactly the opposite is true in the case of the "sudden" approximation which holds only if the external changes are *fast* compared to the *shortest* characteristic period of the system.

In this chapter we focus our attention to another approach which is based on the existence of a constant of motion or an invariant. During the last two decades this approach has been developed extensively. Briefly, the idea is the following. If the Hamiltonian $\hat{H}$ is not a constant of motion, find another operator $\hat{I}$ which is. The eigenfunctions of $\hat{I}$ then play a role similar to that of energy eigenfunctions (when $\hat{H}$ is time-independent). The situation is analogous to that in classical mechanics where the existence of constants of motion greatly simplifies the solution of a problem. Moreover time-dependent classical and quantal invariants are of considerable value in semi-classical formulations of time-dependent quantum mechanical problems. In subsequent sections we intend to discuss the time dependent invariants and their use in path integral formulation of quantum mechanics.

7.2. Classical and Quantal Invariants

7.2.1. Noether invariants

It is well-known that the symmetry properties of the Lagrangian or Hamiltonian imply the existence of conserved quantities. A formal description of the connection between invariance or symmetry properties and conserved quantities is contained in Noether's theorem. The theorem may be stated as follows (for simplicity we restrict ourselves to one-dimension only).

If the functional

$$S[x(t)] = \int L(x, \dot{x}, t) \, dt \qquad (2.1)$$

is invariant under the family of one parameter transformations

$$X = \Psi(x, \dot{x}, t; \varepsilon) \quad ; \quad T = \phi(x, \dot{x}, t; \varepsilon) \qquad (2.2)$$

for arbitrary t' and t'', then

$$I(x, \dot{x}, t) = (\xi \dot{x} - \eta) \frac{\partial L}{\partial x} - \xi L - f = \text{const.} \qquad (2.3)$$

along each extremum of $S[x(t)]$, where ε is a parameter and

$$\xi = (\partial \phi / \partial \varepsilon)_{\varepsilon = 0} \quad , \quad \eta = (\partial \psi / \partial \varepsilon)_{\varepsilon = 0} \qquad (2.4)$$

and $f(x, \dot{x}, t)$ is a function of x, $\dot{x}$, and t. The Noether's theorem already defines the invariant I and also provides a means of obtaining it.

In order to outline a proof of Noether's theorem, we recall that the action functional (2.1) is invariant under a one-parameter family of transformations (2.2) if

$$S = \int_{t'}^{t''} L(x, \dot{x}, t) \, dt = \int_{T'}^{T''} L(X, \dot{X}, T) \, dT \, . \qquad (2.5)$$

This condition implies that the geometry of the problem is invariant and the form of the Eüler equations is preserved under these transformations. Note also that one can add a total time derivative of a function $f(x, \dot{x}, t)$ inside the action integral to obtain a more general condition on the action functionals. The proof of Noether's theorem is as follows. The general variation of the functional (2.1) is given by

$$\delta S = \int_{t'}^{t''} \left[\frac{\partial L}{\partial x} - \frac{d}{dt}\left(\frac{\partial L}{\partial \dot{x}}\right)\right] h(x) \, dt + \left[\left\{L - \dot{x}\left(\frac{\partial L}{\partial \dot{x}}\right)\right\} \delta t + \left(\frac{\partial L}{\partial \dot{x}}\right) \delta x\right]_{t=t'}^{t=t''} \quad (2.6)$$

where,

$$h(x) = \delta x - \dot{x}\, \delta t \quad , \quad \delta x = \eta \varepsilon \quad , \quad \delta t = \xi \varepsilon \, . \quad (2.7)$$

Let the curve $x = x(t)$ be an extremum of $S[x]$. Then

$$\frac{\partial L}{\partial x} - \frac{d}{dt}\left(\frac{\partial L}{\partial \dot{x}}\right) = 0 \quad (2.8)$$

and

$$\delta S = \varepsilon \left\{\left[L - \dot{x}\left(\frac{\partial L}{\partial \dot{x}}\right)\right] \xi + \eta \left(\frac{\partial L}{\partial \dot{x}}\right)\right\} . \quad (2.9)$$

Since by hypothesis $S[x(t)]$ is invariant under the transformation (2.2) δS vanishes, that is,

$$\left[\left\{L - \dot{x}\left(\frac{\partial L}{\partial \dot{x}}\right)\right\} \xi + \eta \left(\frac{\partial L}{\partial \dot{x}}\right) + f(x, \dot{x}, t)\right]_{t=t'}^{t=t''} = 0 \, , \quad (2.10)$$

where the function $f(x, \dot{x}, t)$ is introduced in order to have the general condition of invariance of action functional (2.1). The fact that Eq. (2.10) holds along each extremum now follows from the arbitrariness of t' and t''. Thus we have the Noether invariant

$$I(x, \dot{x}, t) = (\xi \dot{x} - \eta)\left(\frac{\partial L}{\partial \dot{x}}\right) - \xi L - f \quad (2.11)$$

which satisfies the condition

$$\frac{dI}{dt} = \frac{\partial I}{\partial t} + \left(\frac{\partial I}{\partial x}\right)\dot{x} + \left(\frac{\partial I}{\partial \dot{x}}\right)\ddot{x} = 0 \, . \quad (2.12)$$

We now ask the question : Given the Lagrangian

$$L = \frac{\dot{x}^2}{2} - V(x,t) \qquad (2.13)$$

where $V(x,t)$ is the potential function, what is the most general form of $V(x,t)$ so that a Noether invariant exists? In order to answer the question, we need to assume $\xi = \xi(x,t)$, $\eta = \eta(x,t)$ and $f = f(x,t)$, since the potential does not depend explicitly on the velocity $\dot{x}$. The Noether invariant takes the form

$$I(x,\dot{x},t) = \xi \dot{x}^2/2 - \eta \dot{x} + \xi V - f . \qquad (2.14)$$

Inserting the above expression in the defining Eq.(2.12) and equating the coefficients of powers of $\dot{x}$, separately to zero, we arrive at the following set of partial differential equations :

$$\frac{\partial \xi}{\partial x} = 0, \quad \frac{\partial \eta}{\partial x} = \frac{1}{2}\frac{\partial \xi}{\partial t}, \quad \frac{\partial f}{\partial x} = V\frac{\partial \xi}{\partial x} - \frac{\partial \eta}{\partial t}, \qquad (2.15)$$

$$V\frac{\partial \xi}{\partial t} + \xi\frac{\partial V}{\partial t} + \eta\frac{\partial V}{\partial x} - \frac{\partial f}{\partial t} = 0 . \qquad (2.16)$$

From Eqs.(2.15)--(2.16), it follows that

$$\xi(x,t) = a(t) , \quad \eta(x,t) = \frac{\dot{a} x}{2} + b(t) , \qquad (2.17)$$

$$f(x,t) = -\frac{\ddot{a} x^2}{4} - \dot{b} x + c(t) \qquad (2.18)$$

where a, b, and c are arbitrary functions of time and $V(x,t)$ satisfies the partial differential equation

$$a\frac{\partial V}{\partial t} + \left(\frac{\dot{a} x}{2} + b\right)\frac{\partial V}{\partial x} + \dot{a} V + \left(\frac{\dddot{a} x^2}{4} + \ddot{b} x - \dot{c}\right) = 0 . \qquad (2.19)$$

The associated Lagrange system for Eq.(2.19) is

$$\frac{dt}{a} = \frac{2\,dx}{\{\dot{a}x + 2b\}} = \frac{dV}{[-\dot{a}V - \{\dddot{a}(x^2/4) + \ddot{b}x - \dot{c}\}]} . \qquad (2.20)$$

Taking the first and second terms of Eq.(2.20), we have the differential equation

$$a \frac{dx}{dt} - \frac{\dot{a} x}{2} = b , \qquad (2.21)$$

which may be readily solved to obtain the zeroeth-order invariant

$$u = a^{-1/2} x - \int a^{-3/2} b \, dt . \qquad (2.22)$$

From the second and third terms of Eq.(2.20) and making use of Eq.(2.22) we obtain the differential equation

$$\frac{d}{dt}(aV) + \frac{\dot{a}\dot{a}}{4} (u + \int a^{-3/2} b \, dt')^2 + \ddot{b} \left(u + \int a^{-3/2} b \, dt' \right) - \dot{c} = 0.$$

$$(2.23)$$

This equation can be solved easily to obtain the first order invariant

$$v = aV + \left\{ \frac{2a \ddot{a} - \dot{a}^2}{8a} \right\} x^2 - \left\{ \frac{\dot{a}b - 2a\dot{b}}{2a} \right\} x - \left\{ c + \frac{b^2}{2a} \right\}. \qquad (2.24)$$

We now set

$$c = - b^2/2a \; ; \; v = F(u) \qquad (2.25)$$

where F is an arbitrary function of its argument. The admissible potential V is then given by

$$V = \frac{(\dot{a}b - 2 a\dot{b})x}{2a^2} - \frac{(2 a \ddot{a} - \dot{a}^2)x^2}{8a^2} + \frac{1}{a} F\left(a^{-1/2}x - \int a^{-3/2}b \, dt \right). \qquad (2.26)$$

The corresponding invariant is obtained by inserting this form in Eq.(2.14) and reads as

$$I(x,\dot{x},t) = \frac{1}{2}\left[\sqrt{a} \, \dot{x} - \frac{\dot{a}x}{2\sqrt{a}} - \frac{b}{\sqrt{a}} \right]^2 + F\left(a^{-1/2}x - \int a^{-3/2}b \, dt \right) . \qquad (2.27)$$

We may write this in a more elegant form by introducing

$$\rho = \sqrt{a} , \quad \alpha/\rho = \int a^{-3/2} b \, dt . \qquad (2.28)$$

The expressions for the admissible potential V(x,t) and the invariant

$I(x, \dot{x}, t)$ take the following forms

$$V(x,t) = \left(\frac{\dot{\rho}\,\alpha}{\rho} - \ddot{\alpha}\right)x - \frac{\ddot{\rho}\,x^2}{2\,\rho} + \rho^{-2}\,F\!\left(\frac{x-\alpha}{\rho}\right), \qquad (2.29)$$

$$I(x,\dot{x},t) = \frac{1}{2}\Big[\rho(\dot{x}-\dot{\alpha}) - \dot{\rho}(x-\alpha)\Big]^2 + F\!\left(\frac{x-\alpha}{\rho}\right). \qquad (2.30)$$

7.2.2. Derivation of invariant based on Hamiltonian description

We note that the Noether invariant is quadratic in the velocity $\dot{x}$. In Hamiltonian formulation, it will be an invariant quadratic in momentum p. We may paraphrase the previously posed question as follows. Given the Hamiltonian

$$H = \frac{p^2}{2} + V(x,t), \qquad (2.31)$$

what is the permissible form of $V(x,t)$ so that H admits an invariant quadratic in momentum p? It turns out that the form is identical to the one obtained in Eq (2.29). Nevertheless, we outline this derivation here. Assume that the invariant $I(x,p,t)$ is at most quadratic in p, that is, it has the form

$$I(x,p,t) = f_0(x,t) + p\,f_1(x,t) + p^2\,f_2(x,t). \qquad (2.32)$$

Inserting this form in the defining equation for I, viz.,

$$\frac{dI}{dt} = \frac{\partial I}{\partial t} + \{I, H\} = 0 \qquad (2.33)$$

where $\{\,,\,\}$ denotes the classical Poisson bracket, we obtain

$$\left(\frac{\partial f_0}{\partial t} - \frac{\partial V}{\partial x}f_1\right) + p\left(\frac{\partial f_1}{\partial t} + \frac{\partial f_0}{\partial x} - 2\frac{\partial V}{\partial x}f_2\right)$$

$$+ p^2\left(\frac{\partial f_2}{\partial t} + \frac{\partial f_1}{\partial x}\right) + p^3\frac{\partial f_2}{\partial x} = 0. \qquad (2.34)$$

Equating the coefficient of each power of p to zero, we obtain the following set of partial differential equations

$$\frac{\partial f_0}{\partial t} - \frac{\partial V}{\partial x}f_1 = 0,\quad \frac{\partial f_1}{\partial t} + \frac{\partial f_0}{\partial x} - 2\frac{\partial V}{\partial x}f_2 = 0, \qquad (2.35)$$

$$\frac{\partial f_2}{\partial t} + \frac{\partial f_1}{\partial x} = 0 \quad , \quad \frac{\partial f_2}{\partial x} = 0 \ . \tag{2.36}$$

From Eqs. (2.35)--(2.36), it follows that

$$f_2(x,t) = \rho^2/2 \quad , \quad f_1(x,t) = -\rho\dot\rho x + b \ , \tag{2.37}$$

$$f_0(x,t) = \rho^2 V + (\dot\rho\dot\rho + \ddot\rho^2) x^2 - \dot b x + c \ , \tag{2.38}$$

while $V(x,t)$ satisfies the partial differential equation

$$\rho^2 \frac{\partial V}{\partial t} + (\rho\dot\rho x - b)\frac{\partial V}{\partial x} + 2\rho\dot\rho V + (3\dot\rho\ddot\rho + \rho\dddot\rho)\frac{x^2}{2} - \ddot b x + \dot c = 0 \ . \tag{2.39}$$

Here $\rho(t)$, $b(t)$, and $c(t)$ are arbitrary functions of time. As before the partial differential equation can be solved by writing the associated Lagrange system

$$\frac{dt}{\rho^2} = \frac{dx}{(\rho\dot\rho x - b)} = -\frac{dV}{2\rho\dot\rho V + (3\dot\rho\ddot\rho + \rho\dddot\rho)(x^2/2) - \ddot b x + \dot c} \ . \tag{2.40}$$

The first two terms of this equation yield the differential equation

$$\rho^2 \frac{dx}{dt} - \rho\dot\rho x = -b \tag{2.41}$$

and its solution gives the zeroth order invariant

$$u = \frac{x}{\rho} + \int b\,\rho^{-3}\, dt' \ . \tag{2.42}$$

Next, taking the second and third terms of Eq. (2.40) and substituting for x through (2.42) we obtain the differential equation

$$\rho^2 \frac{dV}{dt} + 2\rho\dot\rho V + \left(3\dot\rho\ddot\rho + \rho\dddot\rho\right)\left(u - \int b\,\rho^{-3}\,dt'\right)^2 \frac{\rho^2}{2}$$

$$- \ddot b \left(u - \int b\,\rho^{-3}\,dt'\right)\rho + \dot c = 0 \ . \tag{2.43}$$

This equation when integrated over time yields the first order invariant

$$v = \rho^2 V + \rho \ddot{\rho} \frac{x^2}{2} + (b\dot{\rho} - \dot{b}\rho)\frac{x}{\rho} + b^2 \frac{\rho^{-2}}{2} + c. \qquad (2.44)$$

As before, without loss of generality, we can choose $c = -b^2/2\rho^2$ and set $v = F(u)$ to arrive at the admissible form of $V(x,t)$ given by

$$V(x,t) = \left(-\frac{b\dot{\rho}}{\rho^3} + \frac{\dot{b}}{\rho^2}\right)x - \frac{\ddot{\rho} x^2}{2\rho} + \rho^{-2} F\left(x + \int \frac{b}{\rho^3} dt'\right). \qquad (2.45)$$

Also inserting the expressions (2.37--2.38) for f_0, f_1, and f_2 in Eq.(2.32) we obtain the invariant

$$I(x,p,t) = (\rho p - \dot{\rho} x + b/\rho)^2 + F\left(\frac{x}{\rho} + \int \frac{b}{\rho^3} dt'\right). \qquad (2.46)$$

If we set $\int b\rho^{-3} dt' = -\alpha/\rho$ in the above expressions (2.45) and (2.46) we arrive at Eqs.(2.29) and (2.30).

This theory can be generalized to more than one dimension. In that case the function $\rho(t)$ would still be a scalar but $x(t)$ and $\alpha(t)$ would be replaced by vectors $\vec{x}(t)$ and $\vec{\alpha}(t)$ respectively. However, there could be other invariants present when more than one dimensions are considered. What we desire in the subsequent discussions is an invariant which is closest in form to the Hamiltonian $H(x,p,t)$ and for this the form derived in Eq.(2.30) is adequate.

7.2.3. Examples of invariants

These general considerations allow us to derive several of the invariants discussed in literature. The simplest case is that of a harmonic oscillator with time-dependent frequency $\omega(t)$. Here the equation of motion is

$$\ddot{x} + \omega^2(t) x = 0. \qquad (2.47)$$

Comparing the potential $V(x,t) = \omega(t)x^2/2$ with the general form (2.29), we see that $\alpha = F = 0$ and $\rho(t)$ satisfies the equation

$$\ddot{\rho} + \omega^2(t) \rho = 0. \qquad (2.48)$$

The invariant takes the form

$$\dot{I} = \frac{1}{2}(\rho \dot{x} - \dot{\rho} x)^2 . \tag{2.49}$$

We could have also chosen $F = \frac{1}{2}\omega_0^2 x^2 \rho^{-2}$ where ω_0 is certain constant, in which case $\rho(t)$ satisfies the equation of Pinney,

$$\ddot{\rho} + \omega_0^2 \rho = \omega_0^2 \rho^{-3} \tag{2.50}$$

and the invariant reads as

$$I = \frac{1}{2}[(\rho \dot{x} - \dot{\rho} x)^2 + \omega_0^2 x^2 \rho^{-2}] . \tag{2.51}$$

Equations (2.47) and (2.48) [or (2.50)] together are known as Ermakov pairs. The invariant of (2.51) was first derived by Ermakov by eliminating $\omega^2(t)$ between these two equations. Subsequently Lewis rederived this invariant by assuming, *ab initio*, a quadratic form in x and p for I and evaluating the coefficients from the invariance condition (2.12).

Another known result is for the time-dependent oscillator with an inverse square potential. Here the Lagrangian takes the form

$$L = \frac{1}{2}(\dot{x}^2 - \omega^2(t) x^2) - g x^{-2} . \tag{2.52}$$

Here $\alpha = 0$, but $F = \frac{1}{2}\omega_0^2 x^2 \rho^{-2} + g \rho^2 x^{-2}$ and the invariant takes the form

$$I = \frac{1}{2}\left[\left(\rho \dot{x} - \dot{\rho} x\right)^2 + \omega_0^2 \left(\frac{x}{\rho}\right)^2\right] + g \left(\frac{\rho}{x}\right)^2 , \tag{2.53}$$

while $\rho(t)$ satisfies the same equation (2.50) as before. In fact, an useful generalization of (2.52) is the Lagrangian

$$L = \frac{1}{2}(\dot{x}^2 - \omega^2(t) x^2) - \rho^{-2} F\left(\frac{x}{\rho}\right) , \tag{2.54}$$

where F is an arbitrary function of its argument and ρ satisfies Eq.(2.50). The equation of motion and the invariant I read as

$$\ddot{x} + \omega^2(t) x + \rho^{-2} \frac{\partial F}{\partial x} = 0 , \tag{2.55}$$

$$I(t) = \frac{1}{2}\left[(x\dot{\rho} - \rho\dot{x})^2 + \omega_0^2 \left(\frac{x}{\rho}\right)^2\right] + F\left(\frac{x}{\rho}\right) \quad . \quad (2.56)$$

We might add here that Noether's theorem can be applied to a more general Lagrangian of the form

$$L = \frac{1}{2} a(t) [\dot{x}^2 - \omega^2(t)x^2] - P(x,t) \quad . \quad (2.57)$$

Physically $a(t) > 0$ represents either a variable mass of the particle or a frictional force depending linearly on particle velocity. It is left as an exercise to show that the admissible form for $P(x,t)$ is given by

$$P(x,t) = \rho^{-2} F\left(\frac{\sqrt{a}\,x}{\rho}\right) \quad (2.58)$$

where F is again an arbitrary function of its argument and $\rho(t)$ obeys the equation

$$\ddot{\rho} + \Omega^2(t)\rho = \omega_0^2 \rho^{-3} \quad (2.59)$$

where

$$\Omega^2(t) = \omega^2(t) + \frac{(\dot{a}^2 - 2a\ddot{a})}{4a^2} \quad . \quad (2.60)$$

The invariant I and the classical equations of motion are given by

$$I = \frac{1}{2}(z^2 + \omega_0^2 y^2) + F(y) \quad , \quad \frac{d}{dt}(a\dot{x}) + \omega^2(t)ax + \sqrt{a}\,\rho^{-3}\frac{\partial F}{\partial y} = 0 \quad (2.61)$$

where, $y = \sqrt{a}\,x/\rho$, $z = \rho^2 \frac{dy}{dt}$.

As a final remark, we mention that classical Lagrangians involving velocity-dependent potentials may also admit an invariant. Although these velocity-dependent systems are of considerable interest in classical mechanics, their physical interpretation in the quantum context is difficult and requires a much deeper study. Nevertheless, quantum mechanical considerations for such systems may be best introduced through Feynman path integral approach. For this reason, we consider here the system described by the Lagrangian

$$L = \frac{1}{2} a(t) [\dot{x}^2 - \omega^2(t) x^2] - Q(x, \dot{x}, t) \quad . \tag{2.62}$$

It is tedious but straightforward to show that a Noether invariant for the problem exists if

$$Q(x, \dot{x}, t) = \rho^{-2} F(y, z) \quad , \tag{2.63}$$

where y and z are as defined earlier and $\rho(t)$ satisfies the same auxiliary equation as (2.59) The classical equation of motion and the invariant have the form

$$\frac{d}{dt} \left(a \dot{x} - \frac{1}{\sigma} \frac{\partial F}{\partial z} \right) + \omega^2(t) a x + \frac{1}{\rho^2 \sigma} \frac{\partial F}{\partial y} - \frac{\dot{\sigma}}{\sigma^2} \frac{\partial F}{\partial z} = 0 \quad , \tag{2.64}$$

$$I(t) = \frac{1}{2} (z^2 + \omega_0^2 y^2) + F(y, z) - z \frac{\partial F}{\partial z} \quad , \tag{2.65}$$

with $\sigma = \rho/\sqrt{a}$.

Going beyond the time-dependent harmonic oscillator and all its variations, we may now consider the interesting case of non-stationary Coulomb potential. Applying our general criterion that $V(\vec{q}, t)$ should have the standard admissible form (2.29), we find that for Coulomb potential $V(\vec{q}, t) = -Z(t)/r$, where $r = |\vec{q}|$, the "time-dependent charge", $Z(t)$ should have the form $Z(t) = Z_0/\rho(t)$, where $\rho(t)$ satisfies the equation $\ddot{\rho} = 0$ and Z_0 is a constant ($\vec{\alpha}(t) = 0$). This yields that the admissible time-dependence $\rho(t) = (1 + \beta t)$ with β, an arbitrary constant. The required invariant then reads as

$$I(\vec{q}, \dot{\vec{q}}, t) = \frac{1}{2} (\rho \dot{\vec{q}} - \dot{\rho} \vec{q})^2 - Z_0 \frac{\rho}{r} \quad . \tag{2.66}$$

Finally we remark that it is sufficient to know the classical invariant. It turns out that the classical invariant also becomes the quantum invariant when the canonical momentum p is replaced by the quantum mechanical operator $(\hbar/i)\partial/\partial x$ with the auxiliary functions $\alpha(t)$ and $\rho(t)$ as c-numbers. Care is to be taken to preserve the order of the operator product $\hat{x}\hat{p}$ and $\hat{p}\hat{x}$. In fact, the derivation presented in the Hamiltonian formulation may be done quantum mechanically using the defining relation

$$i\hbar \frac{d\hat{I}}{dt} = i\hbar \frac{\partial \hat{I}}{\partial t} + [\hat{I}, \hat{H}] = 0 \tag{2.67}$$

and the form

$$\hat{I} = \frac{1}{4}(\hat{p}^2 f_2 + 2\hat{p} f_2 \hat{p} + f_2 \hat{p}^2) + (\hat{p} f_1 + f_1 \hat{p}) + f_0 , \qquad (2.68)$$

where the symbol [,] denotes the commutator bracket and $f_i = f_i(x, t)$, ($i = 0, 1, 2$) are functions of x and t only.

7.3. Schrödinger Equation and Invariants

Consider a quantum problem in which the Hamiltonian $\hat{H}(t)$ depends explicitly on the time t. Assume that $\hat{H}(t)$ admits a time-dependent Hermitian invariant operator $\hat{I}(t)$. This implies

$$i\hbar \frac{\partial \hat{I}}{\partial t} + [\hat{I}, \hat{H}] = 0 . \qquad (3.1)$$

The quantum mechanical problem is described by the Schrödinger equation

$$i\hbar \frac{\partial \psi}{\partial t} = \hat{H} \psi . \qquad (3.2)$$

It follows from Eq. (3.1) that

$$i\hbar \left(\frac{\partial \hat{I}}{\partial t} \right) \psi + (\hat{I}\hat{H} - \hat{H}\hat{I}) \psi = 0$$

and after using (3.2)

$$i\hbar \frac{\partial}{\partial t}(\hat{I}\psi) = \hat{H}(\hat{I}\psi) . \qquad (3.3)$$

This means that action of the invariant operator on a Schrödinger wavefunction produces another solution $\hat{I}\psi$ of the Schrödinger equation. Now suppose that the invariant operator $\hat{I}$ admits a complete orthogonal set of eigenstates $\phi_\lambda(x,t)$ corresponding to eigenvalues λ :

$$\hat{I} \phi_\lambda(x, t) = \lambda \phi_\lambda(x, t) \; ; \; (\phi_\lambda, \phi_{\lambda'}) = \delta_{\lambda \lambda'} . \qquad (3.4)$$

Since $\hat{I}$ is Hermitian, λ's are real. We also assume that $\hat{I}$ does not contain any operator of the form $\partial/\partial t$. Next differentiate both sides of the eigenvalue equation (3.4) with respect to t

$$\frac{\partial \hat{I}}{\partial t} \phi_\lambda + \hat{I} \frac{\partial \phi_\lambda}{\partial t} = \frac{\partial \lambda}{\partial t} \phi_\lambda + \lambda \frac{\partial \phi_\lambda}{\partial t} . \qquad (3.5a)$$

On the other hand from the defining equation (3.1)

$$\frac{\partial \hat{I}}{\partial t} \phi_\lambda + \frac{1}{i\hbar} (\hat{I} - \lambda) \hat{H} \phi_\lambda = 0 . \qquad (3.5b)$$

Eliminating $(\partial \hat{I}/\partial t)\phi_\lambda$ from these two equations, we have

$$(\hat{I} - \lambda)\left(i\hbar \frac{\partial \phi_\lambda}{\partial t} - \hat{H}\phi_\lambda\right) = i\hbar \frac{\partial \lambda}{\partial t} \phi_\lambda . \qquad (3.6)$$

Now take the scalar product with $\phi_{\lambda'}$. We obtain

$$i\hbar (\phi_{\lambda'}, (\hat{I} - \lambda)\frac{\partial \phi_\lambda}{\partial t}) - (\phi_{\lambda'}, (\hat{I} - \lambda)\hat{H}\phi_\lambda) = i\hbar \frac{\partial \lambda}{\partial t} (\phi_{\lambda'}, \phi_\lambda)$$

that is,

$$(\lambda' - \lambda)(\phi_{\lambda'}, (i\hbar \frac{\partial \phi_\lambda}{\partial t} - \hat{H}) \phi_\lambda) = i\hbar \frac{\partial \lambda'}{\partial t} \delta_{\lambda'\lambda} . \qquad (3.7)$$

There are two implications of this equation : (i) $\partial \lambda/\partial t = 0$, that is, the eigenvalues λ are time-independent and the time-dependence is relegated to the eigenstate ϕ_λ ; (ii) for $\lambda' \neq \lambda$,

$$(\phi_{\lambda'}, (i\hbar \frac{\partial}{\partial t} - \hat{H}) \phi_\lambda) = 0 . \qquad (3.8)$$

Since this equation does not hold for $\lambda' = \lambda$, we cannot conclude immediately that ϕ_λ also satisfies Schrödinger equation. We notice, however, that there is still some freedom left to fix the phase of ϕ_λ. We introduce a new function ψ_λ, such that

$$\psi_\lambda = \exp(i\alpha_\lambda(t)) \phi_\lambda . \qquad (3.9)$$

Since we have assumed that $\hat{I}(t)$ does not contain an operator of the form $\partial/\partial t$, ψ_λ is also an eigenfunction of $\hat{I}$ corresponding to the eigenvalue λ. Thus the derivation of Eq.(3.8) is still valid when ϕ_λ is replaced by ψ_λ. The idea is to choose α_λ so that Eq.(3.8) holds even for $\lambda' = \lambda$, that is,

$$(\psi_\lambda, (i\hbar \frac{\partial}{\partial t} - \hat{H}) \psi_\lambda) = 0 . \qquad (3.10)$$

It is clear from (3.10) that α obeys the differential equation

$$\hbar \frac{d\alpha_\lambda}{dt} = (\phi_\lambda, (i\hbar \frac{\partial}{\partial t} - \hat{H}) \phi_\lambda) . \qquad (3.11)$$

Thus if $\phi_\lambda(x,t)$ are normalized eigenfunctions of the invariant operator, the functions $\psi_\lambda = \exp(i\alpha_\lambda(t))\phi_\lambda(x,t)$ are the solutions of the Schrödinger equation. Moreover the general solutions of the Schrödinger equation can be written as

$$\psi(x,t) = \sum_\lambda C_\lambda \exp[i\alpha_\lambda(t)] \phi_\lambda(x,t) \qquad (3.12)$$

where $\alpha_\lambda(t)$ are determined from (3.11) and the expansion coefficients are given by

$$C_\lambda = \exp[-i\alpha_\lambda(0)] (\phi_\lambda(x,0), \psi(x,0)) . \qquad (3.13)$$

Illustrative example

As an example, consider the interesting case of a time-dependent Coulomb potential. According to Eq (2.66) the quantal invariant is

$$I(\vec{q}, \vec{p}, t) = \frac{1}{2} (\rho \vec{p} - \dot{\rho} \vec{q})^2 - Z_0 \frac{\rho}{r} \quad ; \quad r = |\vec{q}| \qquad (3.14)$$

and we have to obtain λ and ϕ_λ according to $I \phi_\lambda = \lambda \phi_\lambda$. Applying a unitary transformation

$$U = \exp\left[-\frac{i\dot{\rho} \vec{q}\cdot\vec{q}}{2\hbar\rho}\right] \qquad (3.15)$$

to define $\phi'_\lambda = U \phi_\lambda$, we can write

$$U \vec{p} U^+ = \vec{p} + \dot{\rho} \vec{q}/\rho \quad , \quad \hat{I}' = U \hat{I} U^+ = \frac{\vec{P}^2}{2} - Z_0/R , \qquad (3.16)$$

where $\vec{R} = \vec{q}/\rho$ and $\vec{P} = \rho\vec{p} = -i\hbar \nabla_R$. Thus, we see that $\hat{I}'$ is identical to the Hamiltonian $\hat{H}(\vec{P},\vec{R})$ of the usual time-independent Coulomb potential. The eigenvalues and the eigenfunctions ϕ'_λ would then be related to the standard wavefunction $\psi_\lambda(\vec{R})$ of the hydrogen-like problem. The relation is

$$\psi_\lambda(\vec{R}) = \rho^{1/2} \phi'_\lambda(\vec{q}, t) . \qquad (3.17)$$

We now determine the phases $\alpha_\lambda(t)$. For this purpose consider the transformed Hamiltonian:

$$U \hat{H} U^+ = U \left[\frac{\vec{p}^2}{2} - \frac{Z_0}{r\rho} \right] U^+ = \frac{\hat{I}'}{\rho^2} + \frac{\dot{\rho}}{\rho} \vec{q}\cdot\vec{p} + \left(\frac{\dot{\rho}}{\rho} \right)^2 \frac{\vec{q}^2}{2} - \frac{i\hbar}{2} \frac{\dot{\rho}}{\rho} \qquad (3.18)$$

and transform $i\hbar\partial/\partial t$ operator:

$$U \left(i\hbar \frac{\partial}{\partial t} U^+ \right) = i\hbar \frac{\partial}{\partial t} + \left(\frac{\dot{\rho}}{\rho} \right)^2 \frac{\vec{q}^2}{2} . \qquad (3.19)$$

Therefore, it is easy to see that

$$U \left(i\hbar \frac{\partial}{\partial t} - \hat{H} \right) U^+ = i\hbar \frac{\partial}{\partial t} - \frac{\hat{I}'}{\rho^2} - \frac{\dot{\rho}}{\rho} \vec{q}\cdot\vec{p} + \frac{i\hbar}{2} \frac{\dot{\rho}}{\rho} , \qquad (3.20)$$

$$i\hbar \frac{\partial}{\partial t} \phi'_\lambda(\vec{q}, t) = i\hbar \frac{\partial}{\partial t} (\psi_\lambda(\vec{R}) \rho^{-1/2}) = \left[- i\hbar \frac{\dot{\rho}}{2\rho} + \frac{\dot{\rho}}{\rho} (\vec{q}\cdot\vec{p}) \right] \phi'_\lambda . \qquad (3.21)$$

Combining all these results the equation determining α_λ becomes

$$\hbar \frac{d\alpha_\lambda}{dt} = (\phi_\lambda, (i\hbar \frac{\partial}{\partial t} - \hat{H}) \phi_\lambda) = (\phi'_\lambda, U(i\hbar \frac{\partial}{\partial t} - \hat{H}) U^+ \phi'_\lambda)$$

$$= (\phi'_\lambda, (-\hat{I}' \bar{\rho}^{-2}) \phi'_\lambda) = -\lambda \bar{\rho}^{-2} . \qquad (3.22)$$

The phases $\alpha_\lambda(t)$ are then determined as

$$\alpha_\lambda(t) = - \frac{\lambda}{\hbar} \int dt \, \rho^{-2} . \qquad (3.23)$$

The complete wavefunction which satisfies the time-dependent Schrödinger equation corresponding to the non-stationary Coulomb problem can be constructed by combining all the above results with the standard solution of the time-independent case.

Problems

3.1. Show that for the time-dependent harmonic oscillator of Eq.(2.47)

$$\lambda_n = (n + \frac{1}{2})\hbar\omega_0 \; ; \quad \hbar \alpha_n(t) = - \lambda_n \int^t dt \, \rho^{-2} \qquad (3.24a)$$

$$\phi_n(x,t) = (\omega_0/\pi)^{1/4}(2^n n!\rho)^{-1/2} \exp\left[(i\rho\dot{\rho} - \hbar\omega_0)\frac{x^2\rho^2}{2\hbar}\right] H_n(\sqrt{\omega_0}\, x/\rho),$$

(3.24b)

where $H_n(y)$ is the Hermite polynomial.

3.2. In problem 3.1, show that the invariant operator may be written as

$$\hat{I} = (a^+ a + \tfrac{1}{2}) \quad ; \quad [a, a^+] = 1.$$ (3.25)

3.3. Generalize the results of Problems 3.1 and 3.2 to the cases of i) Time-dependent harmonic oscillator (TDHO) with a linear force, ii) TDHO with a damping $\gamma\dot{x}$ and a linear force, iii) TDHO with an additional inverse square potential g/x^2, $(g > -\hbar^2/8)$.

7.4. Feynman Propagator

Let us examine the relevance of this theory to Feynman propagator. For $t'' > t'$, Eq. (3.12) may be rewritten as

$$\psi(\vec{q}'',t'') = \sum_\lambda C_\lambda \exp[i\alpha_\lambda(t'')]\, \phi_\lambda(\vec{q}'',t'')$$

$$= \sum_\lambda e^{-i\alpha_\lambda(t')} \int \phi_\lambda^*(\vec{q}',t')\psi(\vec{q}',t')d\vec{q}'\, e^{i\alpha_\lambda(t'')}\phi_\lambda(\vec{q}'',t'')$$

$$= \int K(\vec{q}'',t'';\vec{q}',t')\, \psi(\vec{q}',t')\, d\vec{q}' .$$ (4.1)

Thus the propagator K takes the form

$$K(\vec{q}'',t'';\vec{q}',t') = \sum_\lambda \exp[i(\alpha_\lambda(t'') - \alpha_\lambda(t'))]\phi_\lambda^*(\vec{q}',t')\phi_\lambda(\vec{q}'',t'') .$$ (4.2)

Notice that this is a generalization of the usual Feynman-Hibbs expansion formula for time-independent Hamiltonian. For time-dependent problems, the conserved quantity is the invariant $\hat{I}$ rather than the Hamiltonian $\hat{H}$ and the propagator admits a natural expansion in terms of the eigenfunctions of the invariant operator.

Problem

Show that for various cases of TDHO, in Problems 3.1 and 3.3 of Sec. 7.3, the above series may be summed into a closed form.

One expects that the existence of an invariant may simplify the derivation of the propagator. As an example, consider the time-dependent oscillator with the classical invariant given in Eq (2.51). It is clear that the propagator is given by the usual Van Vleck--Pauli formula. A simplification arises in the computation of S_{cl} in the following way. We define a canonical transformation

$$P = I(x,p,t)/\omega_0 \quad ; \quad Q = \cot^{-1}\{\rho\,(p\,\rho - x\,\dot{\rho})/\omega_0 x\} \tag{4.3}$$

with $\{Q,P\} = 1$. The generating function $F(x,Q,t)$ of this transformation

$$F = \tfrac{1}{2}\omega_0\left(\frac{x}{\rho}\right)^2 \cot Q + x^2 \frac{\dot{\rho}}{2\rho} \tag{4.3a}$$

$$p = \frac{\partial F}{\partial x} \quad , \quad P = -\frac{\partial F}{\partial Q} \quad . \tag{4.4b}$$

The new Hamiltonian $\tilde{H}(Q,P,t)$ is obtained as

$$\tilde{H} = H + \frac{\partial F}{\partial Q} = \omega_0 P/\rho^2 \, , \tag{4.5}$$

with the equations of motion

$$\dot{Q} = \frac{\partial \tilde{H}}{\partial P} = \omega_0/\rho^2 \quad ; \quad \dot{P} = -\frac{\partial \tilde{H}}{\partial Q} = 0 \quad . \tag{4.6}$$

The new Lagrangian is

$$\tilde{L}(Q, \dot{Q}, t) = \dot{Q} P - \tilde{H} \equiv 0 \tag{4.7}$$

But the new Lagrangian is related to the old one by

$$\tilde{L} = L - \frac{dF}{dt} \quad . \tag{4.8}$$

Hence $\tilde{L} = 0$ implies $L = dF/dt$ and

$$S_{cl}(x'', t''; x', t')$$

$$= \int_{t'}^{t''} L\,dt = F(t'') - F(t')$$

$$= \tfrac{1}{2}\left[\omega_0\left[\left(\frac{x''}{\rho''}\right)^2 \cot Q'' - \left(\frac{x'}{\rho'}\right)^2 \cot Q'\right] + \frac{\dot{\rho}''}{\rho''}x''^2 - \frac{\dot{\rho}'}{\rho'}x'^2\right]. \tag{4.9}$$

Note here that the use of the invariant has helped us to reduce the original Lagrangian to a total time derivative of a function $F(t)$. It is now a simple matter to use the following results obtained from Eqs. (4.6):

$$Q(t) = \omega_0 \int_{t'}^{t} \rho^{-2} \, dt = \phi(t,t') \; ; \qquad P'' = P' \qquad (4.10)$$

to prove the identities

$$\frac{x'}{\rho'} \cot Q' = \frac{x''}{\rho''} \operatorname{cosec} \phi(t'',t') - \frac{x'}{\rho'} \cot \phi(t'',t') ,$$

$$\frac{x''}{\rho''} \cot Q'' = \frac{x''}{\rho''} \cot \phi(t'',t') - \frac{x'}{\rho'} \operatorname{cosec} \phi(t'',t') . \qquad (4.11)$$

Inserting these identities in the expression for S_{cl}, we have

$$S_{cl}(x'',t'';x',t') = \frac{1}{2}\left[\frac{\dot\rho''}{\rho''} x''^2 - \frac{\dot\rho'}{\rho'} x'^2\right] +$$

$$\omega_0 \left\{\left[\left(\frac{x''}{\rho''}\right)^2 + \left(\frac{x'}{\rho'}\right)^2\right] \cot \phi(t'',t') - 2 \frac{x''x'}{\rho''\rho'} \operatorname{cosec} \phi(t'',t')\right\} , \qquad (4.12a)$$

$$\frac{\partial^2 S_{cl}}{\partial x' \partial x''} = - \frac{\omega_0 \operatorname{cosec} \phi(t'',t')}{\rho''\rho'} . \qquad (4.12b)$$

By Van Vleck--Pauli formula we have

$$K(x'',t'';x',t') = \left[\frac{\omega_0}{-2\pi i\hbar \rho'\rho'' \sin \phi(t'',t')}\right]^{1/2} \exp\left\{\frac{i}{2\hbar}\left[\frac{x''^2 \dot\rho''}{\rho''} - \frac{x'^2 \dot\rho'}{\rho'}\right]\right\}$$

$$\times \exp\left\{\frac{i\omega_0}{2\hbar \sin \phi(t'',t')}\left[\left(\frac{x''^2}{\rho''^2} + \frac{x'^2}{\rho'^2}\right)\cos \phi(t'',t') - \frac{2x'x''}{\rho'\rho''}\right]\right\} \qquad (4.13)$$

As a check, it is easy to see that when $\omega^2(t) = \omega_0^2$ (independent of time), $\rho = 1$ and the expression reduces to the propagator of the harmonic oscillator with frequency ω_0.

Problem

For an harmonic oscillator (HO) with constant frequency ω_0, the invariant I may be taken as H itself. Use the above method to derive the propagator.

The above example merely illustrates the fact that the existence of invariant indeed simplifies the derivation of the propagator. This prompts the general questions: Given a classical Lagrangian that admits an invariant, what form does the propagator take? In particular, what is the role played by the invariant in this approach to quantization? Clearly the existence of an invariant imposes certain conditions on the admissible forms of the potential $V(x,t)$ which may simplify the derivation of the propagator. We have no answers to these general questions for the moment. However there is an amusing simplification in the derivation of the propagator, if the Lagrangian admits an invariant which is quadratic in momentum. This result is quite sufficient for us since invariants which are quadratic in momentum are closest in form to the energy.

7.5. Invariants Quadratic in Momentum and the Propagator

Consider a system described by a classical Lagrangian

$$L(\vec{q}, \dot{\vec{q}}, t) = \frac{1}{2} \dot{\vec{q}}^2 - V(\vec{q}, t) . \tag{5.1}$$

Now, it has been shown in Sec.7.2 that this Lagrangian admits an invariant, quadratic in momentum, if and only if the potential is of the form

$$V(\vec{q}, t) = (\frac{\rho \ddot{\vec{\alpha}}}{\rho} - \dot{\vec{\alpha}}) \vec{q} - \frac{\rho \ddot{\vec{q}}^2}{2 \rho} + \rho^{-2} F\left(\frac{\vec{q} - \vec{\alpha}}{\rho}\right) , \tag{5.2}$$

where $\vec{\alpha}(t)$ and $\rho(t)$ are arbitrary functions of time and $F(\vec{x})$ is an arbitrary function of $\vec{x}$. The associated invariant has the form

$$I(\vec{q}, \vec{p}, t) = [\rho(\vec{p} - \dot{\vec{\alpha}}) - \dot{\rho}(\vec{q} - \vec{\alpha})]^2 + F\left(\frac{\vec{q} - \vec{\alpha}}{\rho}\right). \tag{5.3}$$

Now, we do a little trick. Insert the potential $V(\vec{q}, t)$ in the Lagrangian and after some trivial algebra rewrite the Lagrangian as

$$L = \frac{d\chi}{dt} + L_0 \quad ; \quad L_0 = \frac{1}{2} \rho^2 \left[\frac{d}{dt}\left(\frac{\vec{q} - \vec{\alpha}}{\rho}\right)\right]^2 - \rho^{-2} F\left(\frac{\vec{q} - \vec{\alpha}}{\rho}\right) , \quad (5.4)$$

where, χ is defined as

$$\chi = \frac{\dot{\rho} \vec{q}^2}{2 \rho} + \frac{\vec{W}\cdot\vec{q}}{\rho} - G \quad ; \quad W = \vec{\alpha} \rho - \vec{\alpha} \dot{\rho} , \quad G = \int^t dt \, \frac{\vec{W}\cdot\vec{W}}{2\rho^2} . \quad (5.5)$$

The dx/dt term contributes only a constant phase factor and the propagator takes the form

$$K(\vec{q}'', t''; \vec{q}', t') = \exp\left[(\chi(t'') - \chi(t'))\right] K_0 . \quad (5.6)$$

Here K_0 is the new propagator corresponding to the new Lagrangian L_0. Since $\vec{\alpha}(t)$ is a given function of time, it is possible to introduce a new variable $\vec{Q} = \vec{q} - \vec{\alpha}$ in the path integration. As a result the end points will shift from $\vec{q}'$ to $\vec{Q}' = \vec{q}' - \vec{\alpha}'$ and $\vec{q}''$ to $\vec{Q}'' = \vec{q}'' - \vec{\alpha}''$. Thus

$$K_0 = \int \exp\left(\frac{i}{\hbar} \int_{t'}^{t''} L_0 \, dt\right) \mathcal{D}[q(t)], \quad L_0 = \frac{\rho^2}{2}\left[\frac{d}{dt}\left(\frac{\vec{Q}}{\rho}\right)\right]^2 - F\left(\frac{\vec{Q}}{\rho}\right) . \quad (5.7)$$

We now introduce a new parameter τ related to time t by

$$\tau(t) = \int^t ds \, \rho^{-2}(s) , \quad (5.8)$$

and define $\vec{\xi} = \vec{Q}/\rho$ so that

$$\int_{t'}^{t''} dt \, L_0 = \int_{\tau'}^{\tau''} d\tau \, \bar{L}_0 \quad ; \quad \bar{L}_0 = \frac{1}{2}\left(\frac{d\vec{\xi}}{dt}\right)^2 - F(\vec{\xi}) . \quad (5.9)$$

The parameter τ would in turn induce a transformation in the path differential measure

$$\mathcal{D}[\vec{Q}(t)] = (\rho'' \rho')^{-3/2} \mathcal{D}[\vec{\xi}(\tau)] . \quad (5.10)$$

Proof of result (5.10)

In polygonal approximation, we partition the time interval in N equal subintervals, $\Delta t_k = t_{k+1} - t_k = \varepsilon$ ($k = 0, 1, \ldots, N$), with $t_0 = t'$, $t_N = t''$ and $\vec{Q}_k = \vec{Q}(t_k)$. The path differential measure then takes the form

$$\mathcal{D}\,[Q(t)] \longrightarrow (2\pi i\hbar\varepsilon)^{-3N/2} \prod_{k=1}^{N-1} d\vec{Q}_k \ . \tag{5.11}$$

Writing for convenience, $\beta = 1/\rho$, $\beta_k = \beta(t_k)$, we obtain

$$\Delta\tau_k = \int_{t_k}^{t_{k+1}} ds\,\beta^2(s) = \int_0^\varepsilon ds\,\beta^2(t_k + s) = \varepsilon\,[\beta_k^2 + 2\varepsilon\,\dot\beta_k\,\beta_k + O(\varepsilon^2)]$$

$$= \varepsilon\,\beta_k\,\beta_{k+1}\,[\,1 + O(\varepsilon^2)\,] \ . \tag{5.12}$$

Hence,

$$(2\pi i\hbar\varepsilon)^{-N/2} = \prod_{k=0}^{N-1}[(\beta_k\beta_{k+1}/2\pi i\hbar\Delta\tau_k)^{1/2}(1 + O(\varepsilon^2))]$$

$$= [1 + O(\varepsilon)]\,[\beta_N\beta_0/2\pi i\hbar\Delta\tau_0]^{1/2} \prod_{k=1}^{N-1} \beta_k(1/2\pi i\hbar\Delta\tau_k)^{1/2} \ .$$

Therefore,

$$(2\pi i\hbar\varepsilon)^{-3N/2} \prod_{k=1}^{N-1} d\vec{Q}_k = (\rho''\rho')^{-3/2} \prod_{k=0}^{N-1} (2\pi i\hbar\Delta\tau_k)^{-3/2} \prod_{k=1}^{N-1} d\vec{\xi}_k \ .$$

In the limit $\varepsilon \to 0$, $\Delta\tau_k \to 0$, we arrive at Eq. (5.10).

We have therefore shown that

$$K = (\rho''\rho')^{-3/2} \exp\left[\frac{i}{\hbar}(\chi'' - \chi')\right] \bar{K}_0(\vec{\xi}'',\tau'';\vec{\xi}',\tau') \ , \tag{5.13}$$

where

$$\bar{K}_0(\vec{\xi}\,'',\tau\,'';\vec{\xi}\,',\tau') = \int \exp\left[\frac{i}{\hbar}\int_{\tau'}^{\tau''} d\tau\,\bar{L}_0\right] \mathcal{D}\,[\vec{\xi}(\tau)] \ . \tag{5.14}$$

This important result shows that the propagator for the original time-dependent problem has been reduced to that for an associated time-independent problem with the Lagrangian $\bar{L}_0$.

One can also generalize this result if

$$L = a(t)\,\vec{\dot q}^2 - V(\vec{q},t) \tag{5.15}$$

with $V(q, t)$ of the form given earlier.

7.5.1. Illustrative examples

A. A forced TDHO :

The Lagrangian reads as

$$L = \frac{1}{2}\left[\dot{x}^2 - \omega^2(t)x^2\right] + f(t)x \ . \tag{5.16}$$

It is easy to see from the general form (5.2) of $V(x, t)$ that ρ and α satisfy the equations

$$\ddot{\rho} + \omega^2(t)\rho = 0 \ ; \quad \ddot{\alpha} + \omega^2(t)\alpha = f(t) \ , \tag{5.17}$$

while $F = 0$. The propagator is therefore related to a free particle propagator

$$\bar{K}_0(\xi'',\tau'';\xi',\tau') = (2\pi i\hbar(\tau'' - \tau'))^{-1/2} \exp\left(\frac{i(\xi'' - \xi')^2}{2\hbar(\tau'' - \tau')}\right) \ . \tag{5.18}$$

The complete propagator reads as

$$K(x'',t'';x',t') = \left(\frac{c}{2\pi i}\right)^{1/2} \exp\left[\frac{i}{2}\left(ax''^2 + bx'^2 - 2cx''x' + dx'' + ex' + g\right)\right]$$

$$a = (\dot{\rho}'' + c\rho')/\rho'' \ , \quad b = (-\dot{\rho}' + c\rho'')/\rho' \ , \quad c = [\rho'\rho''(\tau''-\tau')]^{-1} \ ,$$

$$d = \frac{2}{\rho''}[W(t'') - c(\alpha''\rho' - \alpha'\rho'')] \ , \quad e = \frac{2}{\rho'}[c(\alpha''\rho' - \alpha'\rho'') - W(t')] \ ,$$

$$g = \frac{c(\alpha''\rho' - \alpha'\rho'')^2}{\rho'\rho''} - \int_{t'}^{t''} (W/\rho)^2 \, dt \ .$$

In order to evaluate these coefficients we need to use any solution of the differential equations for ρ and α. We choose the solution

$$\rho(t) = \sigma(t) \cos\beta(t) \ , \quad \ddot{\sigma} + \omega^2(t)\sigma = \omega_0^2/\sigma^3 \ , \quad \dot{\beta} = \omega_0/\sigma^2$$

where ω_0 is a constant.
The corresponding solution for $\alpha(t)$ is

$$\alpha(t) = \frac{\sigma(t)}{\omega} \int^t H(s) \sin\phi(t,s) \, ds \ , \quad \phi(t,s) = \beta(t) - \beta(s)$$

where $H(s) = f(s)\sigma(s)$. After some algebra we obtain

$$a = \frac{\ddot{\sigma}''}{\sigma''} + \frac{\omega_0}{\sigma''^2} \cot \phi(t'',t') \quad ; \quad b = -\frac{\dot{\sigma}'}{\sigma'} + \frac{\omega_0}{\sigma'^2} \cot \phi(t'',t') ,$$

$$c = \frac{\omega_0}{\sigma'\sigma'' \sin \phi(t'',t')} \quad , \quad d = \frac{2}{\sigma'' \sin \phi(t'',t')} \int_{t'}^{t''} H(t) \sin \phi(t,t') dt,$$

$$e = \frac{2}{\sigma' \sin \phi(t'',t')} \int_{t'}^{t''} H(t) \sin \phi(t'',t) dt,$$

$$g = -\frac{2}{\omega_0 \sin \phi(t'',t')} \int_{t'}^{t} dt \int_{t'}^{t} ds\, H(t)\, H(s) \sin \phi(t'',t) \sin \phi(s,t') .$$

Combining these results we arrive at the complete expression for the propagator

$$K(x'',t'';x',t') = \left(\frac{\omega_0}{2\pi i\hbar \sin \phi(t'',t')}\right)^{1/2} \exp\left[\frac{i}{2\hbar}\left(\frac{\dot{\sigma}''}{\sigma''} x''^2 - \frac{\dot{\sigma}'}{\sigma'} x'^2\right)\right]$$

$$\times \exp\left[\frac{i\omega_0}{2\hbar \sin \phi(t'',t')}\left\{\left(\frac{x''^2}{\sigma''^2} + \frac{x'^2}{\sigma'^2}\right) \cos \phi(t'',t') - \frac{2x''x'}{\sigma''\sigma'}\right.\right.$$

$$+ \frac{2x''}{\omega_0 \sigma''} \int_{t'}^{t''} H(t) \sin \phi(t,t') dt' + \frac{2x'}{\omega_0 \sigma'} \int_{t'}^{t''} H(t) \sin \phi(t'',t) dt$$

$$\left.\left.- \frac{2}{\omega_0^2} \int_{t'}^{t''} dt \int_{t'}^{t''} ds\, H(t)\, H(s) \sin \phi(t'',t') \sin \phi(s,t')\right\}\right] . \quad (5.19)$$

The formula is very general and with a little bit of patience we recover all special cases, viz., i) HO with constant frequency, ii) Forced HO with constant frequency iii) Free particle acted on by a force $f(t)$.

It is important to note here that the present derivation does not require explicit path integration as the auxiliary time-independent problem involves only a free-particle propagator. A global time transformation given by Eq. (5.8) essentially maps all versions of the oscillator onto a free particle in a coordinate system $\vec{\xi} = [\vec{x} - \vec{\alpha}(t)]/\rho$.

B. TDHO with additional potential g/x^2

In this case $\alpha = 0$ but $F(x) = g/x^2$ and $\rho(t)$ satisfies the same equation as in example A. The propagator for the associated time-independent problem is

$$\bar{K}_0(\xi'',\tau'';\xi',\tau') = \int \exp\left[\frac{i}{\hbar}\int\left[\frac{1}{2}\left(\frac{d\xi}{d\tau}\right)^2 - \frac{g}{\xi^2}\right]d\tau\right] \mathcal{D}[\xi(\tau)] \quad . \quad (5.20)$$

Defining $\gamma = \frac{1}{2}(1+8g/\hbar^2)^{1/2}$, we have, $g = \frac{\hbar^2}{2}(\gamma - \frac{1}{2})(\gamma + \frac{1}{2})$ and the propagator $\bar{K}_0$ corresponds to the radial propagator of a free particle moving in three dimensions with "angular momentum" $(\gamma - \frac{1}{2})$. Therefore $\bar{K}_0$ reads as

$$\bar{K}_0 = \left[\frac{\sqrt{\xi''\xi'}}{i\hbar(\tau''-\tau')}\right] \exp\left[\frac{(\xi''^2+\xi'^2)}{2i\hbar(\tau''-\tau')}\right] I_\gamma\left[\frac{\xi''\xi'}{i\hbar(\tau''-\tau')}\right] . \quad (5.21)$$

Introducing the phase factors the complete propagator takes the form

$$K(x'',t'';x',t') = (c\sqrt{x'x''}/i\hbar) \exp[i(ax''^2 + bx'^2)/\hbar] I_\gamma(c\,x''x'/i\hbar) ,$$

where a, b, c are still given as in example A. Inserting the values of a, b, and c, we obtain the propagator

$$K(x'',t'';x',t') = \left(\frac{\omega_0\sqrt{(x''x')}}{i\hbar\sigma''\sigma'\sin\phi(t'',t')}\right) \exp\left[\frac{i}{2\hbar}\left(\frac{\dot\sigma''}{\sigma''}x''^2 - \frac{\dot\sigma'}{\sigma'}x'^2\right)\right]$$

$$\times \exp\left[\frac{i\omega_0}{2\hbar}\left(\frac{x''^2}{\sigma''^2} + \frac{x'^2}{\sigma'^2}\right)\cot\phi(t'',t')\right] I_\gamma\left(\frac{\omega_0 x''x'}{i\hbar\sigma''\sigma'\sin\phi(t'',t')}\right).$$

$$(5.22)$$

7.6. Role Played by the Invariant I

Let us now discuss the role played by the invariant I in our scheme. The classical Hamiltonian corresponding to the Lagrangian $\bar{L}_0$ of Eq (5.9) is

Time Dependent Invariants and Feynman Propagator 211

$$\bar{H}_0 = \frac{1}{2}\vec{p}_\xi^{\,2} + F(\vec{\xi}) \tag{6.1}$$

where p_ξ is the canonical momentum conjugate to the new variable ξ. On the other hand, it is easy to see that the invariant when expressed in terms of the new variable $\vec{\xi} = (\vec{q} - \vec{\alpha})/\rho$ and τ is identical to $\bar{H}_0$.

The corresponding quantum mechanical Hamiltonian operator $\hat{\bar{H}}_0$ and the invariant operator $\hat{I}_0$ are obtained by using $\vec{p}_\xi = -i\hbar\vec{\nabla}_\xi$ in Eq.(6.1). The result is

$$\hat{\bar{H}}_0 = -(\hbar^2/2)\,\vec{\nabla}_\xi^{\,2} + F(\vec{\xi}) \equiv \hat{I}_0 \,. \tag{6.2}$$

The propagator therefore represents the Green's function of the Schrödinger equation

$$i\hbar\,\frac{\partial}{\partial\tau}\,\psi_n(\vec{\xi},\tau) = \hat{\bar{H}}_0\,\psi_n(\vec{\xi},\tau) \,. \tag{6.3}$$

Hence, if the associated stationary problem

$$\hat{\bar{H}}_0\,\phi_n(\vec{\xi}) = \lambda_n\,\phi_n(\vec{\xi}) \tag{6.4}$$

has a complete set of normalized eigenfunctions corresponding to the eigenvalues λ_n, the propagator $\bar{K}_0$ has the expansion

$$\bar{K}_0(\vec{\xi}'',\tau'';\vec{\xi}',\tau') = \sum_n \exp\left[-(i/\hbar)\lambda_n(\tau''-\tau')\right]\varphi_n^*(\vec{\xi}')\phi_n(\vec{\xi}'') \,. \tag{6.5}$$

The notation $\sum_n$ implies in general a summation over discrete eigenvalues and integration over continuous eigenvalues. Inserting the original variables $(\vec{q},t)$, we see that the propagator has the following expansion in terms of the eigenfunctions of the invariant operator

$$K(\vec{q}'',t'';\vec{q}',t') = (\rho'\rho'')^{-1/2} \exp\left[\frac{-i}{\hbar}\{\chi(t'') - \chi(t')\}\right]$$

$$\times \sum_n \exp\left[-\frac{i}{\hbar}\lambda_n \int_{t'}^{t''} dt\, \rho^{-2}\right] \phi_n\left(\frac{\vec{q}''-\vec{\alpha}''}{\rho''}\right)\phi_n^*\left(\frac{\vec{q}'-\vec{\alpha}'}{\rho'}\right) . \quad (6.6)$$

Lastly, we may compare the path integral approach with Lewis and Riesenfeld theory outlined in Sec. 7.3. Consider the quantal Hamiltonian operator $\hat{H}$ and the invariant corresponding to the potential (5.2). If we perform a unitary transformation

$$\psi_n' = U\psi_n ; \quad U = \exp\left[-\frac{i}{\hbar}\left\{\frac{\dot{\rho}\vec{q}^2}{2\rho} + \frac{(\dot{\vec{\alpha}}\rho - \vec{\alpha}\dot{\rho}).\vec{q}}{\rho}\right\}\right] \quad (6.7)$$

where ψ_n are the eigenfunctions of $\hat{I}$, the equation $\hat{I}\psi_n = \lambda_n\psi_n$ transforms into

$$\hat{I}'\psi_n'(\vec{q},t) = \lambda_n\psi_n'(\vec{q},t) ; \quad \hat{I}' = U\hat{I}U^+. \quad (6.8)$$

This transformed invariant $\hat{I}'$ when expressed in terms of the new variables $\vec{\xi} = (\vec{q} - \vec{\alpha})/\rho$ coincides with $\hat{I}_0$ or $\hat{H}_0$ of Eq. (6.2) and the corresponding normalized eigenfunctions take the form

$$\phi_n(\vec{\xi}) = \sqrt{\rho}\,\psi_n'(\vec{q},t) . \quad (6.9)$$

The phases are determined by the equation

$$\hbar\frac{d\alpha_n}{dt} = \langle\psi_n'|U(i\hbar\frac{\partial}{\partial t} - \hat{H})U^+|\psi_n'\rangle = -\rho^{-2}\langle\psi_n'|[\hat{I}' + \frac{1}{2}W^2]|\psi_n'\rangle$$

$$= -\rho^{-2}[\lambda_n + \frac{1}{2}W^2] \quad (6.10)$$

which leads to expression for the phase

$$\alpha(t) = -\frac{1}{\hbar} \int^{t} [\lambda_n + \frac{1}{2} W^2] \rho^{-2} dt . \qquad (6.11)$$

It is interesting to note here that these phases appear naturally in the expansion (6.6) of the propagator. Also in the Feynman propagator approach, the steps which are essentially equivalent to the above (cf. Eq.(6.7)--(6.10)) are carried out classically on the Lagrangian resulting in a transformation of the path differential measure. The quantum superposition principle is manifest in the reduced propagator $\bar{K}_0$.

7.7. Global Time Transformation in Feynman Path Integral

In the foregoing analysis we introduced essentially a global time transformation $d\tau = dt/\rho^2(t)$ along with the scale transformation $\vec{\xi} = \vec{q}/\rho(t)$. We show here that regardless of the existence of an invariant, such a transformation often simplifies a path integral. The main issue here is to relate the propagator in "old" coordinates and time to that in "new". The required steps are as before, viz., express the action $S(\vec{q},t)$ in terms of the new variables $(\vec{\xi},\tau)$ and make a transformation of the path differential measure. Consider first the kinetic energy term

$$\frac{m}{2} \dot{\vec{q}}^2 = \frac{m}{2\bar{\rho}^2} \left\{ \dot{\vec{\xi}}^2 - \frac{\xi^2}{\bar{\rho}^2} [\bar{\rho} \ddot{\bar{\rho}} - 2 \dot{\bar{\rho}}^2] + \frac{d}{d\tau}\left[\frac{\dot{\bar{\rho}}}{\bar{\rho}} \xi^2\right] \right\} , \qquad (7.1)$$

where we use the notation

$$\bar{\rho}(\tau) = \rho(t) \quad ; \quad \dot{\bar{\rho}} = \frac{d\bar{\rho}}{d\tau} . \qquad (7.2)$$

The potential energy term reads as

$$V(\vec{q}, t) = V(\vec{\xi} \bar{\rho}(\tau), \int^{\tau} \bar{\rho}^2(s) ds) . \qquad (7.3)$$

The action $S(t'',t')$ then takes the form

$$S(t'', t') = \int_{t'}^{t''} L \, dt = \left(\frac{\xi^2 \dot{\bar{\rho}}}{\bar{\rho}} \right)_{\tau'}^{\tau''} + \bar{S}(\tau'', \tau') \tag{7.4}$$

where the new action is

$$\bar{S}(\tau'', \tau') = \int_{\tau'}^{\tau''} \bar{L} \, d\tau = \int_{\tau'}^{\tau''} [\tfrac{1}{2} m \, \dot{\vec{\xi}}^2 - \bar{V}(\vec{\xi}, s)] \, ds \tag{7.5}$$

and the new potential is

$$\bar{V}(\vec{\xi}, \tau) = \bar{\rho}^2 \, V\left(\xi \, \bar{\rho}(\tau), \int^{\tau} ds \, \bar{\rho}^2(s) \right) + \tfrac{1}{2} m \, \Omega^2(\tau) \xi^2 . \tag{7.6}$$

The time transformation has introduced an additional potential corresponding to harmonic oscillator of frequency $\Omega(s)$ given by

$$\Omega^2(\tau) = (\bar{\rho} \, \ddot{\bar{\rho}} - 2 \dot{\bar{\rho}})/ \bar{\rho}^2 = \bar{\rho}^3 \, \ddot{\bar{\rho}} . \tag{7.7}$$

The time transformation brings in the change of path differential measure. As before the measure transforms as

$$\mathcal{D}[\vec{q}(t)] = (\rho' \rho'')^{-3/2} \, \mathcal{D}[\vec{\xi}(\tau)] . \tag{7.8}$$

Combining these results, the Feynman propagator is written as

$$K(\vec{q}'', t''; \vec{q}', t') = \frac{1}{(\rho' \rho'')^{3/2}} \exp\left[\frac{im}{2\hbar} \left\{ \frac{\dot{\bar{\rho}}''}{\bar{\rho}''} \vec{\xi}''^2 - \frac{\dot{\bar{\rho}}'}{\bar{\rho}'} \vec{\xi}'^2 \right\} \right] \bar{K}(\vec{\xi}'', \tau''; \vec{\xi}', \tau') .$$

$$\tag{7.9}$$

Note here that the exponential factor in (7.9) may also be written as

$$\exp\left[\frac{\mathrm{i}m}{2\hbar}\left\{\frac{\dot{\rho}''\vec{q}''^2}{\rho''} - \frac{\dot{\rho}'\vec{q}'^2}{\rho'}\right\}\right] \tag{7.10}$$

and the new propagator has the form

$$\bar{K}(\vec{\xi}'',\tau'';\vec{\xi}',\tau') = \int \mathcal{D}\,[\vec{\xi}(\tau)]\,\exp\,[\,\mathrm{i}S(\tau'',\tau')/\hbar\,]\,. \tag{7.11}$$

The utility of a such a global time transformation and the associated coordinate transformation can be illustrated with the help of following two examples.

Harmonic oscillator with variable frequency

Consider a harmonic oscillator with time-dependent frequency. Here the potential $V(\vec{q},t) = m\,\omega^2(t)q^2/2$ and the transformed action is

$$S(\tau'',\tau') = \frac{m}{2}\int_{\tau'}^{\tau''} d\tau\,[\dot{\vec{\xi}}^2 - (\Omega^2(\tau) + \bar{\omega}^2(\tau)\bar{\rho}^4)\xi^2] \tag{7.12}$$

where $\bar{\omega}(\tau) = \omega(t)$. The idea now is to choose the transformation function $\rho(t)$ or $\bar{\rho}(\tau)$ so that the calculation of the propagator $\bar{K}$ is greatly simplified. To this end, we may write

$$\Omega^2(\tau) + \bar{\omega}^2(\tau)\bar{\rho}^4 = \omega_0^2 = \text{const.} \tag{7.13}$$

which amounts to determining the form of $\rho(t)$. Since $\Omega^2 = \rho^3\ddot{\rho}$, we find from (7.13) that $\rho(t)$ must obey the following equation

$$\ddot{\rho} + \omega^2(t)\rho = \omega_0^2/\rho^3\,. \tag{7.14}$$

In particular, we may even choose $\omega_0 = 0$. In that case, the propagator $\bar{K}$ becomes just the free particle propagator. For $\omega_0 \neq 0$, the propagator $\bar{K}$ corresponds to the propagator of a harmonic oscillator with constant frequency ω_0. In any case, the use of time transformation simplifies matters considerably. Also, the expression for invariant is easy to obtain. The transformed Hamiltonian in the new variable is

$$\bar{H} = \frac{m}{2}\left[\left(\frac{d\vec{\xi}}{ds}\right)^2 + \omega_0^2\,\xi^2\right] \qquad (7.15)$$

which when expressed in term of the old coordinates, is exactly the Lewis invariant

$$I = \frac{m}{2}[(\vec{q}\rho - \dot{\rho}\vec{q})^2 + \omega_0^2\,(q/\rho)^2] \,. \qquad (7.16)$$

Particle in an infinite potential well with a moving boundary

We consider a particle of mass m moving within a square well with two infinite walls. The left boundary is chosen to be the origin and is fixed. The right boundary is allowed to undergo an arbitrary time-dependent shift. At any instant t, the position of the moving wall is $x = L\rho$ where L represents the position of the moving wall at $t = 0$. That is, $\rho(0) = 1$ and $\tau = 0$ if $t = 0$. The potential well is therefore defined by

$$V(x,t) = 0 \,, \quad 0 \le x \le L\,\rho(t), \qquad (7.17)$$

and is infinite elsewhere. Clearly, if x' or x'' are outside the interval $[0, L\rho]$, the propagator vanishes. The propagator is non-zero only when both x' and x'' lie within the interval $[0, L\rho]$. Apply the time transformation $d\tau = dt/\rho^2$ and the coordinate scaling $Q = x/\rho$, we can write the propagator as

$$K(x'',t'';\,x',t') = \frac{1}{\sqrt{\rho'\rho''}}\exp\left[\frac{im}{2\hbar}\left\{\frac{\dot{\rho}''\,Q''^2}{\rho''} - \frac{\dot{\rho}'\,Q'^2}{\rho'}\right\}\right]\bar{K}(Q'',\tau'';Q',\tau')$$

$$(7.18)$$

$$\bar{K}(Q'',t'';Q',t') = \int \exp\left\{\frac{im}{2\hbar}\left[\left(\frac{dQ}{d\tau}\right)^2 - \Omega^2(\tau)Q^2\right]\right\}\mathcal{D}\,[Q(\tau)] \qquad (7.19)$$

where $\Omega^2(\tau)$ is as defined in (7.7). Thus the computation of the propagator for the infinite well with variable width is reduced to the determination of the propagator (7.19). The latter corresponds to the motion of a harmonic oscillator with a time dependent frequency restricted within the internal [0, L]. Analytical calculation of the propagator (7.19) is, however, not possible in general. When $\Omega^2(\tau) = 0$, the auxiliary propagator $\bar{K}$ reduces to that for an ordinary square well with fixed boundaries. When $\Omega^2(\tau) = \omega_0^2$, a constant, $\bar{K}$ corresponds to the propagator of a harmonic oscillator with a constant frequency moving in the internal [0, L].

As a final comment, we mention here that the time transformations may also be used with the Hamiltonian formulation of the path integral. Here the relevant transformations are known as the generalized canonical transformations (GCT) and are defined as

$$d\tau = dt/\rho^2(t) \quad ; \quad \vec{\xi} = \vec{q}/\rho(t), \quad ; \quad \vec{p}_\xi = \vec{p}\,\rho(t) . \tag{7.20}$$

The propagator

$$K(\vec{q}'',t'';\vec{q}',t') = \int \mathcal{D}\,[\vec{q}(t)]\,\mathcal{D}\,[\vec{p}(t)]\,\exp\left[\frac{i}{\hbar}\int_{t'}^{t''} dt\,(\vec{p}.\dot{\vec{q}} - H)\right] \tag{7.21}$$

where,

$$H = \frac{\vec{p}^2}{2m} + V(\vec{q},t) , \tag{7.22}$$

then takes the form

$$K(\vec{q}'',t'';\vec{q}',t') = \frac{1}{\sqrt{\rho'\rho''}}\,\exp\left[\frac{im}{2\hbar}\left\{\frac{\dot{\rho}''\,\vec{\xi}''^2}{\rho''} - \frac{\dot{\rho}'\,\vec{\xi}'^2}{\rho'}\right\}\right]\bar{K}(\vec{\xi}'',\tau'';\vec{\xi}',\tau') \tag{7.23}$$

where the new propagator

$$\bar{K}(\vec{\xi}'',t'';\vec{\xi}',t') = \int \mathcal{D}\,[\vec{\xi}(t)]\mathcal{D}\,[\vec{p}_\xi(t)]\,\exp\left[\frac{i}{\hbar}\int_{\tau'}^{\tau''} d\tau\,(\vec{p}_\xi.\dot{\vec{\xi}} - \bar{H})\right] . \tag{7.24}$$

Here the new Hamiltonian reads as

$$\bar{H} = \frac{\vec{p}_\xi^2}{2m} + V(\vec{\xi}, \tau) \ . \tag{7.27}$$

We might add that use of either Hamiltonian or Lagrangian formulation is more a matter of taste rather than convenience.

Notes and References

The quadratic invariant for the classical harmonic oscillator was first derived by

E. D. Courant and H. D. Snyder, Ann. Phys. **3**, 1 (1958), and rederived by

H. R. Lewis Jr., Phys. Rev. Lett. **18**, 510 (1967); J. Math. Phys. **9**, 510 (1968).

The use of invariant to obtain solution of the Schrödinger equation for a time-dependent Hamiltonian was pointed out in the paper by

H. R. Lewis Jr. and W. B. Riesenfeld, J. Math. Phys. **10**, 1458 (1969) and more recently by

J. G. Hartley and J. R. Ray, Phys. Rev. **A24**, 2873 (1981)

J. G. Hartley and J. R. Ray, Phys. Rev. **A25**, 2388 (1982).

A large number of references discuss the derivation of invariants. These are found in the paper by

H. R. Lewis Jr. and P. G. Leach, J. Math. Phys. **23**, 2371 (1982). This paper also discusses the issue of the form of the potential $V(\vec{x},t)$ that admits an invariant quadratic in momentum.

Use of invariants to simplify the derivation of Feynman propagator is discussed in the following papers.

D. C. Khandekar and S. V. Lawande, Phys. Lett. **A67**, 175 (1978),

D. C. Khandekar and S. V. Lawande, J. Math. Phys. **20**, 1870 (1979),

S. V. Lawande and A. K. Dhara, Phys. Lett. **A99**, 353 (1983),

A. K. Dhara and S. V. Lawande, Phys. Rev. **A30**, 560 (1984),

A. K. Dhara and S. V. Lawande, J. Phys. **A17**, 2423 (1984).

A class of time dependent invariants have been considered by

A. K. Dhara and S. V. Lawande, J. Math. Phys. **27**, 1331 (1986).

Global time transformation in path integrals have been employed by

P. Y. Cai, A. Inomata and P. Wong, Phys. Lett. **A91**, 331 (1982),

G. Junker and A. Inomata, Phys. Lett. **A110**, 195 (1985).

Generalized canonical transformations have been used to treat the quantum mechanical problems of a particle in a square well with a variable width in the paper

A. Munier, J. R. Burgan, M. Feix and E. Fijalkow, "Schrödinger equations with time-dependent boundary conditions", J. Math. Phys. **22**, 1219 (1981).

Hamiltonian formulation of a path integral under generalized canonical transformations is discussed by

L. Chetouani, L. Guechi and T. F. Hamman, Phys. Rev. **A40**, 1157 (1989).

CHAPTER 8

THE CUMULANT APPROXIMATION FOR FEYNMAN PROPAGATORS

8.1. Introduction

In the previous chapters we have presented techniques of evaluating path integrals exactly in a number of situations. However, while investigating problems of practical interest, it is not always possible to evaluate a path integral exactly owing to the complicated nature of the associated Lagrangian (Hamiltonian). Hence exact evaluation of the desired physical quantities is not possible and one has to be content with approximation methods for obtaining the estimates of their values along with the errors involved. One such method, which has been used very extensively is the so-called cumulant expansion method. In this chapter we would discuss this method and subsequently illustrate its use in treating some problems in theoretical physics.

8.2. The Cumulant Approximation

Let us consider a one-dimensional system characterized by the action functional S. The first step in developing cumulant expansion is to write the action as $S = S_0 + \tilde{S}_1$. The action S_0 is usually referred to as the "trial action" and is so chosen that the corresponding propagator $K_0(x'',x';T)$ is exactly known. The path integral representation for the propagator $K(x'',x';T)$ corresponding to the action S is then expressed as

$$K(x'',x';T) = \int \exp\left(\frac{i}{\hbar} S\right) \mathcal{D}\,[x(t)] \equiv \int \exp\left(\frac{i}{\hbar} S_0\right) \exp\left(\frac{i}{\hbar} \tilde{S}_1\right) \mathcal{D}\,[x(t)]. \quad (2.1)$$

In most of the problems, we naturally encounter a parameter (say α) which is small compared to other quantities occurring in S. Let us set $\tilde{S}_1 = \alpha S_1$. The cumulant expansion consists in writing the propagator K in the form

$$K(x'',x';T) = K_0 \exp\left[i\alpha K_1 + i\alpha^2 K_2 + \ldots + i\alpha^n K_n + \ldots \right]. \quad (2.2)$$

The term K_n is called the cumulant of order n. To obtain explicit expressions for $K_1, K_2, \ldots, K_n, \ldots$, we expand K in Eqs.(2.1) and (2.2) in a power series in α and compare the coefficients of various powers of α. The first two terms K_1 and K_2 read as

$$K_1 = \frac{1}{K_0} \int (S-S_0) \exp\left\{ \frac{i}{\hbar} S_0 \right\} \mathcal{D}\,[x(t)], \quad (2.3a)$$

$$K_2 = \frac{1}{2i} \left[K_1^2 - \frac{1}{K_0} \int (S-S_0)^2 \exp\left\{ \frac{i}{\hbar} S_0 \right\} \right] \mathcal{D}\,[x(t)]. \quad (2.3b)$$

In a similar way, the expressions for higher order cumulants can be obtained. However, in most of the applications the first cumulant K_1 itself is enough to arrive at a meaningful approximation. Therefore, we would not give the formal expression for higher order cumulants. Using, the notation for expectation value given in chapter 1, Eqs.(2.3) can be cast in the conventional probabilistic language

$$K_1 = \langle S - S_0 \rangle; \quad K_2 = \frac{1}{2i}\left[K_1^2 - \langle (S - S_0)^2 \rangle \right], \quad (2.4)$$

where the symbol $\langle A \rangle$ stands for the average or the expectation value of A with respect to trial action S_0. Thus within the first cumulant approximation the propagator $K(x'',x';T)$ corresponding to Eq.(2.1) reads as

$$K = K_0 \exp\left[\frac{i}{\hbar} \langle S - S_0 \rangle \right]. \quad (2.5)$$

Due to the oscillatory nature of the terms occurring in the expression for the propagator, it is difficult to make any quantitative statements regarding the nature of approximation. However, one can be more precise if one considers the propagator in "imaginary time" or the problem of evaluating the density matrix $\rho(x'',x';\beta)$. Recall that in chapter 1, we established a formal correspondence between $\rho(x'',x';\beta)$ and the Feynman propagator, namely,

$$\rho(x'',x';\beta) = K(x'',x'; -i\hbar T). \tag{2.6}$$

Hence Eq. (2.2) relating the exact propagator K and the cumulants of various orders modifies to

$$\rho(x'',x';\beta) = \rho_0(x'',x';\beta) \exp[-\alpha\,\rho_1 - \alpha^2 \rho_2 + \ldots] \tag{2.7}$$

where the quantities $\rho_1, \rho_2, \ldots, \rho_n$ are obtained by setting T equal to $-i\hbar\beta$ in the expressions for $K_1, K_2, \ldots, K_n$. Thus, the first cumulant approximation $\tilde{\rho}$ to the density matrix ρ reads as :

$$\tilde{\rho} = \rho_0 \exp(-\langle S - S_0 \rangle). \tag{2.8}$$

In order to investigate the nature of approximation, we shall use the so-called Jensen's inequality stated below.

Jensen's inequality

Let $A[x(t)]$ be a real functional of $x(t)$. Then :

$$\langle e^A \rangle \geq e^{\langle A \rangle} \tag{2.9}$$

The inequality (2.9) can be proved easily. Let us consider a finite set containing positive numbers $X_1, X_2, \ldots, X_n$ each occurring $p_1, p_2, \ldots, p_n$ times respectively, then using the inequality between the arithmetic and geometric means it follows

$$\frac{\sum p_j X_j}{\sum p_j} \geq \left(\prod_{k=1}^{n} X_k^{p_k} \right)^{\frac{1}{\sum p_j}}. \tag{2.10}$$

Setting $X_j = e^{A_j}$, it follows from the above inequality that

$$\frac{\sum p_j e^{A_j}}{\sum p_j} \geq e^{\sum p_j A_j / \sum p_j} \qquad (2.11)$$

which in the language of expectation value implies the inequality (2.9). The continuum analogue of the left hand side of (2.11) reads as

$$< e^A > = \frac{\int p(x) e^{A(x)} dx}{\int p(x) dx} , \qquad (2.12)$$

where A is a function of a random variable x with the associated probability distribution $p(x)$. The inequality (2.9) can be now derived by writing the integrals as Riemann sums and using (2.11) while assuming the usual conditions about the convergence of the two integrals involved. Finally, if A is a functional we can consider it as a function of many variables and use the multi-dimensional analogue of (2.12). As the the number of variables go to infinity we would recover (2.9).

Having seen a very rough sketch of the proof of Jensen's inequality, we return to the question of the nature of cumulant approximation. Setting $A = S - S_0$ in (2.9) yields

$$\rho \geq \tilde{\rho} . \qquad (2.13)$$

Hence, the approximate density matrix $\tilde{\rho}$ provides a lower bound to the exact density matrix ρ. However, similar bounds cannot be obtained for higher cumulants. Therefore, one cannot say *a priori* that the inclusion of higher cumulants will result in a better approximation for ρ. Moreover, the actual computation of higher cumulants is quite tedious. For these reasons, most of the theoretical investigations are restricted only to first cumulant approximation.

Equation (2.13) can be used to obtain a bound on the ground state

224 *Path-Integral Methods and their Applications*

energy E_0 of the system described by the density matrix ρ. The density matrix ρ admits an expansion

$$\rho = \sum_n e^{-\beta E_n} \psi_n(x'') \psi_n(x') , \qquad (2.14)$$

where ψ_n's are orthonormal eigenfunctions of the time-independent Schrödinger equation and E_n's are associated eigenvalues. Similarly

$$\rho_0 = \sum_n e^{-\beta E_n^0} \psi_n^0(x'') \psi_n^0(x') , \qquad (2.15)$$

where the superscript o indicates the corresponding quantities for the system described by the density matrix ρ_0. Since for a finite value of x, both $\psi_0(x)$ and $\psi_0^0(x)$ do not vanish, it is easy to see

$$\lim_{\beta \to \infty} \left(-\frac{1}{\beta} \ln \rho \right) = E_0 ; \qquad (2.16)$$

$$\lim_{\beta \to \infty} \left(-\frac{1}{\beta} \ln \rho_0 \right) = E_0^0 . \qquad (2.17)$$

Next, Jensen's inequality implies

$$\rho \geq \tilde{\rho} = \rho_0 \exp(-\langle S - S_0 \rangle). \qquad (2.18)$$

This leads to

$$\ln \rho > \ln \rho_0 - \langle S - S_0 \rangle \qquad (2.19)$$

and therefore using the result in Eqs. (2.16)--(2.17) we obtain

$$E_0 \leq E_0^0 + \lim_{\beta \to \infty} \beta^{-1} \langle S - S_0 \rangle . \qquad (2.20)$$

Note that, whereas the value of $\langle S - S_0 \rangle$ will depend on the end points x' and x", the limit of $\beta^{-1}<S - S_0>$ as $\beta \rightarrow \infty$ is independent of values of x' and x". Hence for estimating the ground state energy, the density matrix $\tilde{\rho}$ can be evaluated at any convenient values of x' and x" without affecting the estimate. The choice x' = x" = 0 very often leads to a considerable simplification in the algebraic manipulations. We might mention that although we have derived the result (2.20) for a one-dimensional system it remains valid for higher dimensional systems as long as the action functional takes only real values.

8.3. Spectrum of Positionally Disordered Systems

The perfect crystalline substances are characterized by a regular arrangement of atoms on a lattice. The electrons experience a periodic potential. Consequently the electronic levels, in perfect crystals, form a series of bands separated by energy gaps. The electrical properties of these substances are largely governed by electron-density in the top-most energy band known as the conduction band. In metals the conduction band overlaps with the lower energy band and there is no band gap. In perfect insulators the band gap is of the order of 5--8 eV whereas in semiconductors the band gap is of the order of 1 eV.

When the impurities are introduced in a semiconductor lattice, they occupy random sites and give rise to energy levels in the gap region. As the density of impurities increases, the electronic levels of impurity atoms form an energy band within the gap which slowly merges with the conduction band. The presence of this band makes it easier for electrons belonging to lower band to migrate to conduction band. Therefore, the structure of this band and in particular its "tail" is of immense importance.

8.3.1. *Basic formulation*

The mathematical model used in these studies considers an electron in the field of randomly distributed N impurities confined to a volume V. The effect of the presence of other electrons as well as of the periodicity of the lattice is introduced by replacing the bare electron mass with an effective mass. We shall, however, continue to denote it by the same symbol m. In chapter 3, it has been shown that the properties of such a system are governed by an average propagator, $G(\vec{q}'',\vec{q}';T)$, which in the limit of weak and densely distributed scatterers has a functional integral representation

$$G = \int \mathcal{D}\,[\vec{q}(t)]\, \exp\left[\frac{i}{\hbar}\int_0^T \frac{m\dot{\vec{q}}^2}{2} dt - \frac{\rho\eta^2}{2\hbar^2}\int_0^T dt \int_0^T ds\, W(\vec{q}(t)-\vec{q}(s))\right] \quad (3.1)$$

where

$$W(\vec{q}(t)-\vec{q}(s)) = \int v(\vec{q}(t)-\vec{Q})\, v(\vec{q}(s)-\vec{Q})\, d\vec{Q} \quad (3.2)$$

is the autocorrelation function (ACF) associated with the potential v. Further, the quantity ρ denotes the density of the scatterers while η denotes the strength of the potential.

8.3.2. Evaluation of G

Since the electron-impurity potential is weak, we may choose the trial action S_0 to correspond to a free particle. The formal expression for G within the first cumulant approximation can be written using Eq. (2.5) and reads as

$$G(\vec{q}'',\vec{q}';T) = K_0 \exp\left\{\frac{-\rho\eta^2}{2\hbar^2}\tilde{W}\right\} \quad (3.3)$$

where, for brevity we have set

$$\tilde{W} = \int_0^T dt \int_0^T ds\, \langle W(\vec{q}(t)-\vec{q}(s))\rangle. \quad (3.4)$$

The symbol $\langle\ldots\rangle$ in Eq. (3.4) and subsequently denotes the average with respect to the functional S_0. Further K_0 is the free particle propagator.

The quantity characterizing the spectrum adequately is the density of electronic states n(E). It is related to G as

$$n(E) = \frac{1}{2\pi\hbar} \int_0^\infty dT \, \exp\left[\frac{i}{\hbar} ET\right] \text{Tr } G \tag{3.5}$$

where Tr denotes the trace operation.

Since, K_0 as well as W have translational symmetry, Tr G will be equal to $G(0,0;T)$ expect for a volume factor. Hence while evaluating the average $\langle W \rangle$, we can safely substitute $\vec{q}' = \vec{q}'' = 0$. Using the definition of the average (cf. Eq.(2.31) of chapter 1) we may write:

$$\langle W(\vec{q}(t)-\vec{q}(s)) \rangle = \frac{1}{K_0} \int W(\vec{q}(t)-\vec{q}(s)) \, \exp\left(\frac{im}{2\hbar} \int_0^T \dot{\vec{q}}^2 \, dt\right) \mathcal{D} \, [\vec{q}(t)]. \tag{3.6}$$

Next, note that W depends only on the positions $\vec{q}(t)$ and $\vec{q}(s)$ at fixed times t and s respectively. Hence, the path integral in the time interval $(0, t)$, (t, s) and (s, T) can be separately performed and would yield the free particle propagators for the respective intervals. Thus we write

$$\langle W \rangle = \left(\frac{T}{t|t-s||T-s|}\right)^{d/2} \int d\vec{q}(t) \int d\vec{q}(s) \, W(\vec{q}(t)-\vec{q}(s))$$

$$\times \exp\left[\frac{im}{2\hbar} \left\{ \frac{\vec{q}^2(t)}{t} + \frac{(\vec{q}(t)-\vec{q}(s))^2}{|t-s|} + \frac{\vec{q}^2(s)}{T-s} \right\}\right]. \tag{3.7}$$

In Eq.(3.7), d denotes the dimensionality of the system under consideration. Setting $\vec{q}(t) - \vec{q}(s) = \vec{u}$ and performing the integration over $\vec{q}(s)$ it is easy to see that

$$\langle W \rangle = \left(\frac{mT}{2\pi i\hbar|s-t|(T-|s-t|)}\right)^{d/2} \int d\vec{u} \, W(\vec{u}) \, \exp\left[\frac{imTu^2}{2\hbar|t-s|(T-|s-t|)}\right]. \tag{3.8}$$

The next task is to evaluate $\widetilde{W}$, i.e., to integrate $\langle W \rangle$ of Eq.(3.8) with respect to t and s. To this effect we observe that if f is an arbitrary function of $|t-s|(T-|t-s|)$. Then,

$$\int_0^T dt \int_0^T ds\, f(|t-s|(T-|t-s|)) = 2T \int_0^{T/2} f(u(T-u))\, du. \tag{3.9}$$

Hence, we can write

$$\tilde{W} = 2T \int_0^{T/2} d\tau \left(\frac{mT}{2\pi i\hbar \tau(T-\tau)}\right)^{d/2} \int d\vec{u}\, W(\vec{u}) \exp\left[\frac{imTu^2}{2\hbar\tau(T-\tau)}\right]. \tag{3.10}$$

The change of variable from τ to x through the transformation

$$u^2 + x^2 = \frac{u^2 T^2}{4\tau(T-\tau)}$$

simplifies $\tilde{W}$ to

$$\tilde{W}(T) = 2^d\, T^{2-d/2} \left(\frac{m}{2\pi i\hbar}\right)^{d/2} \int W(\vec{u})\, d\vec{u}$$

$$\times \int_0^\infty dx\, \frac{|u|^{2-d}}{(u^2+x^2)^{(3-d)/2}} \exp\left[\frac{2im}{\hbar T}(x^2+u^2)\right]. \tag{3.11}$$

For one, two and three-dimensional systems, $\tilde{W}$ has the following explicit forms :

$$\tilde{W}(T) = \sqrt{2T}\left(\frac{m}{\pi i\hbar}\right) \int d\vec{u}\, W(\vec{u}) \exp\left(\frac{im}{\hbar T} u^2\right) K_{1/2}\left(\frac{-im}{\hbar T} u^2\right), \qquad d=3 \tag{3.12}$$

$$\tilde{W}(T) = \left(\frac{mT}{\pi i\hbar}\right) \int d\vec{u}\, W(\vec{u}) \exp\left(\frac{im}{\hbar T} u^2\right) K_0\left(\frac{-im}{\hbar T} u^2\right), \qquad d=2 \tag{3.13}$$

$$\tilde{W}(T) = 2 \left(\frac{m\pi T^3}{i\hbar}\right)^{1/2} \int [1 - \Phi(u\sqrt{2im/\hbar T})]\, W(u)\, du, \qquad d=1 \tag{3.14}$$

where K_ν denotes the Bessel function of the second kind and Φ denotes the error function.

8.3.3. *The behaviour of density of states*

For arbitrary W, it is not possible to simplify the expression of $\tilde{W}$ any further and hence, in general, the density of states $n(E)$ has to be numerically evaluated using Eq.(3.5). However, we can carry the analysis further if we limit ourselves to knowing the behaviour of $n(E)$ only in the deep interior of the band which corresponds to large negative values of E. It is in this region where a non-zero $n(E)$ changes the properties of semiconductors drastically.

To study the behaviour of $n(E)$ for large negative E ($E \rightarrow -\infty$), we note that $\tilde{W}(T)$ when viewed as a function of the complex variable T does not have a singularity in any finite part of the lower half of complex T plane. As $|T| \rightarrow \infty$, the behaviour of $\tilde{W}(T)$ corresponding to various dimensional systems can be obtained from the respective expressions (3.12)--(3.14). Thus,

$$\tilde{W}(T) \rightarrow T\, (ia_0 + O(1/T)), \qquad d = 3, \qquad (3.15)$$

$$\rightarrow \frac{a_0}{2i} T \ln T + O(T), \qquad d = 2, \qquad (3.16)$$

$$\rightarrow \sqrt{(2m/i\hbar)}\; T^{3/2} \int_0^\infty W_0(x)\, dx + O(T), \qquad d = 1, \qquad (3.17)$$

where a_0 is a constant having the value

$$a_0 = \frac{4m}{\hbar} M_1, \qquad M_1 = \int_0^\infty x\, W_0(x)\, dx, \qquad (3.18)$$

Here $W_0(x)$ denotes the angular average of $W(\vec{u})$, $x = |\vec{u}|$. We are now ready to analyze the behaviour of $n(E)$ as $E \rightarrow -\infty$. Consider the integral appearing in Eq.(3.5) for $n(E)$. For three-dimensional systems if $E < E_0 = -\rho \eta^2 a_0^2/2\hbar$, which is obviously finite if the first moment M_1 is finite. In such a case we can close the contour along a semicircular arc of radius $R \to \infty$, in the lower half of the complex T-plane. Cauchy's theorem can be readily invoked to conclude that $n(E) = 0$ for $E < E_0$. However, in one and two-dimensional cases the point at infinity is a singular point and hence the foregoing considerations do not apply. For one and two-dimensional systems $n(E)$ given by the integral (3.5) has to be evaluated by other methods.

The foregoing discussion can be put in the form of a theorem stated below.

Theorem 3.1

Within first cumulant approximation with free particle trial action, there exists an $E_0 < 0$ such that $n(E) = 0$ for $E < E_0$ provided M_1 the first moment of the ACF is finite.

Problem 8.3.1

Use the theorem 3.1 to show that if the ACF is a Gaussian :

$$W(\vec{q}) = (\pi L^2)^{-3/2} \exp(-q^2/L^2) , \quad (3.19)$$

$$n(E) = 0 \quad \text{for} \quad E \leq - \frac{m\rho\eta^2}{L\hbar^2 \pi^{3/2}} . \quad (3.20)$$

Problem 8.3.2

Show that for the ACF $W(\vec{q}) = \exp[-\lambda|\vec{q}|]/\lambda$, the cut-off energy E_0 of theorem (3.1) is given by $E_0 = -2\rho\eta^2 m/\hbar^2\lambda^3$.

Recall that the result expressed through the problem 8.3.1 is different from the result (5.6) of chapter 3, which implies a non-zero density of states (DOS) for all negative values of energies. As pointed out there the latter result has been due to Bezak and Saya-kanit. The discrepancy can be explained by noting that these authors approximated the Gaussian correlation function by its finite series expansion. This makes M_1 unbounded, and consequently the theorem 3.1 is not applicable.

Having discussed the qualitative aspects, we shall now be more specific and consider the case of a Gaussian correlation function and derive explicit expression for n(E).

8.3.4. The density of states for Gaussian correlation

For a Gaussian correlation function (cf. Eq.(3.19)), the integrals in Eqs.(3.12)--(3.14) can be readily performed. This leads to the following expressions for $\tilde{W}(T)$

$$\tilde{W}(T) = \frac{T^2}{4(\pi)^{3/2}} \frac{1}{L(L^2/4 + i\hbar T/8m)}, \qquad d = 3, \qquad (3.21)$$

$$\tilde{W}(T) = \frac{1}{\sqrt{\pi}} \frac{T^2}{\sqrt{(-i\hbar T/2m)}} \ln\left\{\frac{1+\sqrt{(-i\hbar T/2mL^2)}}{1-\sqrt{(-i\hbar T/2mL^2)}}\right\}, \qquad d = 2, \qquad (3.22)$$

and

$$\tilde{W}(T) = \frac{mT}{i\hbar\pi} \frac{1}{\sqrt{(1+2mL^2/i\hbar T)}} \ln\left\{\frac{\sqrt{(1+2mL^2/i\hbar T)} + 1}{\sqrt{(1+2mL^2/i\hbar T)} - 1}\right\}, \qquad d = 1. \qquad (3.23)$$

This value of $\tilde{W}(T)$ can be used to evaluate the average propagator G (in the appropriate dimension) of Eq.(3.3). The DOS n(E) can subsequently be obtained by using Eq.(3.5).

The three-dimensional system

For d = 3, the expression for n(E) reads as

$$n(E) = \frac{V}{2\pi\hbar} \int_{-\infty}^{\infty} dT \left(\frac{m}{2\pi i\hbar T}\right)^{3/2} \exp\left\{\frac{-\rho\eta^2 T^2}{L\hbar^2(4\pi)^{3/2}}\left(\frac{L^2}{4} + \frac{i\hbar T}{8m}\right)^{-1}\right\}. \qquad (3.24)$$

Using an integral representation

$$\frac{1}{(iT)^{3/2}} = \frac{2}{\sqrt{\pi}} \int_{-\infty}^{\infty} x^2 \exp(-iTx^2) \, dx, \qquad (3.25)$$

the DOS n(E) of Eq.(3.24) can be rewritten as

232 Path-Integral Methods and their Applications

$$n(E) = n_0 e^{-\alpha^2} \int_{-\infty}^{\infty} x^2 \, dx \int_{-\infty}^{\infty} du \, \exp\left[iu(p^2-x^2) - \frac{i\alpha^2}{u-i}\right] \qquad (3.26)$$

where $n_0 = \dfrac{mV}{4\pi^3 L \hbar^2}$, $\varepsilon = \dfrac{2mEL^2}{\hbar^2}$, $\alpha^2 = \dfrac{2\rho\eta^2 m^2 L}{\hbar^4 \pi^{3/2}}$ and $p^2 = \alpha^2 + \varepsilon$.

To simplify the expression for n(E) further, we should evaluate the integral over u. The integral can be easily evaluated by expanding $\exp[-\alpha^2/(u-i)]$ in a power series in $1/(u-i)$ and subsequently integrating each term of the series. The resulting series can be identified with the Bessel function $I_1[2\alpha \sqrt{(p^2-x^2)}]$ if $x^2 < p^2$. For $x^2 > p^2$, it can be shown that each term of the series vanishes. Hence

$$n(E) = n_0 e^{-\alpha^2} \int_{-\infty}^{\infty} dx \, x^2 \, B(p^2-x^2) \quad , \qquad (3.27)$$

where the function $B(\mu)$ is defined as

$$B(\mu) = 2\pi \, \delta(\mu) + \frac{2\pi\alpha \, e^{-\mu}}{\sqrt{\mu}} \, I_1(2\alpha\sqrt{\mu}) \, \theta(\mu)$$

and θ is the Heaviside function. The relation (3.27) implies that

$$n(E) = 0 \qquad p^2 < 0 \, , \qquad (3.28a)$$

while for $p^2 \geq 0$ the DOS is given by

$$n(E) = 2\pi n_0 e^{-\alpha^2}\left[p + 2\alpha p^2 \int_{-p}^{p} dx \, \frac{x^2 \, e^{-(p^2-x^2)}}{\sqrt{(p^2-x^2)}} \, I_1[2\alpha(p^2-x^2)^{1/2}]\right]. \qquad (3.28b)$$

Thus we see that n(E) is zero below $E_0 = -\rho\eta^2 m/(\hbar^2 L \pi^{3/2})$ as predicted by Theorem 3.1. Near $E = E_0$, the parameter p is small in magnitude. Hence to the leading order n(E) can be approximated by the first term of Eq. (3.28b). Thus

$$n(E) \simeq 2n_0\pi\, e^{-\alpha^2}(\varepsilon + \alpha^2)^{1/2}, \qquad E \simeq E_0, \qquad (3.29)$$

which implies that n(E) decays in a power law fashion. The behaviour is depicted in Fig. 8.1.

Problem 8.3.3

Verify that in the limit L → ∞, n(E) of Eq.(3.29) reduces to the DOS corresponding to a free particle.

Two-dimensional system

In two-dimensional system the expression for DOS n(E) reads as :

$$n(E) = \frac{n_0}{2\pi i}\int \frac{du}{u}\exp\left[\varepsilon u + \frac{\mu u}{\sqrt{(1+1/u)}} + \ln\left\{\frac{\sqrt{(1+1/u)}+1}{\sqrt{(1+1/u)}-1}\right\}\right] \quad (3.30)$$

where in writing Eq.(3.30), we have used the expression for G from Eq.(3.3) with $\tilde{W}(T)$ given by Eq.(3.22). Further $\varepsilon = 2mEL^2/\hbar$, $n_0 = V/2\pi\hbar^2$ and $\mu = m^2L^2\rho\eta^2/\pi\hbar^4$. The integral in Eq.(3.30) cannot be evaluated analytically. The numerical evaluation of n(E) is facilitated if one changes u to θ by $u = \sh^2\theta$. With this transformation, the expression for n(E) can be written as $n(E) = n_0(1 + J/\pi)$. The quantity J has the following expression

$$J = \int_0^\infty dx\, \frac{e^{-\varepsilon/2}\, e^{-\mu\theta(x)}}{\ch x}\left[\sin\left(\frac{\varepsilon}{2}\sh x - \Omega(x)\right)\sh x - \cos\left(\frac{\varepsilon}{2}\sh x - \Omega(x)\right)\right].$$

(3.31)

Further, the quantities θ(x) and Ω(x) have the expressions

$$\theta(x) = \frac{x\sh x}{\ch x} + \frac{\pi}{4}\frac{(\sh^2 x - 1)}{\ch x}, \qquad \Omega(x) = \frac{\mu\pi}{2}\frac{\sh x}{\ch x} + \frac{\mu x}{2}\frac{(1-\sh^2 x)}{\ch x}. \qquad (3.32)$$

In Fig. 8.1 we depict the behaviour of n(E). It is clear that n(E) is non-zero for all negative values of E. As expected on physical considerations, the numerical results indeed indicate that J → -π as

$E \to -\infty$, implying thereby that $n(E) \to 0$. However, this has not been shown analytically. For large positive energies $J \to 0$ since the integrand decays exponentially. Thus $n(E) \to n_0$, the free particle DOS in two-dimensions.

System in one dimension

For one-dimensional system, $n(E)$ is described by the integral (3.5) when the expression (3.3) for G is used along with $\widetilde{W}(T)$ of Eq.(3.23).

$$n(E) = \left[\frac{V}{2\pi\hbar}\right] \int_{-\infty}^{\infty} dT \left(\frac{m}{2\pi i \hbar T}\right)^{1/2} \exp\left(\frac{iET}{\hbar}\right) \exp\left[\frac{\rho\eta^2}{2\sqrt{(\pi\hbar^2)}} \frac{T^2}{(-i\hbar T/2m)^{1/2}}\right]$$

$$\times \ln\left[\frac{1 + (-i\hbar T/2mL^2)^{1/2}}{1 - (-i\hbar T/2mL^2)^{1/2}}\right] \qquad (3.33)$$

which can be recast in the following form

$$n(E) = n_1 \, \text{Re}\left(\int_0^{\infty} \frac{dZ}{\sqrt{i}} \exp\left[\frac{-\alpha_1^2 Z^3}{\sqrt{(-i)}}\right] \ln\left[\frac{1 + Z\sqrt{-i}}{1 - Z\sqrt{-i}}\right] + i\varepsilon Z^2\right) \qquad (3.34)$$

with the notations $n_1 = \frac{2mVL}{\hbar^2 \pi^{3/2}}$, $\alpha_1^2 = \frac{2\, m^2 L^3 \rho \eta^2}{\hbar^4 \sqrt{\pi}}$ and $\varepsilon = \frac{2mL^2 E}{\hbar^2}$. Once again, we can analyze the behaviour of $n(E)$ for extreme values of energy, i.e., $E \to \pm \infty$. For this purpose, we split the range of integration in Eq.(3.34) into two parts, $(0, 1)$ and $(1, \infty)$ and denote the respective contributions to the integral by n_I and n_{II} respectively. First consider n_I. This can be approximated as

$$n_I = \text{Re} \int_0^1 \frac{dZ}{\sqrt{i}} \exp\left[i\varepsilon Z^2 - 2\alpha_1^2 Z^4\right] + O(Z^6) \qquad (3.35)$$

whereas in the interval $(1, \infty)$, for large values of E, the integral (3.34), behaves as :

$$n_{II} = \text{Re} \int_1^{\infty} \frac{dZ}{\sqrt{i}} \exp\left[i\varepsilon Z^2 - \pi\alpha_1^2 \, (-2\alpha_1^2/3) \, (\sqrt{iZ})^3\right]. \qquad (3.36)$$

The explicit asymptotic evaluation of integrals in Eqs. (3.35) and (3.36) yields

$$n(E) = 2n_0 \sqrt{(\pi/\varepsilon)}, \quad \varepsilon \rightarrow +\infty,$$

$$= n_0\sqrt{2\pi/|\varepsilon|} \exp(-4|\varepsilon|^3/27a^2) \quad \varepsilon \rightarrow -\infty. \quad (3.37)$$

Equation (3.37) indicates that in one-dimension the DOS indeed displays the exponential tail as is expected from our qualitative considerations. The behaviour of the density of states is depicted in Fig. 8.1.

The results obtained for the behaviour of the DOS for one and two-dimensional systems are in qualitiaive agreement with those obtained by other methods. For example, Halperin and Lax using the Green's function approach also find an exponential tail.

The behaviour of n(E) for three-dimensional system, however, is in marked contrast with the experimental results or the earlier theoretical studies of Kane or Lifshitz. Since in the present analysis, n(E) has been evaluated analytically within the first cumulant approximation, we can attribute the power law type behaviour of n(E) near cutoff to the choice of free particle trial action. This choice of trial action implies that our results are relevant only for weakly disordered systems where the experiments are still inconclusive. For systems which are not weakly disordered, one should choose a different trial action. In that case one cannot proceed analytically and has to rely on numerical results. Saya-kanit has found numerically that n(E) is non-zero for all negative energies with a different choice of trial action. However, the analysis being numerical in nature is not very transparent.

8.4. The Polaron Problem

An electron in an ionic crystal polarizes the lattice in its neighbourhood. As the electron moves the polarization produced by it reacts on the electron. Thus due to the interaction with this polarization field, the electron acquires a self-energy in the field. An electron moving with its accompanying polarization is called a *Polaron*.

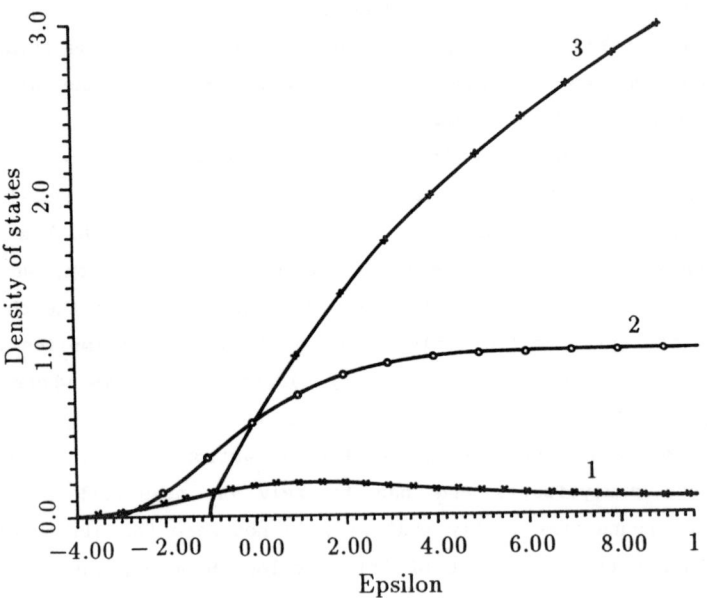

Fig. 8.1 *Qualitative behaviour of the density of states for one, two and three dimensional systems shown by curves marked by numbers 1, 2 and 3 respectively.*

Among the properties which distinguish the polaron from a bare electron, some of the significant ones are the polaron self-energy and its effective mass. The interest in these investigations is two-fold. First, there is an intrinsic interest in the properties of ionic crystals. Secondly, here is an example of a problem in non-relativistic field theory which can be analytically studied using path integral methods. In spite of the fact that several methods have been used to study the problem, the path integral method still remains the best.

Due to the interaction of an electron with phonon field, its ground state energy is lowered. Also, the presence of the phonon field leads to an increase in the inertia of the electron. This is reflected in the fact that the polaron possesses a higher effective mass.

The polaron problem was investigated in detail by Frölich using the conventional first order perturbation theory. He assumed that the crystal lattice acts like a perfect dielectric medium and then proceeded to construct the Hamiltonian for the electron--crystal interaction. He assumed that due to the periodicity of the lattice the electron feels a periodic potential even in absence of any lattice vibrations. This can be accounted for by considering the electron as a free particle but with a different mass. This fictitious particle now interacts with crystal vibrations (phonons). The phonons are modelled as a set of harmonic oscillators. To simplify the problem further, Frölich assumed that the important phonon waves interacting with electrons have the same frequency ω. Under these assumptions he showed that the potential felt by the electron due to polarization of the lattice can be written as :

$$V(\vec{q}) = i(\sqrt{2}\pi\alpha)^{1/2} (\hbar^5 \omega^3/m)^{1/4} \sum_{\vec{k}} \frac{1}{k} \left(\hat{a}^+_{\vec{k}} e^{-i\vec{k}\cdot\vec{q}} - \hat{a}_{\vec{k}} e^{i\vec{k}\cdot\vec{q}} \right) , \quad (4.1)$$

where α is a dimensionless constant depending on static and high frequency dielectric constants of the crystal. As a typical example, for NaCl crystal α ≅ 5, while in general, it ranges between 1 to 20. m is the mass associated with the electron after taking into account the periodicity of the lattice and ω is the frequency of optical phonons assumed to be constant for all vibrational modes $\vec{k}$. Furthermore $\hat{a}^+_{\vec{k}}$ and $\hat{a}_{\vec{k}}$ denote respectively the creation and annihilation operators for a phonon with wave vector $\vec{k}$. Lastly $\vec{q}$ denotes the position vector of the electron.

Next, the Hamiltonian $\hat{H}_0$ for a free electron and the lattice can be written as

$$\hat{H}_0 = \frac{\hat{p}^2}{2m} + \hbar\omega \sum_{\vec{k}} \hat{a}^+_{\vec{k}} \hat{a}_{\vec{k}} . \qquad (4.2)$$

Next we scale the variables by choosing the units such that $\hbar = m = \omega = 1$. The electron-phonon system will be characterized by the Hamiltonian

$$\hat{H} = \frac{\hat{p}^2}{2} + \sum_{\vec{k}} \hat{a}^+_{\vec{k}} \hat{a}_{\vec{k}} + i\,(\alpha\pi\sqrt{2})^{1/2} \sum_{\vec{k}} \frac{1}{k}\left(\hat{a}^+_{\vec{k}} e^{-i\vec{k}\cdot\vec{q}} - \hat{a}_{\vec{k}} e^{i\vec{k}\cdot\vec{q}} \right). \qquad (4.3)$$

However, in order to use path integral methods, we need to construct a classical Hamiltonian. Since the quantum Hamiltonian has been written in the language of second quantization, we must transform it so as to express the Hamiltonian operator in coordinate space. For this purpose, we define the phonon coordinates $\hat{Q}_{\vec{k}}$ and the momenta $\hat{P}_{\vec{k}}$ as

$$\hat{Q}_{\vec{k}} = \frac{1}{\sqrt{2}}\left(\hat{a}^+_{\vec{k}} - \hat{a}_{-\vec{k}} \right), \qquad \hat{P}_{\vec{k}} = \frac{i}{\sqrt{2}}\left(\hat{a}_{\vec{k}} + \hat{a}^+_{-\vec{k}} \right) \qquad (4.4)$$

to arrive at the classical Hamiltonian function

$$H = \frac{p^2}{2} + \sum_k \frac{1}{2}\,(\,P^2_{\vec{k}} + Q^2_{\vec{k}}\,) + \sqrt{2}\,(\sqrt{2}\,\alpha\,\pi)^{1/2} \sum_k \frac{1}{k} Q_{\vec{k}}\, e^{i\vec{k}\cdot\vec{q}} . \qquad (4.5)$$

Further, making use of the fact $\hat{a}^*_{\vec{k}} = \hat{a}_{-\vec{k}}$, Eq. (4.5) simplifies to

$$H = \frac{P^2}{2} + \sum_{\vec{k}} \frac{1}{2}(P^2_{\vec{k}} + Q^2_{\vec{k}})$$

$$+ 2\sqrt{2} \, (\alpha\pi\sqrt{2})^{1/2} \sum_{\vec{k}} \frac{1}{k} \, (\text{Re } Q_{\vec{k}} \cos \vec{k}.\vec{q} - \text{Im } Q_{\vec{k}} \sin \vec{k}.\vec{q}) \qquad (4.6)$$

where Re $Q_{\vec{k}}$ (Im $Q_{\vec{k}}$) refers to real (imaginary) part of $Q_{\vec{k}}$.

8.4.1. Free energy and ground state energy

Having constructed the classical Hamiltonian function, we can now proceed with the evaluation of the two-point density matrix $\rho(\vec{q}'',\vec{q}',Q_{\vec{k}}'',Q_{\vec{k}}';\beta)$ which has the path integral representation

$$\rho(\vec{q}'',\vec{q}',Q_{\vec{k}}'',Q_{\vec{k}}';\beta) = \int \exp\left[-\int_0^\beta H(\tau) \, d\tau\right] \mathcal{D}\,[\vec{q}(\tau)] \, \mathcal{D}\,[Q_{\vec{k}}(\tau)], \qquad (4.7a)$$

where $\mathcal{D}\,[Q_{\vec{k}}]$ denotes the functional integration over both real and imaginary parts of $Q_{\vec{k}}$ and β is related to the inverse of the absolute temperature. The partition function Z is related to the density matrix by the usual relation

$$Z = \int \rho(\vec{q},\vec{q},Q_{\vec{k}},Q_{\vec{k}};\beta) \, d\vec{q} \, dQ_{\vec{k}}. \qquad (4.7b)$$

Once the partition function is known, one can calculate the temperature variation of free energy F by means of the relation $F = -\beta^{-1}\ln Z$. It is well known, that as $\beta \to \infty$ the free energy $F \to E_0$, the ground state energy. Thus E_0 is obtained by computing the limiting expression

$$E_0 = \lim_{\beta \to \infty} \left(-\frac{1}{\beta} \ln Z\right). \qquad (4.8)$$

Next, note that the classical Hamiltonian H (cf. Eq.(4.6)), is quadratic in $Q_{\vec{k}}$. It resembles a set of oscillators perturbed by a constant force $\cos(\vec{k}.\vec{q})$ (or $\sin(\vec{k}.\vec{q})$). Therefore, using the results of chapter 2, we can perform the path integration over $Q_{\vec{k}}$. The resulting expression being Gaussian in $Q_{\vec{k}}''$ and $Q_{\vec{k}}'$ the integration over $Q_{\vec{k}}$ indicated in Eq.(4.7b) can be carried out. This leaves us with the evaluation of the so-called reduced density matrix $\tilde{\rho}(\vec{q}'',\vec{q}';\beta)$,

$$\tilde{\rho}(\vec{q}'',\vec{q}';\beta) = \int \rho(\vec{q}'',\vec{q}',Q_{\vec{k}},Q_{\vec{k}};\beta) \, dQ_{\vec{k}} \quad , \tag{4.9}$$

which can be expressed as a functional integral involving an effective action (in electron coordinates alone) in the form

$$\tilde{\rho}(\vec{q}'',\vec{q}';\beta) = \int \exp(-S_{eff}) \, \mathcal{D}[\vec{q}(t)] \tag{4.10}$$

where the effective action S_{eff} is given by

$$S_{eff} = \frac{1}{2} \int_0^\beta dt \, \dot{\vec{q}}^2 - \alpha\pi\sqrt{2} \int_0^\beta dt \int_0^\beta ds \int \frac{d\vec{k}}{(2\pi)^3} \frac{G(t,s)}{k^2} e^{i\vec{k}.[\vec{q}(t) - \vec{q}(s)]}$$

(4.11a)

with

$$G(t,s) = \frac{\cosh[(\beta/2) - |t-s|]}{\sinh(\beta/2)} \quad . \tag{4.11b}$$

The second term in Eq.(4.11a) arises due to path integration over phonon variables. The double integral is an indication of the memory effects introduced in chapter 3. What is remarkable about path integral method is that we have been able to eliminate the phonon degrees of freedom exactly. Moreover, since we are studying the problem through density matrix, the formulation is capable of studying the temperature variation of physical properties of the polaronic system.

The evaluation of ρ can be carried out exactly only as far as the path integration over phonon coordinates is concerned. For further analysis we shall employ the first cumulant method to evaluate $\tilde{\rho}$. For this purpose, the choice of the trial action S_0 is dictated by the strength α of electron-phonon coupling. For small values of α, S_0 can be chosen to correspond to a free particle and $\tilde{\rho}$ can be approximated as

$$\tilde{\rho} = \rho_0 \exp(-\langle S - S_0 \rangle) \qquad (4.12)$$

where ρ_0 is the free particle density matrix. Next, the average $\langle S-S_0 \rangle$ can be written as

$$\langle S-S_0 \rangle = -\alpha\pi\sqrt{2} \int_0^\beta dt \int_0^\beta ds \int \frac{d^3\vec{k}}{(2\pi)^3} \frac{G(t,s)}{k^2} \langle e^{i\vec{k}\cdot[\vec{q}(t)-\vec{q}(s)]} \rangle. \qquad (4.13)$$

As noted earlier, to obtain the partition function it is enough to evaluate only the diagonal element of $\tilde{\rho}$. Further, due to the translational invariance of S and S_0, the required diagonal element $\tilde{\rho}(\vec{q},\vec{q};\beta) = \tilde{\rho}(0,0;\beta)$. In Sec. 3, we have already evaluated the average $\langle W(\vec{q}(t) - \vec{q}(s)) \rangle$ for an arbitrary function W (cf. Eq.(3.8)). Setting $W = e^{i\vec{k}\cdot[\vec{q}(t)-\vec{q}(s)]}$, we obtain

$$\langle \exp[i\vec{k}\cdot(\vec{q}(t)-\vec{q}(s))] \rangle = e^{-k^2 b(t,s)} \qquad (4.14)$$

where

$$b(t,s) = \frac{1}{2}(|t-s|)(1 - |t-s|/\beta) \qquad (4.15)$$

and hence the complete expression for the diagonal element of the density matrix is given by

$$\tilde{\rho} = \left(\frac{1}{2\pi\beta}\right)^{3/2} \exp\left(\alpha\pi\sqrt{2} \int_0^\beta dt \int_0^\beta ds \int \frac{d^3\vec{k}}{(2\pi)^3} \frac{G\, e^{-k^2 b}}{k^2}\right). \qquad (4.16)$$

The integral over $\vec{k}$ can be performed to arrive at

$$\tilde{\rho} = \left(\frac{1}{2\pi\beta}\right)^{3/2} \exp\left(\frac{\alpha\sqrt{2}}{4\sqrt{\pi}} \int_0^\beta dt \int_0^\beta ds \left(\frac{G(t,s)}{\sqrt{b}}\right)\right). \quad (4.17)$$

Having evaluated $\tilde{\rho}$, we evaluate the free energy F as

$$F = -\frac{1}{\beta}\left\{\ln\left(\frac{1}{2\pi\beta}\right)^{3/2} + \left(\frac{\alpha\sqrt{2}}{4\sqrt{\pi}} \int_0^\beta dt \int_0^\beta ds \frac{G(t,s)}{\sqrt{b}}\right)\right\} \quad (4.18)$$

which can be used to compute the ground state energy E_0 of the polaron by letting $\beta \to \infty$. Hence

$$E_0 = \lim_{\beta \to \infty} F = -\frac{\alpha}{\sqrt{\pi}} \int_0^\infty du \frac{e^{-u}}{\sqrt{u}} = -\alpha. \quad (4.19)$$

Now recall that the density matrix in the first cumulant approximation yields a lower bound to the exact density matrix. Hence it follows that the free energy F of Eq.(4.18) and the ground state energy E_0 of Eq.(4.19) provide the respective upper bounds. This result agrees with the estimate for E_0 obtained earlier by Frölich. However, what we have been able to conclude further is that the estimate of Eq.(4.19) is indeed an upper bound.

For small values of α, the upper bounds in Eqs.(4.18)--(4.19) are very near to actual values of free energy and ground state energy respectively. However, when α is large, the upper bounds given by Eqs.(4.18)--(4.19) deviate considerably from their actual values. The reason for this can be traced in the choice of the trial action. For large α, one should choose a trial action different from the free particle trial action. One such choice used very often in literature is

$$S_0 = \frac{1}{2} \int_0^\beta d\tau\, \vec{q}^2(\tau) + \frac{C}{W} \int_0^\beta dt \int_0^\beta ds\, G_w(t,s)\, [\vec{q}(t)-\vec{q}(s)]^2 \quad (4.20)$$

with the kernel $G_w(t,s) = (W/2)\cosh[W\{(\beta/2) - |t-s|\}]/\sinh(W\beta/2)$. One can evaluate afresh an estimate for $\tilde{\rho}$ using S_0 of Eq.(4.20) and consequently the free energy F. The resulting expression is then minimized with respect to C and W to obtain their optimum values. The estimates for the free energy F and the ground state energy E_0 thus obtained remain valid for a wider range of values of α. The details of the calculations with the trial action S_0 of Eq.(4.20) can be found in a review article cited at the end of this chapter.

8.4.2. *The effective mass*

The dynamics of the polaronic system is usually studied by treating it as a free particle with an effective mass. Before estimating this parameter one must have an adequate definition of effective mass. In this discussion we shall use the definition given by Saitoh which has been found to be adequate. He considers a polaron under the influence of a vanishingly small force $\vec{f}$ and defines its effective mass by the response of the polaron to this external force. If the polaron is to be modelled as a free particle, it is expected that for small $\vec{f}$, the dependence of the polaron density matrix on $\vec{f}$ will be similar to the density matrix associated with a free particle under the influence of a constant external force. The diagonal term of the desired density matrix reads as (cf. chapter 2, Eq.(3.2c))

$$\rho_{free}(0,0;\beta) = \left(\frac{1}{2\pi\beta}\right)^{3/2} \exp\left(\frac{-\beta^3}{24m}\vec{f}^2\right).$$

Guided by these considerations, Saitoh gave the following prescription to obtain the effective mass of the polaron. Couple the polaron to a constant force $\vec{f}$, the inverse of the effective mass is given by the coefficient of $-\beta^3 f^2/24$ in the expression for the polaron density matrix in the limit $\vec{f} \to 0$. When the polaron is coupled to an external force $\vec{f}$, the polaron action S_1 is given by

$$S_1 = S_{eff} + \int_0^\beta \vec{f}.\vec{q}(t) \, dt. \qquad (4.21)$$

To evaluate the density matrix $\tilde{\rho}$ associated with (4.21), we also modify our trial action to $S_0(\vec{f})$ by adding the $\vec{f}$ dependent term of Eq. (4.21) to the free particle action. The new trial action $S_0(\vec{f})$ is of the form

$$S_0(\vec{f}) = \frac{1}{2}\int_0^\beta \dot{\vec{q}}^2(\tau)\,d\tau + \vec{f}\cdot\int_0^\beta \vec{q}(\tau)\,d\tau \tag{4.22}$$

which we shall continue to denote by S_0. As before, we evaluate $\langle S_1 S_0\rangle$ and find

$$\langle S_1 - S_0\rangle = -\frac{\sqrt{2\alpha}}{2\pi}\int_0^\beta dt \int_0^\beta ds\, G(t,s) \int_0^\infty dk\, \frac{\sin k|A|f}{k|A|f} e^{-k^2 b}, \tag{4.23}$$

where $A = (t-s)(\beta-t-s)/2$ and b is as in Eq. (4.15)

In the limit $f \to 0$, Eq. (4.23) assumes the form

$$\langle S_1 - S_0\rangle = -\frac{\sqrt{2}}{\pi}[Z_1 + f^2 Z_2 + O(f^4)] \tag{4.24}$$

where

$$Z_1 = -\frac{\sqrt{\pi}}{2}\int_0^\beta du\, \frac{(\beta-u)\,G(u)}{\sqrt{b(u)}}, \quad Z_2 = \frac{\sqrt{2}}{288\sqrt{\pi}}\int_0^\beta \frac{(\beta-u)^3 u^2 G(u)}{b^{3/2}(u)}. \tag{4.25}$$

It is easy to deduce the temperature dependent effective mass m of the polaron using its definition and the result in Eq. (4.24). It reads as

$$\frac{1}{m(\beta)} = 1 - \frac{\alpha\,\beta^{-3/2}}{3\sqrt{\pi}}\int_0^\beta \sqrt{u}(\beta-u)^{3/2}G(u)du = 1 - \frac{\alpha}{12}\frac{\sqrt{\pi\beta}\,I_1(\beta/2)}{\sinh(\beta/2)}, \tag{4.26}$$

where $I_1(u)$ is the modified Bessel function of the second kind. Equation (4.26) describes the variation of the effective mass of the polaron with temperature in the weak coupling limit. The value of the effective mass m_0 at absolute zero can be obtained by taking limit $\beta \to \infty$ in Eq. (4.26). This yields,

$$\frac{1}{m(\infty)} \equiv \frac{1}{m_0} = (1 - \frac{\alpha}{6}) \qquad (4.27)$$

which is also the estimate obtained by Frölich. Note, however, that through Eq.(4.26) we have obtained the complete temperature dependence of the effective mass $m(\beta)$.

Before we conclude, we must also add that there are several other definitions of effective mass which are prevalent in the literature. For example, Feynman assumed that near absolute zero, the matrix element of the electron-phonon density matrix between the ground state of phonons is similar to the free particle density matrix. Hence he defined the effective mass as the coefficient of $-q^2/2\beta$, in the limit $\beta \rightarrow \infty$ of the polaron density matrix. This definition is physically reasonable. However, if one tries to assert that the polaron density matrix itself is similar to the free-particle density matrix, it leads to erroneous results. In fact the effective mass evaluated using this definition diverges as $\beta \rightarrow \infty$. Also, various estimates of effective mass obtained using other definitions agree with each other only when α is small. For large values of α, every definition gives a different estimate. However, since it is not possible to find any bound for the effective mass, one cannot comment on the accuracy of the methods.

8.5. The Bi-Polaron Problem

The possibility of formation of a bi-polaron structure was first considered by Pekar. He considered two identical electrons (holes) in the presence of electron-phonon coupling in a crystal lattice. On qualitative arguments he showed that the two electrons, which will normally repel each other due to the coulomb interaction, can form a bound pair if the electron-phonon coupling is sufficiently strong. Such a bound pair is called a bi-polaron. It has been conjectured that these bi-polarons play a very important role in the mechanism of high-temperature superconductivity. This has resulted in vigorous efforts in bi-polaron investigation.

For the three-dimensional bi-polaron, the quantum Hamiltonian H is given by the sum of three terms :

i) The Hamiltonian H_0 for the free electrons and the lattice,

$$\hat{H}_0 = \frac{1}{2m}(\hat{p}_1^2 + \hat{p}_2^2) + \hbar\omega \sum_{\vec{k}} \hat{a}_{\vec{k}}^{+} \hat{a}_{\vec{k}}, \qquad (5.1)$$

ii) The electrostatic Coulomb repulsion V_e between two electrons,

$$V_e = e^2/|\vec{q}_1 - \vec{q}_2| \qquad (5.2)$$

iii) The electron-phonon coupling term V_{pe}

$$V_{pe} = \sum_{\vec{k}} \frac{1}{k}\left[\hat{a}_{\vec{k}}^{+}\left(e^{-i\vec{k}\cdot\vec{q}_1} + e^{-i\vec{k}\cdot\vec{q}_2}\right) - \hat{a}_{\vec{k}}\left(e^{i\vec{k}\cdot\vec{q}_1} + e^{i\vec{k}\cdot\vec{q}_2}\right)\right]. \qquad (5.3)$$

Following a procedure similar to the last section, one can construct the classical Hamiltonian function which takes the form

$$H = \frac{1}{2}(p_1^2 + p_2^2) + \sum_{\vec{k}} \frac{1}{2}(P_{\vec{k}}^2 + Q_{\vec{k}}^2) + \frac{\delta}{|\vec{q}_1 - \vec{q}_2|} + 2\sqrt{2}(\alpha\pi\sqrt{2})^{1/2}$$

$$\times \sum_{\vec{k}} \frac{1}{k}\left[\mathrm{Re}(Q_{\vec{k}})[\cos(\vec{k}\cdot\vec{q}_1) + \cos(\vec{k}\cdot\vec{q}_2)] - \mathrm{Im}(Q_{\vec{k}})[\sin \vec{k}\cdot\vec{q}_1 + \sin \vec{k}\cdot\vec{q}_2]\right]$$

$$(5.4)$$

where as before we have chosen the units such that $\hbar = m = \omega = 1$. In these new units, the strength of coulomb repulsion is denoted by δ. α is as in Sec. 4.

Having constructed the classical Hamiltonian function, we can repeat the steps followed in Sec.4.2 to evaluate the partition function Z for the bi-polaron system within the first cumulant approximation. First, as in the last section, the path integration over the phonon coordinates can be exactly performed leaving behind the task of evaluating a functional integral similar to (4.10). Here S_{eff}, the effective action for the bi-polaron system is given by

$$S_{eff} = \int_0^\beta \left\{ \frac{1}{2}(\vec{q}_1^2 + \vec{q}_2^2) + \frac{\delta}{|\vec{q}_1 - \vec{q}_2|} \right\} - S_1 \qquad (5.5a)$$

where

$$S_1 = \pi\alpha\sqrt{2} \int_0^\beta dt \int_0^\beta ds \int \frac{d^3\vec{k}}{(2\pi)^3} G(t,s) \sum_{i,j=1}^{2} \exp[i\vec{k}\cdot\{\vec{q}_j(t) - \vec{q}_i(s)\}] \qquad (5.5b)$$

and $G(t,s)$ is given by Eq. (4.11).

It is convenient to introduce the centre of mass and relative coordinates $\vec{q}_1 + \vec{q}_2 = \sqrt{2}\,\vec{\mu}$ and $\vec{q}_1 - \vec{q}_2 = \sqrt{2}\,\vec{\eta}$ in Eqs. (5.5). This simplifies the expression for S_{eff} to

$$S_{eff} = \int_0^\beta dt \left[\frac{1}{2}(\vec{\mu}^2 + \vec{\eta}^2) + \frac{\delta}{|\vec{\eta}|\sqrt{2}} \right] - S_1, \qquad (5.6a)$$

where S_1 now reads as

$$S_1 = \sqrt{2}\pi\alpha \int_0^\beta dt \int_0^\beta ds \int \frac{d^3\vec{k}}{(2\pi)^3} G(t,s)\, e^{i\vec{k}\cdot[\vec{\mu}(t) - \vec{\mu}(s)]} (b^+_{\vec{k}} + b^-_{\vec{k}}) \qquad (5.6b)$$

with $b^\pm_{\vec{k}}$ being defined as

$$b^\pm_{\vec{k}} = \exp[\pm i\vec{k}\cdot(\vec{\eta}(t) - \vec{\eta}(s))] + \exp[\pm i\vec{k}\cdot(\vec{\eta}(t) + \vec{\eta}(s))]. \qquad (5.6c)$$

At this stage, we introduce the cumulant approximation. For this purpose we must choose a trial action. Since it is expected that the system can form a bound pair only if electron-phonon coupling is very strong, the free particle trial action used in Sec. 4 is no more adequate. For the present problem, we choose a trial action of the form

$$S_0 = \int_0^\beta dt\, [\vec{\mu}^2 + \vec{\eta}^2 + \omega^2\eta^2]/2 + \frac{\omega^2}{4\beta} \int_0^\beta dt \int_0^\beta ds\, [\vec{\mu}(t) - \vec{\mu}(s)]^2. \qquad (5.7)$$

As in Sec. 4, the quantities $\langle S_{eff} \rangle$ and $\langle S_0 \rangle$ should be evaluated afresh using the trial action of Eq. (5.7). Due to a complicated structure of the trial action the computations involve long and tedious but straightforward algebra. We shall leave the explicit evaluation of these averages as an exercise for the reader and only write the final expressions for the various quantities.

First, the trial density matrix ρ_0 (the propagator in imaginary time β) corresponding to S_0 is the product of two factors ρ_0^1 and ρ_0^2. The factor ρ_0^1 corresponds to the propagator associated with a harmonic oscillator of frequency ω (cf. Eqs. (3.3) of chapter 2) while ρ_0^2 corresponds to the Bezak action (cf. Eq. (4.9) of chapter 3). Next it is easy to see that

$$\left\langle \frac{\delta}{\sqrt{2}} \int_0^\beta \frac{dt}{|\eta|} \right\rangle = \frac{\delta}{\sqrt{\pi}} \int_0^\beta dt \left(\frac{\omega \sinh \omega\beta}{\sinh \omega(\beta-t) \sinh \omega t} \right)^{1/2}. \quad (5.8)$$

Further the evaluation of $\langle S_1 \rangle$ leads to

$$\langle S_1 \rangle = \alpha \sqrt{(2/\pi)} \int_0^\beta dt \int_0^\beta ds \, G(t,s) \left[\frac{1}{\sqrt{B_+ + C}} + \frac{1}{\sqrt{B_- + C}} \right]. \quad (5.9)$$

Here the symbols B_+, B_- and C are defined as follows.

$$B_\pm = B(t,t) + B(s,s) \pm 2 B(t,s) \quad (5.10a)$$

with

$$B(t,s) = \frac{\sinh [\omega(\beta-t_+)] \sinh (\omega t_-)}{\omega \sinh (\omega \beta)}. \quad (5.10b)$$

$$C = C(t,t) + C(s,s) - 2C(t,s) \quad (5.11a)$$

with

$$C(t,s) = \frac{2}{\omega} \cosh\left[\frac{\omega}{2}(t-s)\right] \sinh\left(\frac{\omega}{2} t_-\right) \sinh\left[\frac{\omega}{2}(\beta-t_+)\right] \left\{\sinh\left(\frac{\omega\beta}{2}\right)\right\}^{-1}. \quad (5.11b)$$

The quantities t_+ and t_- in Eqs. (5.10) and (5.11) refer to max(t,s) and min(t,s) respectively.

Lastly, the expression for $<S_0>$ reads as

$$<S_0> = <T> + \frac{3\omega\beta}{2} \coth(\beta\omega/2) - \frac{3}{2} ,$$

where the symbol $<T>$ denotes the average of the kinetic energy terms in the expression (5.7) for S_0. This average, however, cancels out in the final computation of $<S_{eff} - S_0>$. The expressions for the various averages obtained above can be used to estimate the ground state energy and effective mass of the bi-polaron. For estimating ground state energy, we use Eq. (2.20) of Sec. 2. Since the explicit expressions for various averages are known, it is easy to obtain the necessary limits needed for the purpose. This yields

$$E_0 \leq \frac{3\omega}{2} + \delta \sqrt{(2\omega/\pi)} - 2\alpha \, I \qquad (5.12)$$

where

$$I = \sqrt{(\omega/\pi)} \left(1 + \int_0^\infty e^{-u} [1 - e^{-\omega u}]^{-1/2} du \right) . \qquad (5.13)$$

To get the best estimate, the right-hand side of the inequality (5.12) must be minimized with respect to ω. An approximate calculation yields

$$\omega = \frac{2}{9\pi} (2\sqrt{2}\,\alpha - \delta)^2 , \qquad \alpha \geq \delta / 2\sqrt{2} \qquad (5.14)$$

$$\omega = 0 , \qquad \alpha \leq \delta / 2\sqrt{2}$$

which implies

$$E_0 \leq \frac{-1}{3\pi} (2\sqrt{2}\,\alpha - \delta)^2 \qquad \alpha \geq \delta / 2\sqrt{2}. \qquad (5.15)$$

In a similar way, an estimate for the effective mass can be obtained. The obvious requirement for the stability of the system is that the ground state energy of the bi-polaron system should be less than that of two free polarons. This yields a minimum value of α needed for the formation of a bi-polaron. The ground state energy of a free polaron E_0^f in the strong coupling limit is given by $E_0^f = -\gamma_p \alpha^2$, γ_p being Pekar's constant. Hence our criterion $2E_0^f - E_0 > 0$ implies

$$\alpha \geq \frac{\delta}{[\sqrt{2}(2 - \sqrt{3\pi\gamma_p})]} \text{, or } \alpha \leq \frac{\delta}{[\sqrt{2}(2 + \sqrt{3\pi\gamma_p})]} \text{,} \qquad (5.16)$$

The first condition in Eq. (5.16) cannot be satisfied. The second condition defines the region of stability.

Problem 8.5.1

Derive an estimate for the effective mass of a bi-polaron using the definition of Sec. 4.

8.6. Polymer Distribution Functions

It has been shown in chapter 3 that the configuration sum $P(\vec{q}'', N; \vec{q}', 0)$ for a system of polymers having their end points at $\vec{q}'$ and $\vec{q}''$ respectively can be expressed as a path integral

$$P(\vec{q}'', N; \vec{q}', 0) = \int \exp(-S) \, \mathcal{D}\,[\vec{q}(\nu)] \qquad (6.1)$$

where

$$S = \int_0^N d\nu \left(\frac{d\vec{q}}{d\nu}\right)^2 + \int_0^N \int_0^N V(\vec{q}(\nu) - \vec{q}(\nu')) \, d\nu \, d\nu' \qquad (6.2)$$

and V is the potential between the monomers at ν and ν'. For a dilute solution of a polyelectrolyte the interaction potential between the two monomers belonging to the same polymer is $V = \alpha |\vec{q}(\nu) - \vec{q}(\nu')|^{-1}$. Since, the solution is dilute, we assume that the situation can be modelled by considering an ensemble of non-interacting polymers. The configuration sum $P(\vec{q}'', N; \vec{q}', 0)$ for this system is given by Eq. (6.1) with the associated action functional S given by Eq. (6.2) with the appropriate form of the potential V.

The action functional of Eq. (6.1) is similar to the effective action (4.11) encountered in the polaron problem, except that the interaction is repulsive. The repulsive nature of the interaction makes the configuration sum (6.1) well defined for all values of α.

A physical quantity of interest in the context of polymers is the end-to-end mean square distance $<q^2>$ defined by Eq. (7.14) of chapter 6. It is a measure of the flexibility (rigidity) of a polymer. Recall, that if the polymer configurations were modelled as free random walks, the average $<q^2> \approx N$, whereas for perfect rod like structures, $<q^2> \approx N^2$. Hence it is interesting to investigate the behaviour of $<q^2>$ for polyelectrolytes.

We shall carry out the evaluation of P within the first cumulant approximation. Further it is assumed that the interaction strength α is small. Hence we choose the trial action S_0 to correspond to a free particle. Thus we write

$$P = P_0 \, e^{-<S-S_0>} \qquad (6.3)$$

where P_0 is the expression corresponding to free particle density matrix (with β replaced by N). Proceeding as in the case of the polaron problem, it is easy to see that

$$<S-S_0> = -\frac{2\alpha}{\pi} \int_0^N d\nu_1 \int_0^N d\nu_2 \int_0^\infty dk \, \frac{\sin ka|\vec{q}''-\vec{q}'| \, e^{-k^2 b}}{ka \, |\vec{q}''-\vec{q}'|} \qquad (6.4)$$

where $a = |\nu_1 - \nu_2|/N$, and b is as defined in Eq. (4.15). With this expression of $<S-S_0>$, one can now write the complete expression for P using Eq. (6.3),

$$P = \exp\left(-\frac{|\vec{q}''-\vec{q}'|^2}{2N} - \frac{2\alpha}{\pi} \int_0^N d\nu_1 \int_0^N d\nu_2 \int_0^\infty dk \, \frac{\sin ka|\vec{q}''-\vec{q}'| \, e^{-k^2 b}}{ka \, |\vec{q}''-\vec{q}'|}\right). \qquad (6.5)$$

Exploiting the symmetry of a and b with respect to ν_1 and ν_2, the integral over ν_2 can be performed. After some rescaling of variables, the expression for P reads as

$$P = \exp\left(\frac{-y^2}{2} - \frac{4\alpha N^{3/2}}{\pi} f(\vec{y})\right), \qquad (6.6)$$

$$f(\vec{y}) = \int_0^1 dv(1-v) \int_0^\infty dk \, \frac{\sin k v y}{kvy} e^{[-k^2 v(1-v)/2]}, \qquad (6.7)$$

where $y = |\vec{q}''-\vec{q}'|/\sqrt{N}$. Consequently, the expression for the mean squared distance $\langle q^2 \rangle$ takes the form,

$$\langle q^2 \rangle = \frac{N \int_0^\infty y^4 \exp\left(\frac{-y^2}{2} - \frac{4\alpha N^{3/2}}{\pi} f(y)\right) dy}{\int_0^\infty y^2 \exp\left(\frac{-y^2}{2} - \frac{4\alpha N^{3/2}}{\pi} f(y)\right) dy}. \qquad (6.8)$$

To study the behaviour of the mean square distance for long polymers, the integrals in Eqs.(6.8) can be estimated using the method of asymptotic analysis. This leads to

$$\langle \vec{q}^2 \rangle = N y_0^2(N) \qquad (6.9)$$

where y_0 is the solution of equation $f'(y) = 0$, with $f''(y) > 0$.

It can be easily seen from the structure of $f(y)$ that there exits a non-zero positive value of y satisfying $f'(y) = 0$, Hence we conclude from (6.9) that in dilute solution the polyelectrolytes behave as free polymers. This conclusion is in conformity with a result obtained by de Gennes.

Notes and References

The cumulant expansion method is discussed in any standard book on Statistical Mechanics. In particular one may consult

R. P. Feynman, "Statistical Mechanics: A set of lecture notes"

Ed. Shahm, (Benjamin, New York, 1975).

Here the reader can also find the derivation of the functional integral representation of the partition function and a sketch of the proof of Jensen's inequality.

The best place to find a qualitative and readable description regarding the problem of positionally disordered systems is contained in

V. Bezak, Proc. Roy. Soc. (London) **A315**, 339 (1970)

and in a recent review article by

V. Saya-kanit and H. R. Glyde, "Path integral approach to the theory of heavily doped semiconductors" in "Path Summation : Achievements and Goals", Ed. Stig Lundqvist, Anedio Ranfagni, Virulh Saya-kanit, and Lawrence S. Schulman (World Scientific, Singapore, 1988). This review also discusses other approaches to study the problem of DOS, namely, those of Halperin and Lax and the semi classical approach of Kane.

The path integral method to study the problem of DOS was developed by

S. F. Edwards and Y. B. Gulyaev, Proc. Phys. Soc. **83**, 495 (1964)

and since then has been extensively used to study this problem. In particular Saya-kanit and his co-workers used this method using harmonic trial action to study a variety of heavily doped semiconductors. They claim that the results for DOS are in agreement with experiments. Various references to Saya-kanit's work can be found in his review article cited earlier.

In a later paper Khandekar *et al* reinvestigated the problem of DOS using free particle trial action and the results of Sec. 3 are based on that work. Their results differ from those of Saya-kanit for three-dimensional systems. These results are contained in a series of three papers.

D. C. Khandekar, Vijay A. Singh, K. V. Bhagwat and S. V. Lawande, Phys. Rev. **B33**, 5482 (1986),

K. V. Bhagwat, D. C. Khandekar and S. V. Lawande, Phys. Rev. **B34**, 8929 (1986),

K. V. Bhagwat, S. V. Lawande and D. C. Khandekar, Int. J. Mod. Phys.

B1, 1321 (1987).

These calculations certainly suggest that the path integral method must be very carefully applied to study the problem of DOS. It appears to us that to a large extent it has to do with the oscillatory nature of Feynman propagator.

Saya-kanit and his co-workers have applied the cumulant approximation method to investigate the problem of broadening of Landau levels in heavily doped semiconductors. This can be found in

Virulh Saya-kanit and Henry R. Glyde, "Path integral approach to the Landau level broadening in the quantum Hall problem", in "Path Integrals From meV to MeV", Ed. Virulh Saya-kanit et al (World Scientific, Singapore 1990).

However, a parallel computation by Leschke and his co-workers suggest that within the first cumulant approximation the DOS may take even negative values for some magnetic field strengths. This is discussed in

K. Broderix, N. Heldt, and H. Leschke, Z. Phys. **B68**, 19 (1987).

The concept of polaron has been nicely explained by Frölich and can be found in

H. Frölich, Adv. Phys. **3**, 325 (1954).

The various properties of polaron and the literature accumulated over the years has been summarized in

"Polarons in Ionic Crystal" Ed. J. T. Devreese (North-Holland, Amsterdam, 1972),

"Polarons and Excitons", Eds. C. G. Kuper and G. D. Whitfield, (Oliver and Boyd, London 1962).

The polaron problem was formulated in terms of a path integral by

R. P. Feynman, Phys. Rev. **97**, 660 (1955).

The treatment followed here is essentially a simplified version of Feynman's approach. We base our calculations through free particle trial action which obviously limit their validity to weak coupling. We also differ from Feynman in the definition of effective mass and follow a definition due to Saitoh. This definition can be found in

M. Saitoh, J. Phys. Soc. (Japan) **49**, 886 (1980).

Saitoh's treatment of the polaron problem can be easily generalized to finite temperatures. Such calculations can be found in

D. P. L. Castrigiano, N. Kokiantonis and H. Stierstorfer, Phys. Lett. **A104**, 364 (1984).

Several workers have attempted to improve upon Feynman's estimate of the ground state energy by introducing more parameters in the trial action. However, the conclusion which emerges from these studies is that even after employing infinite number of variational parameters, the estimate for the ground state energy differs only by 2 % from that of Feynman. See, for example,

J. Adamowski, B. Gerlach and H. Leschke, in "Functional integration, theory and applications", Eds. J. P. Antonie and E. Tirapegui, (Plenum, New York 1980).

Several definitions of the effective mass of the polaron prevalent in the literature are contained in

F. M. Peters and J. T. Devreese, in "Solid State Physics", Eds. H. Ehrenreich, F. Seitz and D. Turnbull, (Academic Press, New York 1984).

It was pointed out by Khandekar et al that within first cumulant approximation most of the definitions of the effective mass lead to erroneous results. This discussion forms the subject matter of

D. C. Khandekar, K. V. Bhagwat and S. V. Lawande, Phys. Rev. **B37**, 3085 (1988).

The first cumulant approximation cannot be applied to arrive at a bound on the effective mass. However, functional integral methods have been applied to obtain bounds on this parameter. This is discussed in

B. Gerlach, H. Löwen and H. Schliffke, Phys. Rev. **B36**, 6320 (1987).

The path integral has also been exploited to obtain an exact estimate for the ground state energy of the polaron in strong coupling limit. This can be found in

Janusz Adamowski, Bernd Gerlach and Hajo Leschke, Phys. Lett. **A79**, 249 (1980).

The cumulant approximation scheme has been used by Devreese and his co-workers to study the polaron system in magnetic fields. However, these studies are able to provide bound on the ground state energy only for weak magnetic fields. This work appears in

J. T. Devreese and F. Brosens, "Path integral application to a polaron in a magnetic field". in " Path Integrals From meV to MeV", Ed. Virulh Saya-kanit *et al* (World Scientific, Singapore 1990).

The limitations on these results are related to the applicability of Jensen's inequality. For a discussion of this aspect refer to

K. Broderix, N. Heldt and H. Leschke, Z. Phys. **B66**, 507 (1987).

The bi-polaron literature is again growing fast primarily because of a conjecture about its applicability in explaining the phenomenon of high temperature superconductivity. This concept was introduced by

S. I. Pekar, "Research on electron theory in crystals" (U.S. A.E.C. Report, Washington D.C. 1963).

More accessible report on the formation of bi-polaron is due to

T. D. Schultz, in "Polarons and Excitons", Eds. C. G. Kuper and G. D. Whitfield, (Oliver and Boyd, London 1963).

The connection of the bi-polaron in explaining the phenomenon of high temperature superconductivity has been discussed by

J. G. Bednorz and K. A. Müller, Z. Phys. **B64**, 189 (1986).

This problem was formulated in terms of a path integral by

H. Hiramoto and Y. Toyozawa, J. Phys. Soc. (Japan) **54**, 245 (1985).

However, they used a form of the interaction Hamiltonian which was cut off at a finite value of the phonon wave vector. This makes analytical evaluation of various parameters very difficult. Khandekar considered a modified model where the interaction Hamiltonian involved all wave vectors and the treatment presented here is essentially based on

D. C. Khandekar, Mod. Phys. Lett. **B4**, 1201 (1990),

D. C. Khandekar, S. V. Lawande and D. Biswas, Phys. Rev. **B43**, 9750 (1991).

The best place to find a description of the path integral formulation to evaluate various parameters relating to polymer physics is a book by

F. W. Wiegel, "Introduction to Path Integral Methods in Physics and Polymer Sciences", (World Scientific, Singapore 1986).

Our formulation of the problem of polyelectrolytes is based on an early work by

G. J. Papadopoulos and J. Thomchick, J. Phys. **A10**, 115 (1977)
and the results presented here form the subject matter of the article

D. C. Khandekar, K. V. Bhagwat and F. W. Wiegel, Mod. Phys. Lett. **B1**, 19 (1987).

The first cumulant method has been used in a variety of other problems. Most of them use the harmonic trial action to evaluate Feynman propagator. In particular, if the frequency of phonon modes in the polaron problem is assumed to depend on the wave vector k one can easily map this problem to study the properties of plasmaron which is discussed by

V. Saya-kanit, M. Nithisoontorn and W. Sritrakool, Physica Scripta **32**, 334 (1985).

Similarly, in the bi-polaron problem if the two particles forming a bound pair happen to possess opposite charge the problem is referred to as exciton-phonon problem and has been discussed by

Janusz Adamowski, Bernd Gerlach and Hajo Leschke, Phys. Rev. **B23**, 2943 (1981),

Janusz Adamowski, Bernd Gerlach and Hajo Leschke, Physica **B117**, 287 (1983).

CHAPTER 9

THE PERTURBATION APPROACH

9.1. Introduction

In standard quantum mechanics based on Schrödinger or Heisenberg methods, an important approximation method for solving problems is the perturbation theory. It turns out that perturbation approach can be easily incorporated in the Feynman path integral with the gain of much formal simplicity. In fact, the early applications of path integral method were based on the perturbation expansion of the propagator. Feynman introduced his famous diagram technique to provide a lucid physical interpretation of the perturbation series. Indeed the concept of the propagator and the associated Feynman diagrams proved to be paradigms for computations in quantum electrodynamics and modern field theory.

A general account of the perturbation theory can be found in the books listed in the references. We take a slightly different view in this chapter. We examine the possible use of perturbation series to derive exact propagators. There are several examples where the perturbation series can be summed yielding a closed form of either the time-dependent propagator or the energy-dependent Green's function. We discuss these examples first on an ad hoc basis and later develop a systematic theory to explain why the theory works.

9.2. The Perturbation Series

Consider a system characterized by a local Lagrangian L given by

$$L = L_0 - \alpha V(\vec{q}) \qquad (2.1)$$

where L_0 is the Lagrangian of the system for which the associated propagator $K_0(\vec{q}'',t'';\vec{q}',t')$ is known. The strength α of the interaction $V(\vec{q})$ is assumed to be small compared to any other interaction strength. We now write the propagator $K(\vec{q}'',t'';\vec{q}',t')$ for the system described by the Lagrangian (2.1) as

$$K(\vec{q}'',t'';\vec{q}',t') = \int \exp\left[\frac{i}{\hbar}\int_{t'}^{t''} L_0 \, dt\right] \exp\left[-\frac{i\alpha}{\hbar}\int_{t'}^{t''} V \, dt\right] \mathcal{D}[\vec{q}(t)]. \quad (2.2)$$

Next, we expand the second exponential in Eq.(2.2) as a power series in α as

$$K(\vec{q}'',t'';\vec{q}',t') = K_0(\vec{q}'',t'';\vec{q}',t') + \sum_{n=1}^{\infty} \frac{(-i\alpha/\hbar)^n}{n!} K_n(\vec{q}'',t'';\vec{q}',t'), \quad (2.3)$$

where the term $K_n(\vec{q}'',t'';\vec{q}',t')$ is given by the functional integral

$$K_n(\vec{q}'',t'';\vec{q}',t') = \int \exp\left[\frac{i}{\hbar}\int_{t'}^{t''} L_0 dt\right]\left(\int_{t'}^{t''} V dt\right)^n \mathcal{D}[\vec{q}(t)]. \quad (2.4)$$

Next, we note that

$$\left(\int_{t'}^{t''} V \, dt\right)^n = n!\int_{t'}^{t''} dt_1 V(\vec{q}_1) \int_{t'}^{t_1} dt_2 V(\vec{q}_2) \ldots \int_{t'}^{t_{n-1}} dt_n V(\vec{q}_n), \quad (2.5)$$

where $\vec{q}_k = \vec{q}(t_k)$, $k = 1, 2, \ldots, n$. Using the result of (2.5) followed by an interchange of the functional integration and n-fold integration over $t_1, t_2, \ldots, t_n$ we can write

$$K_n = \int_{t'}^{t''} dt_1 \int_{t'}^{t_1} dt_2 \ldots \int_{t'}^{t_{n-1}} dt_n \int V(\vec{q}_1)V(\vec{q}_2)\ldots V(\vec{q}_n) \exp\left[\frac{i}{\hbar}\int_{t'}^{t''} L_0 dt\right] \mathcal{D}[\vec{q}(t)], \quad (2.6)$$

where we have used a short-hand notation K_n for $K_n(\vec{q}'',t'';\vec{q}',t')$. Since L_0 is a local Lagrangian, it follows that

$$\int_{t'}^{t''} L_0 \, dt = \sum_{i=0}^{N} \int_{t_i}^{t_{i+1}} L_0 \, dt \qquad (2.7)$$

with the notation $t_0 = t'$ and $t_{N+1} = t''$.

Since the V's occurring in Eq.(2.6) depend only on the positions at $t = t_1, t_2, \ldots, t_n$, we can perform the functional integration over the domains (t', t_n), (t_n, t_{n-1}), ..., (t_1, t'') independently. This yields

$$K_n = \int_{t'}^{t''} dt_n \cdots \int_{t'}^{t_3} dt_2 \int_{t'}^{t_2} dt_1 \int \prod_{j=1}^{n} d\vec{q}_j \, K_0(0,1) \, V(\vec{q}_1) \, K_0(1,2)$$

$$\times V(\vec{q}_2) \ldots K_0(2,3) \, K_0(n-1,n) \, V(\vec{q}_n) \, K_0(n,n+1) \qquad (2.8)$$

where the notation $K_0(i,j)$ denotes the propagator $K_0(\vec{q}_i, t_i; \vec{q}_j, t_j)$ along with the identification $0 \equiv (\vec{q}', t')$ and $n+1 \equiv (\vec{q}'', t'')$. Thus the nth order term of the series can be expressed as an n-fold integral over coordinates q and time t involving the product of unperturbed propagators K_0 and the potentials V.

Each term of the expansion can be given a physical interpretation. The zeroth order term denotes an unperturbed evolution where the effects of the potential $V(\vec{q})$ are not felt. Next consider K_1. The particle travels from $t = t'$ to $t = t_1$, without feeling the influence of the potential. Then it encounters the interaction instantaneously at $t = t_1$. Thereafter in the interval $t_1 < t < t''$ it again has free evolution. Similarly the second order term K_2 denotes free evolution up to time t_1. At $t = t_1$ it encounters the potential and continues to evolve freely until it encounters the potential second time at time $t = t_2$. It then evolves freely up to time $t = t''$. In an analogous way nth order term represents n encounters with the potential. Figure 9.1 gives a schematic account of the first three terms of the perturbation series.

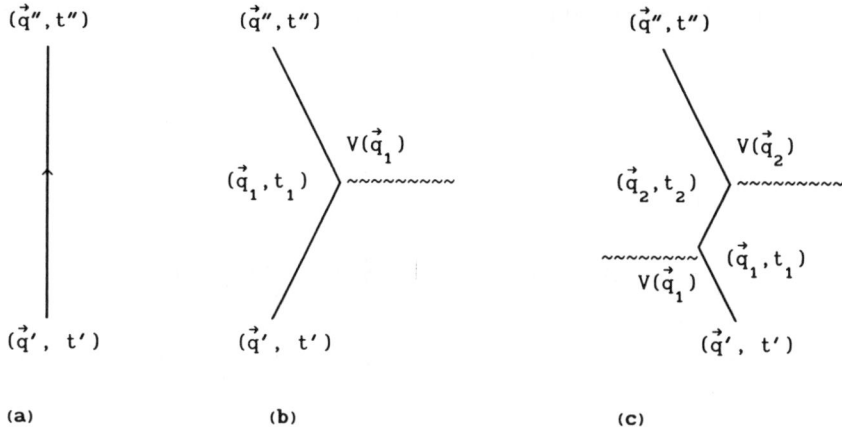

Fig. 9.1. The physical picture of the first three terms of the perturbation series; (a) the zeroth order, (b) the first order, and (c) the second order terms. The solid line denotes free evolution, while the encounter with the potential is indicated by a wavy line.

The perturbation approximation and its various applications have been discussed extensively in several books. However, an important aspect which is not mentioned in the literature is the possibility of obtaining exact propagators by summing the perturbation series. This is perhaps due to the fact that traditionally the perturbation method is looked upon as a tool merely to obtain an approximate expression for the propagator. Hence we shall devote the remaining part of this chapter to discuss this new aspect. For this purpose we confine ourselves to Lagrangians which do not explicitly depend on time. Before considering the general case we would demonstrate the procedure by considering some specific cases.

A considerable simplification occurs in the derivation of exact propagators if one first evaluates the energy-dependent Green's function $G(\vec{q}'',\vec{q}';E)$ instead of the propagator $K(\vec{q}'',\vec{q}';T)$. The two are related by the usual relation

$$G(\vec{q}'',\vec{q}';E) = \int_0^\infty K(\vec{q}'',\vec{q}';T)\, e^{iET/\hbar}\, dT . \qquad (2.9)$$

To set up the perturbation series for $G(\vec{q}'',\vec{q}';E)$, we take the Fourier transform of Eq. (2.3) to obtain

$$G(\vec{q}'',\vec{q}';E) = G_0(\vec{q}'',\vec{q}';E) + \sum_{n=1}^{\infty} \left(-\frac{i\alpha}{\hbar}\right)^n G_n(\vec{q}'',\vec{q}';E) , \qquad (2.10)$$

where $G_n(\vec{q}'',\vec{q}';E)$ has the expression

$$G_n(\vec{q}'',\vec{q}';E) = \int \prod_{j=0}^{n} G_0(\vec{q}_{j+1},\vec{q}_j;E) \prod_{j=1}^{n} V(\vec{q}_j) \, d\vec{q}_j \qquad (2.11)$$

where as before, $\vec{q}_0 = \vec{q}'$, $\vec{q}_{n+1} = \vec{q}''$. The advantage of going over to the Fourier representation is now distinctly seen. Equation (2.11) involves integrations only over coordinates, whereas the perturbation series for the propagator in Eq. (2.8) involved integrations over both coordinates and time. Once $G_n(\vec{q}'',\vec{q}';E)$ is evaluated Eq. (2.10) can be used to obtain the expression for the complete Green's function. The propagator then can be obtained by taking the inverse Fourier transform of the Green's function. With these remarks we proceed to evaluate propagators for some specific cases.

9.3. One-Dimensional Delta-function Potential

For the case of a particle moving under the influence of a one-dimensional δ-function potential, the L_0 of Eq. (2.1) corresponds to a free particle Lagrangian and the potential is given by $V(x) = \delta(x)$. For this specific choice, $K_0(x'',x';T)$ corresponds to the free particle propagator (cf. Eq. (3.1) of Ch. 2) and the Green's function $G_0(x'',x';E)$ is obtained by taking its Fourier transform

$$G_0(x'',x';E) = \left(\frac{m}{2E}\right)^{1/2} \exp[\,ik|x''-x'|\,] \qquad (3.1)$$

where $k^2 = 2m\,E/\hbar^2$. To evaluate $G_n(x'',x';E)$, we use Eq. (2.11) and carry out integration over x_j. This yields

$$G_n(x'',x';E) = G_0(x'',0;E) G_0(0,x';E) \, [G_0(0,0;E)]^{n-1} . \qquad (3.2)$$

Substituting the value of G_0 from Eq.(3.1) in the above equation, the expression for $G_n(x'',x';E)$ reads as

$$G_n(x'',x';E) = \left(\frac{m}{2E}\right)^{1+(n-1)/2} \exp(ik\xi) , \qquad \xi = |x''| + |x'|. \qquad (3.3)$$

The complete Green's function $G(x'',x';E)$ is obtained by summing the series in Eq.(2.10). This gives

$$G(x'',x';E) = G_0(x'',x';E) + \frac{m\alpha}{2i\hbar E}\left[1 + i\left(\frac{m\alpha^2}{2\hbar^2 E}\right)^{1/2}\right]^{-1} \exp(ik\xi). \qquad (3.4)$$

It is easy to see from Eq.(3.4) that if α is negative the Green's function has only one pole at $E = -m\alpha^2/2\hbar^2$. This corresponds to the only bound state for the attractive delta function potential. The bound state wave function can be easily obtained form Eq.(3.4) and is given by

$$\psi(x) = \left(\frac{m|\alpha|}{\hbar}\right)^{1/2} \exp\left(\frac{m\alpha|x|}{\hbar}\right) .$$

In order to obtain the expression for the propagator $K(x'',x';T)$, we take the inverse Fourier transform of $G(x'',x';E)$ given by Eq.(3.4). Thus we get

$$K(x'',x';T) = K_0(x'',x';T) - K^1(x'',x';T) \qquad (3.5a)$$

where

$$K^1(x'',x';T) = \frac{i\alpha m}{2\pi\hbar} \int_{-\infty+i0}^{\infty+i0} \left(\frac{dE}{2\sqrt{E}}\right) \frac{\exp[-(iET/\hbar) + ik\xi]}{\sqrt{E} + (i\alpha/\hbar)\sqrt{(m/2)}} . \qquad (3.5b)$$

To simplify the integral in Eq.(3.5b), we change the variable of integration from E to z by writing $\sqrt{E} = z$. This modifies the expression for $K^1(x'',x';T)$ as

$$K^1(x'',x';T) = \frac{i\alpha m}{2\pi\hbar} \int_\Gamma dz \, \frac{\exp[i\{-z^2 T + \sqrt{(2m)}z\xi\}/\hbar]}{\{z + (i\alpha/\hbar)\sqrt{(m/2)}\}} \qquad (3.6)$$

where Γ is a contour in the complex z plane defined by the rectangular hyperbola $xy = \gamma$, $(\gamma \rightarrow 0+)$.

The evaluation of the integral in Eq. (3.6) would be simpler if one is able to deform the contour Γ into a straight line parallel to the real axis. With this motivation in mind we note that in the case of attractive potential α is negative and the integrand has a singularity on the positive imaginary axis. Taking the contribution of the singularity into account the integral can be written as

$$K^1(x'',x';T) = K^2(x'',x';T) + \frac{i\alpha m}{2\pi\hbar} \int_\Gamma dz \frac{e^{i\{-z^2 T + \sqrt{(2m)} z\xi\}/\hbar}}{z + (i\alpha/\hbar)\sqrt{(m/2)}} , \qquad (3.7a)$$

where $K^2(x'',x';T)$ denotes the contribution from the singularity and is given by

$$K^2(x'',x';T) = \frac{m\alpha}{\hbar} \exp\left[\frac{m\alpha\xi}{\hbar^2} + \frac{im\,\alpha^2 T}{2\hbar^3}\right] \theta(-\alpha) \qquad (3.7b)$$

with $\theta(x)$ denoting the step function. The contour Γ can now be deformed into a line parallel to the real axis such that Im z is positive. Further, we use the integral representation

$$\zeta^{-1} = -i \int_0^\infty dx\, e^{i\zeta x} , \qquad \text{Im } \zeta > 0 \qquad (3.8)$$

and rewrite the expression for K^1 as

$$K^1 = K^2 + \frac{i\alpha m}{2\pi\hbar} \int_\Gamma dz \int_0^\infty d\tau \, \exp\left[\frac{i}{\hbar}\left\{\left(-z^2 T + \sqrt{2m}\,\xi\,z\right) + \tau\left(-i\hbar\,z + \alpha\sqrt{\frac{m}{2}}\right)\right\}\right].$$

$$(3.9)$$

Next the integration over the variable z can be performed readily and the expression for K^1 takes the form

$$K^1 = K^2 + \frac{\alpha m}{2\pi\hbar} \int_0^\infty d\tau \, e^{-(m\alpha/\hbar^2)\tau} K_0(\xi+\tau,0;T). \qquad (3.10)$$

Hence the complete expression for the propagator is given by

$$K(x'',x';T) = K_0(x'',x';T) - \frac{\alpha m}{\hbar^2} \int_0^\infty d\tau \, e^{-(m\alpha/\hbar^2)\tau} K_0(\xi+\tau,0;T) - K^2.$$

(3.11)

We can perform the integration over τ in the second term of right-hand side of Eq.(3.11) to rewrite the propagator as

$$K(x'',x';T) = K_0(x'',x';T) - \frac{\alpha m}{\hbar^2} \sqrt{2} \exp\left(\frac{imT\alpha^2}{2\hbar} - \frac{m^2\alpha^2\xi^2}{4\hbar}\right)$$

$$\times \text{erfc}\left[\xi \left(\frac{m}{2i\hbar T}\right)^{1/2} - \alpha \left(\frac{imT}{2\hbar}\right)^{1/2}\right] - K^2(x'',x';T), \quad (3.12)$$

where erfc(z) denotes the complementary error function with a complex argument.

The last term in Eq.(3.12) denotes the contribution from the bound state which occurs in the case of an attractive delta-function. The other two terms denote the contribution from the continuous spectrum. The derivation sketched above can be easily generalized to take into account a potential described by a lattice of δ-function potentials.

The fact that we are able to obtain the exact expression of the propagator for a one-dimensional δ-function potential raises the question about its applicability to higher dimensional cases. It is well known that δ-function potentials in higher dimensions are too singular in the sense that the corresponding Hamiltonian operators cannot be made self-adjoint. To discuss the problem of point interactions in higher dimensions one usually considers the free motion in space excluding the singular point. This leads to an effective potential which is operator-valued and cannot be discussed by the conventional techniques of path integration.

9.4. Inverse Square Potential

Consider a particle moving under the influence of an inverse square potential. The Lagrangian L_0 of Eq.(2.1) corresponds to the free particle constrained to move on the half line $0 < x < \infty$, while the potential $V(x)$ is given by

$$V(x) = 1/x^2, \quad x > 0; \quad V(x) = \infty, \quad x < 0. \tag{4.1}$$

The corresponding expression for the propagator $K_0(x'',x';T)$ can be found in chapter 4. The expression for the Green's $G_0(x'',x';E)$ corresponding to K_0 is obtained by taking its Fourier transform

$$G_0(x'',x';E) = (x'x'')^{1/2} \left(\frac{m\pi}{i\hbar}\right) J_{1/2}(kr_<) H^{(1)}_{1/2}(kr_>), \tag{4.2}$$

where k is related to E as defined earlier. The symbols $r_<$ and $r_>$ denote $\min(x'',x')$ and $\max(x'',x')$ respectively and $J_\nu(z)$ and $H^{(1)}_\nu(z)$ refer to the Bessel and Hankel functions respectively.

Our next task is to use this expression for $G_0(x'',x';E)$ in Eq.(2.11) to perform the n-fold coordinate integrations for evaluating $G_n(x'',x';E)$. However, the expression (4.2) for $G_0(x'',x';E)$ is not suitable for this purpose. The difficulty in carrying out the n-fold coordinate integrations is due to the coupling between x' and x'' in the expression for $G_0(x'',x';E)$. Hence it is desirable to re-express $G_0(x'',x';E)$ in another suitable form such that the terms in x' and x'' are completely separated. Such a representation for $G_0(x'',x';E)$ has the following form

$$G_0(x'',x';E) = \frac{m}{i\hbar}(x'x'')^{1/2} \int_0^\infty d\nu \, \frac{\nu \sinh \nu\pi}{\nu^2 + \frac{1}{4}} H^{(1)}_{i\nu}(kx'') H^{(1)*}_{i\nu}(kx'). \tag{4.3}$$

When the above expression for $G_0(x'',x';E)$ is inserted in Eq.(2.11) the n-fold coordinate integrations can be performed by using the (orthogonality) relation

$$\int_0^\infty \frac{dx}{x} H^{(1)*}_{i\nu}(kx) H^{(1)}_{i\nu'}(kx) = \frac{2\,\delta(\nu-\nu')}{\nu \sinh \nu\pi}, \tag{4.4}$$

and we obtain

$$G_n(x'',x';E) = \left(\frac{2m}{i\hbar}\right)^{n+1} \frac{(x'x'')^{1/2}}{2} \int_0^\infty d\nu \, \frac{\nu \sinh \nu\pi}{(\nu^2 + \frac{1}{4})^{n+1}} H^{(1)}_{i\nu}(kx'') H^{(1)*}_{i\nu}(kx').$$

$$\tag{4.5}$$

The expression (4.5) for $G_n(x'',x';E)$ is now substituted in Eq. (2.10) and the resulting geometric series is summed to obtain the complete Green's function $G(x'',x';E)$:

$$G(x'',x';E) = \frac{m}{i\hbar} (x'x'')^{1/2} \int_0^\infty d\nu \; \frac{\nu \sinh \nu\pi}{\nu^2 + \frac{1}{4} + \frac{2m\alpha}{\hbar^2}} \; H_{i\nu}^{(1)}(kx'') H_{i\nu}^{(1)*}(kx').$$

(4.6)

This is the correct expression for the energy dependent Green's function for $E > 0$. For $E < 0$, k is pure imaginary and the expression for G can be obtained by analytic continuation of (4.6).

The resulting expression has a similar appearance except for the fact that the Hankel function $H_\nu^{(1)}$ is now replaced by the modified Bessel function K_ν. The expression for the propagator is obtained by taking the inverse Fourier transform of the Green's function. However, an explicit evaluation of the inverse Fourier transform is unnecessary. In comparing the expression (4.6) for G with the expression (4.3) for G_0, we observe that they have identical structure. Therefore the expression for $K(x'',x';T)$ can be immediately obtained from that of $K_0(x'',x';T)$ by appropriate substitutions and reads as

$$K(x'',x';T) = \frac{m}{i\hbar T} (x'x'')^{1/2} \exp\left[\frac{i}{2\hbar T} \frac{m}{}(x'^2+x''^2)\right] I_\gamma\left(\frac{mx'x''}{i\hbar T}\right) \qquad (4.7)$$

where $\gamma^2 = 2m\alpha/\hbar^2 + 1/4$.

The derivation outlined above is valid if the geometric series in Eq. (2.10) converges. This requires $|2m\alpha/\hbar^2| < 1/4$. In other words, $-\hbar^2/8m < \alpha < \hbar^2/8m$. The condition $\alpha > -\hbar^2/8m$ is necessary to avoid the "fall to the centre". The other condition $\alpha < \hbar^2/8m$ can be relaxed by analytic continuation.

9.5. Coulomb Potential

For the Coulomb problem, the Lagrangian L_0 of Eq. (2.1) corresponds to that for a free particle in three-dimensions, while the potential $V(\vec{q}) = 1/r$, $r = |\vec{q}|$. The expression for the propagator $K_0(\vec{q}'',\vec{q}';T)$ has been given in chapter 2. The associated energy dependent Green's function is given as usual by Fourier transform of the propagator. The Green's function admits an expansion

$$G_0(\vec{q}'',\vec{q}';E) = \sum_{l,m} Y_{lm}^*(\hat{q}') Y_{lm}(\hat{q}'') G_{0l}(r'',r';E) ,\qquad(5.1)$$

where $Y_{lm}(\hat{x})$ denotes the spherical harmonic. The symbol $G_{0l}(r'',r';E)$ is the radial Green's function associated with the lth partial wave. Explicitly, it reads as

$$G_{0l} = \frac{m}{i\hbar} (r'r'')^{-1/2} \int \exp\left[\frac{i}{\hbar}\left(\frac{m(r'^2+r''^2)}{2T} + E\,T\right)\right] I_{l+1/2}\left(\frac{mr'r''}{i\hbar T}\right) dT$$

$$= \frac{m\pi}{\hbar} (r'r'')^{-1/2} J_{l+1/2}(kr_<) H^{(1)}_{l+1/2}(kr_>) .\qquad(5.2)$$

The complete Green's function $G(\vec{q}'',\vec{q}';E)$ also admits an expansion similar to that of (5.1) and we denote the expansion coefficient (the radial Green's function) by $G_l(r'',r';E)$. The perturbation series for $G_l(r'',r';E)$ can be obtained by substituting expansion (5.1) in Eq. (2.11) and carrying out the angular integrations. The series representation has the form

$$G_l(r'',r',E) = \sum_{n=0}^{\infty} \left(\frac{-i\alpha}{\hbar}\right)^n G_{nl}(r'',r',E)\qquad(5.3a)$$

with

$$G_{nl}(r'',r';E) = \int_0^\infty \left(\prod_{j=1}^n r_j dr_j\right) \prod_{j=0}^n G_{0l}(r_{j+1},r_j;E)\qquad(5.3b)$$

and we identify $r_0 = r'$. Once again the expression (5.2) for G_{0l} is not suitable for carrying out the radial integrations involved in (5.3b). A convenient representation for this purpose is found to be

$$G_{0l}(r'',r';E) = \frac{2m}{i\hbar}(r'r'')^{-1/2} \int_0^\infty g_l(r'',r',\omega)\,d\omega ,\qquad(5.4a)$$

where the function $g_l(r'',r',\omega)$ is given by

$$g_l(r'',r',\omega) = \frac{e^{ik(r'+r'')\coth\omega}}{\sinh\omega} I_{2l+1}\left(\frac{2k(r'r'')^{1/2}}{i\sinh\omega}\right) .\qquad(5.4b)$$

The advantage of the integral representation is distinctly seen if we consider the integral

$$G_{11}(r'',r';E) = \int_0^\infty r\, dr\, G_{01}(r'',r;E)\, G_{01}(r,r';E). \qquad (5.5)$$

The integral in the above equation can be explicitly performed to obtain

$$G_{11}(r'',r';E) = \left(\frac{2m}{i\hbar}\right)(r'r'')^{-1/2}\left(\frac{2m}{\hbar k}\right)\int_0^\infty \omega\, g_1(r'',r',\omega)\, d\omega. \qquad (5.6)$$

The expression for $G_{nl}(r'',r',E)$ can be obtained by induction on n, and we may write

$$G_{nl}(r'',r',E) = \left(\frac{2m}{i\hbar}\right)(r'r'')^{-1/2}\left(\frac{2m}{\hbar k}\right)^n \frac{1}{n!}\int_0^\infty \omega^n g_1(r'',r',\omega)\, d\omega. \qquad (5.7)$$

When the result (5.7) for G_{nl} is substituted in Eq. (5.3a), the resulting series can be summed to obtain

$$G_1(r'',r',E) = \left(\frac{2m}{i\hbar}\right)(r'r'')^{-1/2}\int_0^\infty \exp\left[\frac{2m\alpha\omega}{i\hbar^2 k}\right] g_1(r'',r',\omega)\, d\omega. \qquad (5.8)$$

Inserting the expression for $g_1(\omega)$ from Eq. (5.4b) in the above equation the integration over ω can be performed to obtain an expression for G_1 in closed form.

$$G_1(r'',r',E) = \frac{m}{\hbar k}\frac{\Gamma(1+1-\nu)}{r''r'(2l+1)!} W_{\nu, l+1/2}(-2ikr'')\, M_{\nu, l+1/2}(-2ikr') \qquad (5.9)$$

for $r'' > r'$.

The symbols W and M refer to the Whittaker functions. In the expression (5.9), we have introduced the notation $\nu = i(m\alpha^2/2\hbar E)^{1/2}$. It may be noted that the result (5.9) is derived for $E > 0$. However, it can be analytically continued so as to obtain the radial Green's function over the entire complex E-plane. The poles of the Green's function are determined from the relation $l + 1 - \nu = -n$, $n = 1, 2, \ldots$, and yield the discrete energy eigenvalues, E_n, which are given by the familiar expression:

$$E_n = -m\alpha^2/2\hbar^2 N^2, \qquad N = n + l + 1.$$

9.6. General Formulation

In the previous sections, we were successful in summing the perturbation series to obtain exact propagators corresponding to three potentials. This suggests that it might be possible to arrive at a general scheme for summing the Feynman-Dyson series. In this section, we develop a systematic formulation for this purpose.

First we recall that the energy dependent Green's function $G(\vec{q}'',\vec{q}';E)$ associated with the Feynman propagator $K(\vec{q}'',\vec{q}';T)$ can be represented by the perturbation series (2.10). Since G_0 is the Green's function for the free particle, it satisfies the differential equation

$$[\nabla_q^2 + k^2]\, G_0(\vec{q},\vec{q}';E) = \frac{2im}{\hbar}\, \delta(\vec{q} - \vec{q}') \qquad (6.1)$$

where $k^2 = 2mE/\hbar^2$. Next consider the integral operator

$$B(\vec{q},\vec{q}') = \frac{-i}{\hbar}\, [V(\vec{q})\, V(\vec{q}')]^{1/2}\, G_0(\vec{q},\vec{q}';E) \qquad (6.2)$$

and assume that $V(\vec{q}) > 0$ for all $\vec{q}$. With this notation the Feynman-Dyson series in (2.10) can be rewritten as

$$G(\vec{q}'',\vec{q}';E) = \frac{i\hbar}{\alpha}\, [V(\vec{q})\, V(\vec{q}')]^{-1/2} \sum_{n=1}^{\infty} \alpha^n\, B^n, \qquad (6.3)$$

which can be formally summed to yield

$$G(\vec{q},\vec{q}') = i\hbar\, [V(\vec{q})\, V(\vec{q}')]^{-1/2}\, \frac{B}{(I - \alpha B)}. \qquad (6.4)$$

Next for $E < 0$, B is a self-adjoint operator, it admits an expansion

$$B(\vec{q},\vec{q}') = \sum_{\{\mu\}} \mu\, \varphi_\mu^*(\vec{q}')\, \varphi_\mu(\vec{q}) \qquad (6.5)$$

where $\varphi_\mu(\vec{q})$ are the normalized eigenfunctions of the Kernel B corresponding to eigenvalue μ. (It is tacitly assumed that the summation goes over to an integral when the μ-spectrum is continuous). Using the representation (6.5), the expression (6.4) for G ($E < 0$) takes the form

$$G(\vec{q}'',\vec{q}';E) = \frac{i\hbar}{\alpha} [V(\vec{q}'')V(\vec{q}')]^{-1/2} \sum_{\mu} \left(\frac{\alpha\mu}{1-\alpha\mu}\right) \varphi_{\mu}(\vec{q}'') \varphi_{\mu}^{*}(\vec{q}').$$

(6.6)

The eigenvalue equation for the Kernel B reads as

$$\int B(\vec{q},\vec{q}') \varphi_{\mu}(\vec{q}') d\vec{q}' = \mu \varphi_{\mu}(\vec{q}).$$ (6.7a)

Moreover, since φ_{μ}s' are orthonormal

$$\int \varphi_{\mu}^{*}(\vec{q}) \varphi_{\mu'}(\vec{q}) d\vec{q} = \delta_{\mu\mu'}.$$ (6.7b)

Defining $\psi(\vec{q}) = \varphi(\vec{q})/\sqrt{V(\vec{q})}$ and using the explicit form for the kernel B from (6.2), the eigenvalue equation (6.7) takes the form

$$\frac{-i}{\hbar} \int d\vec{q}\, G_{0}(\vec{q}'',\vec{q};E) V(q) \psi_{\mu}(\vec{q}) = \mu \psi_{\mu}(\vec{q}'').$$ (6.8)

Similarly, the expression for the Green's function modifies to:

$$G(\vec{q},\vec{q}';E) = \frac{i\hbar}{\alpha} \sum_{\mu} \frac{\alpha\mu}{1-\alpha\mu} \psi_{\mu}^{*}(\vec{q}') \psi_{\mu}(\vec{q}).$$ (6.9)

Since $G_0(\vec{q}'',\vec{q}';E)$ is the Green's function of the Schrödinger equation and satisfies Eq.(6.1), Eq.(6.8) can be transformed into a differential equation

$$\left[\nabla_{\vec{q}}^{2} + k^{2} - \frac{2m}{\hbar^{2}\mu} V(\vec{q})\right] \psi_{\mu}(\vec{q}) = 0.$$ (6.10)

Note that Eq.(6.10) is very similar to the Schrödinger equation. However, eigenfunctions of the Schrödinger equation correspond to different energy eigenvalues, whereas in Eq.(6.10) the solutions ψ_{μ} correspond to a fixed energy. It is this fact, which makes the subsequent calculations different from those involved in solving the conventional Schrödinger equation.

The advantage of transforming the integral equation (6.8) to a differential equation (6.10) lies in the fact that powerful methods are available for solving differential equations while relatively very few methods exist for solving integral equations.

272   *Path-Integral Methods and their Applications*

The above derivation is strictly valid for the case E < 0 where the operator B is self-adjoint. However, in the examples considered here it turns out that the operator B has a complete set of orthonormal eigenfunctions even for E > 0. Moreover, B is symmetric and hence the expansion (6.5) can be used for all E.

9.6.1. *Inverse-square potential (1-dimensional)*

As an illustration of the eigenfunction-expansion technique formulated above, we shall re-derive the expression for the Green's function for the potential $V(x) = 1/x^2$. The equation (6.10) takes the form

$$\left[\frac{d^2}{dx^2} + k^2 - \frac{2m}{\hbar^2\mu}\frac{1}{x^2}\right]\psi_\mu(x) = 0. \qquad (6.11)$$

The solution of Eq. (6.11) can be expressed in terms of Hankel functions $H_{i\nu}^{(1)}$ as

$$\psi_\mu(y) = A_\nu \sqrt{y}\, H_{i\nu}^{(1)}(y), \qquad (6.12)$$

where A_ν is the normalization constant, $y = kx$ and $-\nu^2 = 2m/\hbar^2\mu + \frac{1}{4}$. Acceptable solutions correspond to real values of ν.

The orthogonality relation (6.7b) implies

$$A_\nu^* A_{\nu'} \int (dy/y)\, H_{i\nu}^{(1)*}(y)\, H_{i\nu'}^{(1)}(y) = \delta(\nu-\nu'). \qquad (6.13)$$

Explicit evaluation of the above integral leads to the following expression for A_ν : $A_\nu^2 = (\nu/2)\, \text{sh}\, \nu\pi$. Using the expression for ψ_μ from Eq. (6.12) and the normalization constant A_ν we indeed obtain the expression (5.7) for Green's function.

Thus we see that the unified formulation presented in this section does yield the correct results. One can use this formulation to arrive at the propagators for other cases.

9.6.2. *Harmonic oscillator (1-dimensional)*

For the harmonic oscillator the potential $V(x) = x^2$. The eigenvalue equation (6.10) assumes the form

$$\left[\frac{d^2}{dx^2} + k^2 - \frac{2m\, x^2}{\hbar^2 \mu}\right] \psi_\mu(x) = 0 \ . \tag{6.14}$$

For $E > 0$, the solutions of the above equation exist both for discrete positive as well as a continuum of negative values of μ. The set of normalized eigenfunctions ψ_n and the corresponding discrete eigenvalues $\mu_n > 0$ is given by

$$\psi_n(x) = A_n \exp\left[-\frac{1}{2}\alpha_n^2 x^2\right] H_n(\alpha_n x) \ , \tag{6.15}$$

where $\alpha_n^2 = (m/\hbar)\,\Omega_n$, $\Omega_n = [2/m\mu_n]^{1/2} = E/(n+\frac{1}{2})\hbar$ and H_n is the Hermite polynomial. The normalization constant A_n is given by

$$A_n^2 = \alpha_n^3 \, [2^n n!\,(n+1/2)\,\sqrt{\pi}\,]^{-1} \ . \tag{6.16}$$

According to Eq(6.7b), ψ_n's obey the orthogonality relation

$$A_n A_{n'} \int dx\, x^2\, e^{-(\alpha_n^2 + \alpha_{n'}^2)\frac{x^2}{2}} H_n(\alpha_n x)\, H_{n'}(\alpha_{n'} x) = \delta_{nn'} \ . \tag{6.17}$$

Note, that in the usual orthogonality relation between H_ns', the argument of H_n does not depend on n, while the above orthogonality relation involves Hermite polynomials with their arguments scaled with respect to the order. This is due to the fact that the eigenfunctions ψ_n now represent solutions corresponding to a fixed energy.

In the case when μ is real and negative, say $\mu^{-1} = -m\omega^2/2$, the two linearly independent eigenfunctions can be expressed in terms of the parabolic cylindrical functions

$$\psi_\mu^{(1)}(x) = A_\mu D_\nu ([2im\omega/\hbar]^{1/2} x) \tag{6.18}$$

and

$$\psi_\mu^{(2)}(x) = B_\mu D_\nu (-[2im\omega/\hbar]^{1/2} x) \ , \tag{6.19}$$

where $\nu = -\frac{1}{2} - \frac{iE}{\hbar\omega}$. The continuum analogue of the orthogonality relation (6.7b) assumes the form

$$\int_{-\infty}^{\infty} x^2 \, dx \, D_\nu ([2im\omega/\hbar]^{1/2} x) \, D_{-\nu'-1}([-2im\omega'/\hbar]^{1/2} x)$$

$$= (2\hbar/m)^{3/2} \omega^{-1/2} \pi \sin(-\nu\pi) \exp[-i(\nu + \tfrac{1}{2})] \, \delta(\omega - \omega') \qquad (6.20a)$$

where $\nu' = -\tfrac{1}{2} - \tfrac{iE}{\hbar\omega'}$. This relation enables us to determine the normalization constants A_μ and B_μ,

$$A_\mu = B_\mu = [(2\hbar/m)^{3/2} \omega^{-1/2} \pi \sin(-\nu\pi) \exp\{-i(\nu + \tfrac{1}{2})\}]^{-1/2}. \qquad (6.20b)$$

The set of eigenfunctions (6.15), (6.18) and (6.19) together form a complete set. They satisfy the closure relation

$$\sum_0^\infty \psi_n(x') \psi_n(x) + \int_0^\infty d\omega \, [\psi_\mu^{(1)*}(x') \psi_\mu^{(1)}(x) + \psi_\mu^{(2)*}(x') \psi_\mu^{(2)}(x)]$$

$$= (1/xx') \delta(x - x') . \qquad (6.21)$$

Incidently when $E < 0$, the eigenvalue equation has solutions for a continuum of negative values of μ. These solutions are of the same form as in Eq. (6.19) with $\nu = -\tfrac{1}{2} + \tfrac{iE}{\hbar\omega}$.

The complete expression for the Green's function for $E > 0$ has the form

$$G(x'',x';E) = i\hbar \left[\sum_{n=0}^\infty \frac{2[(n+\tfrac{1}{2})\hbar]^2 e^{-\alpha_n^2 (\frac{x''^2 + x'^2}{2})}}{m[E^2 - \tfrac{2\alpha}{m}\{(n+\tfrac{1}{2})\hbar\}^2]} H_n(\alpha_n x'') H_n(\alpha_n x') \right.$$

$$\left. - \frac{2}{m} \int_0^\infty \frac{d\omega}{\omega^2 + \tfrac{2\alpha}{m}} \left[\psi_\mu^{(1)*}(x') \psi_\mu^{(1)}(x'') + \psi_\mu^{(2)*}(x') \psi_\mu^{(2)}(x'') \right] \right] . \qquad (6.22)$$

It is clear that the poles of the Green's function at $E = E_n = (n+\tfrac{1}{2})\hbar\omega_0$ determine the energy eignvalues, $\omega_0^2 = 2\alpha/m$, being the frequency of the oscillator. The residue at the pole determines the corresponding normalized eigenfunction.

9.6.3. *Coulomb potential*

We shall now consider a three-dimensional case to illustrate the use of the present formulation by applying it to the repulsive potential $V(\vec{q}) = 1/r$, with $r = |\vec{q}|$ and $\alpha > 0$.

Since the potential is spherically symmetric, the solution $\psi_\mu(\vec{q})$ of Eq. (6.10) can be separated into radial and angular parts. Thus we write

$$\psi_\mu(\vec{q}) = \mathcal{R}_\mu(r)\, Y_{lm}(\theta, \phi). \tag{6.23}$$

For convenience the dependence of the function $\mathcal{R}_\mu(r)$ on the indices l and m has been suppressed. It follows that the function $\mathcal{R}_\mu(r)$ satisfies the equation

$$\left[\frac{d^2}{dr^2} + k^2 - \frac{l(l+1)}{r^2} - \frac{2m}{\hbar^2 \mu} \frac{1}{r} \right] [r\, \mathcal{R}_\mu(r)] = 0. \tag{6.24}$$

It is sufficient to consider the solutions of Eq. (6.24) for a negative value of energy and hence we set $k = iK$. These solutions correspond to a set of discrete negative values of μ, viz.,

$$\mu_n = -\frac{m}{\hbar^2 K} \frac{1}{(n + l + 1)}, \quad n = 0, 1, 2, \ldots, \tag{6.25}$$

and can be written in terms of the Laguerre polynomials

$$\mathcal{R}_n(r) = A_n\, r^l\, e^{-Kr}\, L_n^{2l+1}(2Kr). \tag{6.26}$$

The polynomials satisfy the orthogonality relation

$$A_n A_{n'} \int_0^\infty dr\, r^{2l+1}\, e^{-2Kr}\, L_n^{2l+1}(2Kr) L_{n'}^{2l+1}(2Kr) = \delta_{nn'}. \tag{6.27}$$

Explicit evaluation of the above integral leads to the expression for the normalization constant,

$$A_n = \left[\frac{(2K)^{2l+2}\, n!}{(n+2l+1)!} \right]^{1/2}. \tag{6.28}$$

Substituting the value of $\mathcal{R}_n(r)$ from Eq. (6.26) in Eq. (6.23) and making use of the expansion (6.9), we arrive at the expression for the Green's function. It has the form

$$G(\vec{q}'',\vec{q}';E) = \sum_{lm} G_1(r'',r';E) \, Y_{lm}(\theta'',\phi'') Y_{lm}^*(\theta',\phi') \qquad (6.29)$$

where the radial Green's function $G_1(r'',r';E)$ can be represented as

$$G_1(r'',r';E) = \frac{i\hbar m}{Kr''r'} \sum_{n=0}^{\infty} \frac{(2K)^{2l+2} \, n!}{(n+2l+1)!} (r''r')^{l+1} e^{-K(r''+r')}$$

$$\times \frac{L_n^{2l+1}(2Kr) \, L_n^{2l+1}(2Kr)}{\left(n + l + 1 + \dfrac{m\alpha}{\hbar^2 K}\right)} \, . \qquad (6.30)$$

We may cast Eq. (6.30) in the familiar form (5.8) with the help of the standard representation of $(1/z)$ in the form of an integral of an exponential function and the following identity involving the product of Laguerre polynomials

$$\sum_{n=0}^{\infty} \frac{n! \, L_n^\alpha(x) \, L_n^\alpha(y) \, z^n}{\Gamma(n + \alpha + 1)} = \frac{(xyz)^{-\alpha/2}}{(1-z)} \exp\left[\frac{(x+y)z}{z-1}\right] I_\alpha\left(\frac{2\sqrt{xyz}}{1-z}\right) \, . \qquad (6.31)$$

9.6.4. The singular potentials

For strongly localized potentials like for example $\delta(x)$, it is not possible to express the kernel B as in Eq. (6.4) owing to the fact that the condition $V(x) > 0$ is manifestly violated. Hence the B given by the relation (6.2) is ill-defined. This difficulty may be circumvented by redefining the zero of the energy scale. Mathematically, this means the potential $V(x)$ is written as $(C + \delta(x))$ with $C > 0$. However, we shall follow an alternative procedure by defining another operator

$$B_1(x,y) = -(i/\hbar) \, G_0(x,y;E) \, \delta(y) \, . \qquad (6.32)$$

In terms of this operator, the Green's function $G(x'',x';E)$ can be rewritten as

The Perturbation Approach 277

$$G(x'',x';E) = G_0(x'',x';E) + \sum_{n=1}^{\infty} \alpha^n \int_{-\infty}^{\infty} B_1^n(x'',y) \, G_0(y,x';E) \, dy \qquad (6.33)$$

which can be summed to yield

$$G(x'',x';E) = G_0(x'',x';E) + \int_{-\infty}^{\infty} A_1(x'',y) \, G_0(y,x';E) \, dy \, , \qquad (6.34)$$

where the integral operator A_1 has the form

$$A_1(x,y) = \{\alpha \, B_1 [I - \alpha \, B_1]^{-1}\}(x,y) \, . \qquad (6.35)$$

The next difficulty arises from the fact that the operator B_1 and hence A_1 is not self-adjoint. The expansion formula similar to (6.5) requires the knowledge of eigenfunctions of both B_1 and its adjoint. However, the integral equation for B_1 can be easily solved. The operator B_1 has only one nonzero eigenvalue. Thus the explicit expression for A_1 reads as

$$A_1(x,y) = -\left(\frac{i \, \alpha}{\hbar}\right) \frac{G_0(x,0;E) \, \delta(y)}{1 + (i\alpha/\hbar) \, G_0(0,0;E)} \qquad (6.36)$$

which when substituted in Eq. (6.34) yields the expression for the energy dependent Green's function associated with the δ-function potential. These results can be immediately generalized to obtain Green's function associated with the potentials

$$V(x) = V_0(x) + \delta(x) \, , \qquad (6.37)$$

where $V_0(x)$ is a potential such that the corresponding Green's function G_0 is known. The energy dependent Green's function is still given by Eq. (6.34) except that G_0 now corresponds to the potential V_0.

Problem 9.6.1

Consider a three-dimensional spherically symmetric oscillator for which $V(r) = r^2$. Set up the eigenvalue problem (6.10) for this case. Separate the equation in the radial and angular parts. Show that the eigenfunctions $\psi_{nlm}(r)$ corresponding to the discrete spectrum are given by

$$\psi_{nlm}(r) = A_{nlm} \exp[-\rho_n^2/2](\rho_n)^l L_k^\lambda(\rho_n^2)$$

where $\rho_n = (m\omega_n/\hbar)^{1/2} r$, $\omega_n = (2/m\mu_n)^{1/2} = E_n/[(n + \frac{3}{2})\hbar]$, $k = (n-1)/2$, $\lambda = l+1/2$ and A_{nlm} is the normalization constant. Using Eq.(6.7b), we obtain the orthogonality relation satisfied by ψ_{nlm}.

$$A_{nlm} A_{n'lm} \int dr\, r^{2l+2} \exp[-(\rho_n^2 + \rho_{n'}^2)/2](\rho_n)^l L_k^\lambda(\rho_n^2) L_k^\lambda(\rho_{n'}^2) = \delta_{nn'}.$$

Notes and References

A treatment of the perturbation theory and its applications can be found in almost any standard text book on quantum mechanics. The physical visualization of various terms of the perturbation series has been explained nicely in Feynman's book.

R. P. Feynman and A. R. Hibbs, "Quantum Mechanics and Path-Integral" (McGraw-Hill, New York, 1965).

The possibility of summing the perturbation series to solve the Coulomb problem was first considered by

M. J. Goovaerts and J. T. Devreese, J. Maths. Phys. **13**, 1070 (1972).

M. J. Goovaerts, A. Baceno and J. T. Devreese, J. Maths. Phys. **14**, 554 (1973)

consider the problem of δ-function potential in this context. However, little could be achieved because their efforts were aimed at evaluating the time-dependent propagator directly.

It was only Lawande and Bhagwat who observed that considerable simplification arises in summing the perturbation series if one attempts to calculate the Green's function rather than the propagator. Their results are contained in the following papers :

S. V. Lawande and K. V. Bhagwat, Phys. Lett. **A131**, 8 (1988),

K. V. Bhagwat and S. V. Lawande, Phys. Lett. **A135**, 417 (1988),

K. V. Bhagwat and S. V. Lawande, Phys. Lett. **A141**, 321 (1989).

The perturbation series for the δ-function potential was also summed by

D. Bauch, Nuovo Cimento **B85**, 118 (1985).

The unified approach for summing the perturbation series and the discussion of the resulting orthogonality relations are contained in a recent work by

K. V. Bhagwat, Sunil Datta, and D. C. Khandekar (unpublished)

and is discussed by

S. V. Lawande, "Some Techniques of Path Integration", in *Adratico Workshop on Path Integral Methods.* ed. V. Saya-Kanit *et al* (World Scientific, Singapore, 1991)

CHAPTER 10

SEMICLASSICAL PROPAGATOR

10.1. Introduction

One of the extensively used approximation in the context of path-integrals is the "Semiclassical Approximation". The seeds of such an approximation can be seen even in the original work of Feynman. While visualizing the propagator as Σ exp[iS/$\hbar$] over all trajectories, he argues that due to the oscillatory nature of the integrand, the contribution exp[iS/$\hbar$] from two neighbouring trajectories $x(t)$ and $x(t) + \varepsilon\eta(t)$ will cancel out if the difference

$$\delta(S) = S[x(t) + \varepsilon\eta(t)] - S[x(t)]$$

is very large compared to $\hbar$. However, for trajectories $x(t)$ satisfying $\delta(S)/\varepsilon \rightarrow 0$ as $\varepsilon \rightarrow 0$, there is no change in the value of S to the order ε. Hence the major contribution to the propagator is expected from the trajectories satisfying the condition $\delta S = 0$. But this is merely the variational principle used to arrive at the Euler-Lagrange equations of motion governing the classical trajectories of a particle. Now imagine the small parameter ε to be replaced by $\hbar$, then the major contribution to the propagator can be thought of as emerging from the limit of small $\hbar$ ($\hbar \rightarrow 0$) of the exact propagator.

Note that what has been said so far is essentially the subject matter of asymptotic analysis. In other words our motivation is to derive the asymptotic behaviour of the propagator as $\hbar \rightarrow 0$. To set the ideas clear, it is therefore preferable to familiarize ourselves with the basic results pertaining to asymptotic analysis of **ordinary** integrals.

10.2. Asymptotic Analysis

Consider a one-dimensional integral

$$I = \int_a^b e^{f(x)/\lambda} \, dx, \qquad b > a > 0. \tag{2.1}$$

We want to analyze the behaviour of this integral as $\lambda \to 0$. Assume, that $f(x)$ has a single maxima at a point $x = x_0$, $a < x_0 < b$. Further, in the vicinity of the point x_0, $f(x)$ admits a Taylor series expansion. Retaining only the first two terms of this series and noting $f''(x_0) < 0$, we can rewrite (2.1) as

$$I \simeq e^{f(x_0)/\lambda} \int_{x_0-\varepsilon}^{\infty} \exp\left[\frac{1}{2\lambda} |f''(x_0)| (x-x_0)^2\right] dx. \tag{2.2a}$$

The justification of the change in limits in the right-hand side of Eq. (2.2a) can be intuitively understood as follows. Since $f(x)$ is maximum at $x = x_0$, the function $e^{f(x)/\lambda}$ peaks very sharply near $x = x_0$. Moreover, as $\lambda \to 0$ the peak becomes sharper and sharper. Thus it is reasonable to assume that the major contribution to the integral (2.1) arises primarily from a small neighbourhood of the point x_0, say from $x_0 - \varepsilon$ to $x_0 + \varepsilon$. Hence, the error commited by changing the range of integration from $x_0 - \varepsilon$ to ∞ is expected to be small. Next changing the variable of integration from x to $x - (x_0 - \varepsilon)$ (again denoted by x) we get

$$I \simeq e^{f(x_0)/\lambda} \int_0^\infty \exp\left[\frac{1}{2\lambda} |f''(x_0)| x^2\right] dx = \sqrt{\frac{\pi\lambda}{2|f''(x_0)|}} \, e^{f(x_0)/\lambda}. \tag{2.2b}$$

The subject of asymptotic analysis also enables us to get an estimate of these errors. To obtain some rough estimate of the error, we assume that the remainder in the Taylor series of $f(x)$ is of the order of $C x^3$ where C is a negative constant. Now, we know

$$e^{f(x_0)/\lambda} \int_0^\infty x^3 e^{-[|f''(x_0)|x^2]/2\lambda} \, dx \simeq \frac{\lambda^2}{|f''|^2} \, e^{f(x_0)/\lambda}. \tag{2.3}$$

Expanding the remainder term in the exponential function appearing in (2.1) and retaining only the first two terms we obtain the corrections to the approximation (2.2) for the integral I. These turn out to be of the order of λ. Next, we examine whether it is possible to relax the condition on $f(x)$ namely, $f(x)$ had a single maxima. Let $f(x)$ be maximum at the points $x = x_i$, $i = 0, 1, \ldots$. The result (2.2) can be generalized to

$$I \simeq \sum_i \sqrt{\frac{\pi \lambda}{2|f''(x_i)|}} \, e^{f(x_i)/\lambda} . \qquad (2.4)$$

All the above arguments can be made mathematically rigorous by explicitly accounting for the analytical bahaviour of $f(x)$. For example, in the simplest case where $f(x)$ belongs to C^∞ class, one first uses a transformation $f(x) \rightarrow u$ and then uses Watson's lemma to obtain the results (2.2)--(2.4). For details, the reader is referred to the excellent book by Copson cited at the end of this chapter.

The next difficulty in this analysis arises when $f(x)$ has no maxima for real values of x. Assume, that $f(z)$ attains an extrema at a point z_0 in the complex plane so that $f'(z_0) = 0$. In this case one tries to deform the original line of integration (in this case the real line) to another curve passing through z_0 such that along this curve $[\partial^2 f/\partial z^2]_{z=z_0} < 0$. Cauchy's theorem can then be invoked to obtain the results mentioned in (2.2) except for the fact that the various quantities in (2.2) must be evaluated along the deformed contour.

The ideas presented above can be generalized to the case when λ is pure imaginary and $|\lambda| \rightarrow 0$. In this case the extremum points of $f(x)$ would correspond to values of x where the phase is stationary. For this reason this approximation method is also referred to as stationary phase approximation in the context of oscillatory integrals. Of course when λ is pure imaginary we have no reason to restrict to maxima of $f(x)$ and all extrema should be considered.

So far our discussion was centered about the evaluation of a one-dimensional integral in the asymptotic limit. It is obvious that the evaluation of higher dimensional integrals in the asymptotic limit would be straightforward as long as the integrand is a product function. Let us then consider an integral of the type

$$I = \int e^{f(x_1,x_2,\ldots,x_n)/\lambda} \, dx_1 dx_2 \ldots dx_n \, . \tag{2.5}$$

As in the one-dimensional case let us assume that there exists a point $x_0 = (x_{10}, x_{20}, \ldots, x_{n0})$ where $\partial f/\partial x_{i0} = 0$, $i = 1, 2, \ldots, n$. Further assume that the matrix F with elements $F_{ij} = \partial^2 f/\partial x_i dx_j$, evaluated at the point x_0 is negative definite. Then the Taylor series for f can be written as

$$f(x_1, x_2 \ldots) = f(x_{10}, x_{20} \ldots) + x^T.F.x + O(\{x_{i0}\}^3) \tag{2.6}$$

where x is a column vector with components $(x - x_{10})$, $(x - x_{20})$, ... The use of Eq.(2.6) in Eq.(2.5) reduces the integral I to a multidimentional Gaussian integral. Such an integral has already been evaluated in chapter 2. The resulting expression is

$$I \propto \left(\frac{\pi\lambda}{2}\right)^{d/2} |\text{Det } F|^{-1/2} \exp\left[\frac{1}{\lambda} f(x_0)\right] \, . \tag{2.7}$$

The various steps indicated in the previous paragraph can be made precise by stating rigorous conditions on the analytical nature of the function f and on the domain of integration. However, this lies beyond the scope of this book and will not be pursued.

At this point we can correlate the foregoing discussion with the semiclassical approximation for the propagator. In the polygonal definition, the propagator is represented as an N fold integral, $N \to \infty$. Thus the semiclassical approximation can be viewed as an asymptotic evaluation of an integral of dimension $N \to \infty$. It is this aspect which makes the evaluation of semiclassical propagator different as well as difficult from a finite dimensional integral.

We shall now proceed to obtain the semiclassical approximation for the propagator. Our subsequent discussion will be mostly intuitive in nature. However, most of the steps can be rigorously justified.

10.3. Semiclassical Approximation

Consider a system described by a Lagrangian $L = \frac{1}{2} m\dot{x}^2 - V(x)$. The polygonal representation of the propagator $K_N(x'',x';T)$ has the form

$$K_N(x'',x';T) = A_N \int \prod_{j=1}^{N} \exp\left[\frac{i}{\hbar}\left(\frac{m}{2\varepsilon}(x_j - x_{j-1})^2 - \varepsilon V(x_j)\right)\right] \prod_{j=1}^{N-1} dx_j \quad (3.1)$$

where the various quantities are already defined in chapter 1. Next, for evaluating the asymptotic behaviour of $K_N(x'',x';T)$, we look for the maxima of the discretized form of the action S_N

$$S_N = \sum_{j=1}^{N}\left[\frac{m}{2\varepsilon}(x_j - x_{j-1})^2 - \varepsilon V(x_j)\right] \quad (3.2)$$

which are obtained by solving $\partial S_N/\partial x_i = 0$, $i = 1, 2, \ldots, N-1$. We do not differentiate with respect to x_0 and x_N because of the conditions $x_0 = x'$ and $x_N = x''$. This implies that

$$\frac{\partial S_N}{\partial x_j} = \frac{m}{\varepsilon}\left[-(x_{j+1} - x_j) + (x_j - x_{j-1})\right] - \varepsilon \frac{\partial V}{\partial x_j} = 0,$$

$$j = 1, 2, \ldots, N-1 \quad (3.3)$$

which in the limit $\varepsilon \to 0$ reduces to Euler-Lagrange equation

$$m\ddot{x} + \frac{\partial V}{\partial x} = 0, \qquad x(0) = x' \; ; \; x(T) = x'' \; . \quad (3.4)$$

Let x_j^c be the solution of Eq.(3.3). In the limit $\varepsilon \to 0$, x_j^c describes a continuous function $x_c(t)$, namely solution of equation of motion (3.4). Our next step consists of expanding S_N around x_j^c. Hence we set $x_j = x_j^c + \eta_j$. This implies $\eta_0 = \eta_N = 0$. It can be easily verified that the linear term in η in the expansion of S_N is identically equal to zero when we use (3.3) along with end-point conditions on η. Separating out the first two nonvanishing terms of this expansion we can write

$$S_N = \sum_{j=1}^{N}\left[\frac{m}{2\varepsilon}(x_j^c - x_{j-1}^c)^2 - \varepsilon V(x_j^c)\right] + \sum_{j=1}^{N}\left[\frac{m}{2\varepsilon}(\eta_j - \eta_{j-1})^2\right]$$

$$- \frac{\varepsilon}{2}\eta_j^2 \frac{\partial^2 V}{\partial x_j^{c2}}\right] - \varepsilon \sum_{n=3}^{\infty} \frac{1}{n!}\eta_j^n \frac{\partial^n V}{\partial x_j^n} \quad . \quad (3.5)$$

In the limit $\varepsilon \to 0$. The first term on the right-hand side of Eq.(3.5) is simply the value of the action functional S along the classical trajectory x_c, viz., S_{cl}. In the same limit the second term describes the action functional associated with an harmonic oscillator of frequency $\Omega^2(t) = \frac{1}{m}\left[\partial^2 V/\partial x^2\right]_{x=x_c}$. Ignoring higher order terms in Eq.(3.5) the semiclassical expression for the propagator $K(x'',x';T)$ can be written as :

$$K(x'',x';T) = \lim_{N \to \infty} K_N(x'',x';T) = \exp[iS_{cl}/\hbar] \int_{\eta(0)=0}^{\eta(T)=0} \exp[iS_1/\hbar]\, \mathcal{D}\,[\eta(t)]$$

(3.6)

with S_1 defined as

$$S_1 = \int_0^T dt\, \frac{m}{2}\left(\dot{\eta}^2 - \Omega^2(t)\,\eta^2\right). \tag{3.7}$$

The path-integral associated with the action (3.7) has already been evaluated. Borrowing the results of chapter 2 (cf. Eqs.(2.22)--(2.28)) the complete expression for the propagator can be written as

$$K(x'',x';T) = \sqrt{1/2\pi\hbar}\, D^{-1/2}(T)\, \exp[iS_{cl}/\hbar] \tag{3.8}$$

where the function $D(t)$ satisfies the differential equation :

$$\frac{d^2 D}{dt^2} + \Omega^2(t)\, D = 0 \quad ; \quad D(0) = 0,\ \dot{D}(0) = \frac{1}{m}. \tag{3.9}$$

The propagator given by Eq.(3.8) yields the leading contribution to the exact propagator. We shall now explore the possibility of obtaining, a relation between $D(T)$ and S_{cl}. To proceed further we return to the solution of the classical equations of motion. So far we were solving these equations subject to the boundary conditions $x(0) = x'$ and $x(T) = x''$. The classical equation of motion (3.4) along with these conditions completely specifies the motion for all time t. However, since the position of the particle is fixed at two times t' and t'', the momentum p of the particle will be a function of the end-points and time t. In particular, recall that the initial momentum p is related to S_{cl} by the relation $p = \partial S/\partial x'$. Now consider the quantity

$$\frac{1}{J} = \frac{\partial p}{\partial x''} = \frac{\partial^2 S}{\partial x'' \partial x'} \ . \tag{3.10}$$

Alternatively, we can consider the position x of the particle as a function of its initial momentum p and time t. Thus we can write $x = x(p,t)$ satisfying the conditions $x(p,t') = x'$ and $x(p,t'') = x''$. Now x(t) is a solution of the equation of motion (3.4). Differentiating both sides of Eq.(3.4) with respect to p, we obtain

$$m \frac{d^2}{dt^2}\left(\frac{\partial x}{\partial p}\right) + \frac{\partial^2 V}{\partial x^2}\left(\frac{\partial x}{\partial p}\right) = 0 \ . \tag{3.11}$$

Further, the initial conditions imply

$$\frac{\partial x}{\partial p}\bigg|_{t=t'} = \frac{\partial x'}{\partial p} = 0, \quad \frac{d}{dt}\left(\frac{\partial x}{\partial p}\bigg|_{t=t'}\right) = \frac{\partial (\dot x)}{\partial p}\bigg|_{t=t'} = \frac{1}{m} \ . \tag{3.12}$$

Next it is clear from Eq.(3.10) that the quantity J is the value of $\partial x/\partial p\big|_{t=t''}$. Thus we observe that the quantity $\partial x/\partial p$ and D(t) satisfy the same differential equation and identical boundary conditions. Hence D(T) can be immediately obtained from Eq.(3.10) as inverse of J.

Combining the result of Eq.(3.8) and (3.10) we arrive at the familiar expression for the propagator in the semiclassical approximation

$$K(x'',x';T) = \sqrt{\frac{1}{2\pi i \hbar}\frac{\partial^2 S_{cl}}{\partial x'' \partial x'}} \exp\left[\frac{i}{\hbar} S_{cl}\right] \tag{3.13}$$

and as before the expression is valid for time interval before we encounter the first zero of J. In case the classical equation admits more than one solution the semiclassical propagator is obtained by adding the contributions from all the classical trajectories. This still leaves us with the task of obtaining the errors in the approximation (3.13). By analogy with finite dimensional integral, it is expected that these errors would be of the order $\hbar$. However, in view of the oscillatory nature of the path-integral, the exact evaluation of such an estimate is very difficult and will not be considered here.

So far we have been assuming that J does not vanish for any value of T. Let us now attempt to relax this condition. Again the guidelines for this purpose are provided by the corresponding analysis of finite dimensional integrals. Let us assume that while analyzing the integral in (2.1), the point x_0 is such that not only $f'(x_0) = 0$ but also $f''(x_0) = 0$. The obvious choice in this case would be to include the cubic term in the Taylor series expansion of $f(x)$. In other words we approximate I of (2.1) as

$$I \approx \exp(.f(x_0)/\lambda) \int_a^b \exp\left[-\frac{1}{3!}|\partial^3 f/\partial x_0^3|\frac{x^3}{\lambda}\right] dx \qquad (3.14)$$

where we have assumed that $\partial^3 f/\partial x_0^3 < 0$. This restriction can be relaxed by retaining enough terms of the Taylor series expansion of $f(x)$) to ensure that the integrand has the right type of behaviour in the region of integration.

Our interest really lies in the case when λ is pure imaginary. For this case the integral in (3.14) can be identified with an appropriate Airy function $A_i(z)$ having the definition

$$A_i(z) = \frac{1}{2\pi i} \int_\Gamma \exp\left[tz - \frac{t^3}{3}\right] dt$$

where Γ is a contour in the left half plane and goes to infinity asymptotically at angles more than $\pi/6$ to the negative real axis. For higher dimensional integrals similar corrections can be introduced if Det F in Eq. (2.7) vanishes.

Now let us consider what changes in our approach are needed if the quantity $\partial^2 S/\partial x'' \partial x'$ is singular. Though in chapter 2, we have performed such a calculation for quadratic Lagrangians it is not of much help. In the case of quadratic Lagrangians the Taylor series expansion of S_N did not contain any terms of order higher than η_j^2. Hence as soon as we encounter a singularity of Van Vleck-Pauli Determinant, the propagator approaches a delta function of an appropriate argument. In the present case the Taylor series of S_N around the extremum contains all powers of η_j. The terms containing η_j^3 and higher are entirely contributed by the expansion of the potential. Hence by analogy our goal should be to calculate the integral where S_N is approximated as :

$$S_N = S_{Nc} - \frac{1}{3!} \sum_{j=1}^{N} \frac{\partial^3 V}{\partial x_j^3} \eta_j^3 \qquad (3.16)$$

where S_{Nc} denotes the first two terms on the right-hand side of equation (3.5). Notice that in S_{Nc} the η_j are coupled due to the kinetic energy term. Therefore, the first task in the analysis consists of decoupling the entire quadratic term. For this purpose we must transform η_j to u_j by writing $\eta_j = \sum_k c_{jk} u_k$ where c_{jk} are the elements of the matrix C formed by the normalized eigenvectors $(c_0, c_1, \ldots, c_N)$ of a matrix A. The matrix A corresponds to the quadratic form of second term on right-hand side of Eq. (3.5). The non-zero elements of the matrix A are

$$A_{jk} = -\frac{m}{2\varepsilon}, \qquad j = k \pm 1,$$

$$A_{jj} = \left[\frac{m}{\varepsilon} - \frac{\varepsilon}{2} \frac{\partial^2 V}{\partial x_j^2} \right]. \qquad (3.17)$$

Under this transformation the action (3.16) can be formally written as

$$S_N = \sum \lambda_i u_i^2 + \frac{1}{3!} \sum_{j=1}^{N} \frac{\partial^3 V}{\partial x_j^3} \sum_{l\,m\,n} c_{jl} c_{jm} c_{jn} u_l u_m u_n.$$

Since the matrix C can be chosen to be orthogonal the Jacobian of the transformation from η to u is unity. Next consider the integrations over $\{u\}$. If all the eigenvalues λ_i are nonzero we may ignore the contribution due to the cubic terms in the expression for S_N while performing the integrations over u. If however one or more eigenvalues of A vanish the corresponding quadratic term vanishes. And one is forced to consider the cubic term while performing the integration over the corresponding u. Let us assume for simplicity that $\lambda_1 = 0$. In such a case integrations over other variable, viz., $u_2, u_3, \ldots, u_{N-1}$ can be performed by ignoring the higher order term in these variables. The first leading term in the variable u_1 will involve a cubic term in u_1 and would lead to the Airy function.

The foregoing discussion has been very qualitative in nature and various steps indicated above indeed require very careful analysis and an extensive use of the results from asymptotic analysis. For this reason we do not pursue this issue here. In any case the semiclassical approximation is most useful when the simple form as in Eq.(3.13) is valid. For details the reader is referred to Schluman's book which deals with the topic of semiclassical analysis very extensively.

10.4. A Particle in an Inverse Square Potential

To illustrate the evaluation of the propagator in semiclassical approximation let us consider the case of a particle of mass m moving under the influence of an inverse square potential. The Lagrangian for the particle is given by

$$L = \frac{m}{2} \dot{x}^2 - \frac{g^2}{2x^2} . \qquad (4.1)$$

The associated Eüler-Lagrange equations can be easily solved to obtain the solution $x(t)$

$$x^2(t) = x_0^2 + g^2(t - t_0)^2 x_0^{-2} \qquad (4.2)$$

where x_0 and t_0 are two arbitrary constants which should be determined subjecting to the conditions $x(0) = x'$ and $x(T) = x''$.

For further algebraic manipulations, it is convenient to introduce a variable $\cos \theta(t) = x_0/x(t)$. In this new variable Eq.(4.2) and the boundary conditions can be expressed as

$$g(t - t_0) = x_0^2 \tan \theta(t) , \quad - g t_0 = x_0^2 \tan \theta(0) ,$$

$$g(T - t_0) = x_0^2 \tan \theta(T). \qquad (4.3)$$

Using the defining relation for $\cos \theta$ and Eq.(4.3) we can obtain the value of x_0. It is given by

$$x_0^2 = g^2 T^2 [x'^2 + x''^2 - 2x'x'' \cos \{\theta(T) - \theta(0)\}]^{-1}. \qquad (4.4)$$

The relations (4.3) also imply that $\sin\{\theta(T) - \theta(0)\} = gT/x'x''$ which yields two values for $\cos\{\theta(T) - \theta(0)\}$, viz., $\pm[1 - (gT/x'x'')^2]^{1/2}$. The two values of this parameter lead us to two real values of x_0 provided $gT/x'x'' < 1$. This implies the existence of two trajectories. For $gT/x'x'' > 1$, there is no real value of x_0 and hence there is no classical trajectory. Along each trajectory the action functional can be calculated by using the value of $x(t)$ from Eq. (4.2) and we may write

$$S_{cl} = mg[\tan\theta(T) - \tan\theta(0) - 2\{\theta(T) - \theta(0)\}]/2, \quad (4.5)$$

Along each of the two trajectories $\tan\theta$ has a different value which can be calculated easily. When this value is inserted in Eq. (4.5), we obtain the action functional. Denoting these two values by $S_{CL}^{\pm}$ we may write

$$S_{cl}^{\pm} = \frac{m}{2T}\left[x'^2 + x''^2 \pm 2x'x''\left\{1 - \left(\frac{gT}{x''x'}\right)^2\right\}^{1/2}\right] - mg\sin^{-1}\left(\frac{gT}{x'x''}\right). \quad (4.6)$$

Further a straightforward differentiation yields

$$\frac{\partial^2 S_{cl}^{\pm}}{\partial x'\partial x''} = \pm\frac{m}{T}\left\{1 - \left(\frac{gT}{x''x'}\right)^2\right\}^{1/2}. \quad (4.7)$$

Combining the results of Eqs. (4.6) and (4.7) the semiclassical propagator $\tilde{K}(x'',x';T)$ takes the form

$$\tilde{K}(x'',x';T) = -\left[\frac{im}{2\pi\hbar T\sqrt{[1 - (gT/x'x'')^2]}}\right]^{1/2} \exp\left[\frac{im}{2\hbar T}(x'^2 + x''^2)\right]$$

$$\times \{e^{\mathcal{A}(x'x'') - \mathcal{B}(x'x'')} + e^{-\mathcal{A}(x'x'') - \mathcal{B}(x'x'') - i\pi/2}\}$$

where the quantities $\mathcal{A}(u)$ and $\mathcal{B}(u)$ are defined as

$$\mathcal{A}(u) = (imu/\hbar T)[1 - (gT/u)^2]^{1/2}, \quad \mathcal{B}(u) = \sin^{-1}(gT/u).$$

One can immediately compare this result with the exact expression for the propagator obtained in Chapter 4 (cf. Eq. (3.37)). The asymptotic limit of the exact result coincides with the above result except for the fact that the term $mg/\hbar$ is replaced by $[(mg/\hbar)^2 + 1/4]^{1/2}$. However, the difference between the two expressions is of the order $\hbar$ and is indeed to be expected.

Notes and References

The topic of semiclassical approximation has been extensively discussed in

L. S. Schulman, "The Techniques and Applications of Path Integration", John Wiley, New York, (1981).

The subject of asymptotic analysis is contained in

E. T. Copson, "Asymptotic Expansions", (Cambridge University Press, 1965).

CHAPTER 11

NUMERICAL METHODS OF SUMMING OVER PATHS

11.1. Introduction

The intuitive appeal of a path integral as a universal key for formulating and solving problems in diverse areas of physics has inspired the development of several approximate analytical as well as purely numerical methods of evaluating a path integral. In the preceding chapters we examined some of the analytical techniques used so far in literature. In the present chapter, we consider some of the purely numerical ones. These methods rely on the lattice definition of a path integral where the propagator is written as the limit $N \rightarrow \infty$ of an N-dimensional Riemann integral. The idea here is to evaluate this multiple integral numerically. At the outset, there are two difficulties in this task. First the integrals are oscillatory and second the dimensionality N of the integral is very large. The convergence problems associated with the numerical evaluation of oscillatory integrals are well known. The high dimensionality of the path integral makes the use of conventional methods like the trapezoidal and Simpson's rules prohibitively time consuming even with the fastest among the present generation computers.

The first difficulty may be overcome by going over to the imaginary time, that is, by letting $t = -i\tau$. In non-relativistic quantum mechanics, this amounts to replacing the time-dependent Schrödinger equation by diffusion equation. This interpretation is, in principle, useful to derive the energy eigenvalues and the corresponding eigenfunctions of a quantal system from the standard Feynman-Hibbs expansion formula. If we further let $\tau = \hbar\beta$, the path integral represents the density matrix of equilibrium statistical mechanics provided that we identify $\beta = 1/k_B T$ (where T is temperature and k_B is the Boltzmann constant). The main point here is that the use of imaginary time is analogous to use of Wiener measure rather than the Feynman measure. Nevertheless, a large class of physical problems may be solved by numerically evaluating the path integral in imaginary time.

The second difficulty concerning the large dimension of the path integral is overcome by taking recourse to random sampling techniques or Monte Carlo (MC) methods. Indeed, the path integral MC methods have been considerably developed during the last three decades. There are now powerful techniques like Metropolis algorithm and Fourier methods to carry out the "sum over paths" efficiently on a computer. Such computations are now routinely performed in statistical mechanics and more recently in quantum chromodynamics.

11.2. Monte Carlo Method

As a technique in numerical analysis the Monte Carlo (MC) method has gained widespread applicability over the past three decades. There does not seem to be any branch of science and technology, where the method has not been used at one time or the other. The advent of high speed computers has been mainly responsible for the continual development of this computer oriented method. It is thus worthwhile to discuss briefly some basic elements of the MC method. Incidentally, the name Monte carlo comes from the name of a city in Monaco, famous for its casino.

What is Monte Carlo? There is no useful definition which characterizes the method accurately, completely and concisely. It is, however, sufficient to recognize that Monte Carlo is a scheme of studying a real or an artificial stochastic model of physical or mathematical process. Random sampling techniques are used, so that "calculation" involves playing a game of chance. Associated with each "try" in the game of chance is a numerical result or score. The game is so designed that the expected value of the score is that physical or mathematical quantity which the player wishes to obtain. Then according to the central limit theorem (the law of large numbers) in probability theory, if the game is played enough times, the average score approximates the expected value, which is the quantity to be calculated.

Used properly, Monte Carlo can give quick "first cuts" at difficult problems, that is, problems which are intractable by the traditional analytical or numerical techniques. Blind application of crude Monte Carlo estimates, however, produces results, which are inferior to those obtained by using the standard methods of numerical analysis. Direct rigorous estimation of the error range in the answer produced by MC is usually difficult or at times impossible unless the answer is known by other means. The *efficiency* of the MC method, therefore, depends on the ingenuity of the user. It must be emphasized here that MC is an experimental tool and not an analytical tool in mathematics. Thus, one is forced to infer conclusions from observations; not deduce conclusions from postulates.

The computations in MC are concerned with random numbers and we have to decide what we mean by random and how far we are willing to stretch the definition for particular cases. If we assume for the moment, that we have a method for selecting random numbers, we can directly simulate the process in question. For example, the neutrons in a nuclear reactor undergo several events; such as absorption, scattering and fission, which are probabilistic in nature. Hence, we can simulate the performance of a nuclear reactor by choosing random number, which represent random motions of neutrons in it. An important point here is that such a simulation may be carried out on a computer without incurring the cost, in money, time, and safety of an actual reactor. Deterministic problems may also be attacked by Monte Carlo with the help of a stochastic analogue. For example, the diffusion equation has the same solution as one form of the drunkard's walk problem.

Historically, we must credit MC to Lord Kelvin, who mentioned the technique in a discussion of the Boltzmann equation in 1901. However, it took the development of the high speed digital computers to make the method practical. Indeed a systematic development of the MC method was done by the same people who developed the digital computer : Fermi, Metropolis, Ulam, Von Neumann and Kahn. Apparently, the method seems to have been rediscovered by Von Neumann and Ulam during the development of atomic weapons. To give the ultimate credit, we must go back not just ninety years, but more than two centuries to find the earliest example of a stochastic determination of a parameter. It appears that, apart from crafty gamblers, the first MC experiments were performed by Buffon. His stochastic evaluation of π is described in his 1777 "Essaid Arithmetique Morale". We describe this rather amusing method to bring out the basic principles of Monte Carlo.

Buffon's needle

Consider a set of parallel lines separated by a distance d on a plane (Fig. 10.1). A needle of length ℓ (ℓ < d) is thrown on this plane. It is assumed that performance of the experimenter has no control over the final position of the needle. We can characterize the location of the needle by the position of the midpoint x and the angle α the needle makes with the normal to the parallel lines. It is also assumed that the distribution of both x and α are uniform, that is, all values of x in the range 0 ≤ x ≤ d and α in the range $-\frac{\pi}{2} \le \alpha \le \frac{\pi}{2}$ are equally probable.

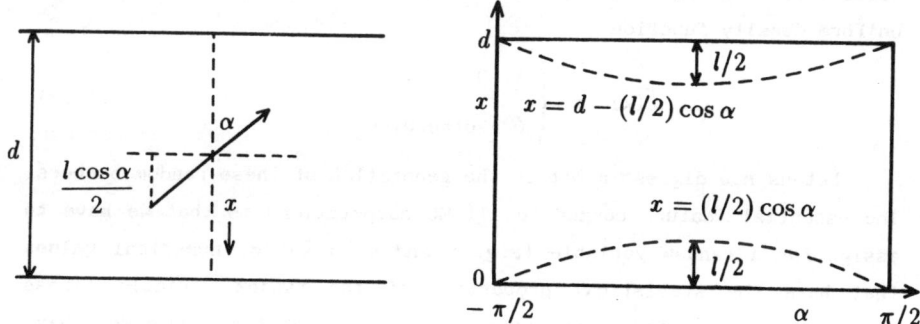

Fig. 11.1. *Buffon's needle experiment*.

It is clear that the needle does not intersect the upper line, if $x < d - \frac{\ell}{2} \cos \alpha$ and the lower line if $x > \frac{\ell}{2} \cos \alpha$. This is shown by the (x, α) diagram in the figure where the region R is the region where no intersections occur. Now the probability that a toss produces no intersection is the ratio of the area D to the total area πd of all possible tosses. A simple calculation yields

$$q = \frac{1}{\pi d} \iint_R dx \, d\alpha = \frac{1}{\pi d} \int_{-\pi/2}^{\pi/2} d\alpha \int_{1/2 \cos \alpha}^{d-1/2 \cos \alpha} dx = 1 - \frac{2\ell}{\pi d}. \qquad (2.1)$$

Hence, the probability of an intersection is

$$p = 1 - q = 2\ell/\pi d. \qquad (2.2)$$

In particular, for $\ell = d/2$ we have $p = 1/\pi$. The actual technique is now clear. One rules the field of parallel lines, selects the needle, tosses and records whether there is an intersection. If the total number of tosses is n and r of them are "successful" then the probability $\tilde{p}$ of success is estimated by $\tilde{p} = r/n$.

Random numbers :

How do we simulate this experiment on a computer? Most computers provide a routine for generating random numbers in the range (0,1). Basically, these numbers represent samples drawn independently from the uniform density function

$$f(x) = \begin{cases} 1 & 0 \leq x \leq 1 \\ 0 & \text{otherwise} \end{cases}. \qquad (2.3)$$

Let us now digress a bit on the generation of these random numbers. The essential feature common to all MC computations is that we have to assign for a random variable (e.g. x and α) a set of numerical values that have the statistical properties of the random variable. These values are called random numbers on the ground that they could well have been obtained by chance by an appropriate random process. If we want truly random numbers, we can extract them from the noise of an electron tube or by the decay of a radioactive element. However, what is desirable for practical MC calculations, is a long enough sequence of random numbers generated on a computer, which satisfy certain statistical tests for randomness. The set of numbers must be reproducible, so that we can repeat our calculation with confidence that the same set of values will be generated. Such a set of numbers that are "random" so far as the computation is concerned but are reproducible are called pseudo random numbers.

Three methods that have been used for the generation of long sequences of pseudo random numbers are: (i) The mid-square method, (ii) the congruence method, and (iii) Fibonacci series. In the mid-square method invented by Metropolis and Von Neumann, each number x_{i+1} is generated by squaring its predecessor x_i and by taking the approximate number of digits from the middle of the square. The main point here is that, although the entire sequence of number is completely determined once the first number is specified, the numbers behave statistically as though they were drawn at random. That is, the numbers are nearly uniform and more or less uncorrelated so that the sampling errors made in using them are small.

A variation of the mid-square method was suggested by Lehmer in 1949. This method known as the multiplicative congruential algorithm has the form

$$X_i \equiv b X_{i-1} \pmod{m} \quad (i = 1, 2, \ldots) \qquad (2.4)$$

where X_0 and b are positive integers and m is a large positive integer. The statement $X \equiv Y \pmod{m}$ is read "X is congruent to Y modulo m" and (2.4) implies that X_i is the remainder when $b X_{i-1}$ is divided by m. The sequence $\{X_i/m\}$ will be the required sequence of pseudo-random numbers. The algorithm is expected to produce no more than m distinct numbers X_i; there is also the question of the statistical behaviour of the sequence $\{X_i/m\}$ with respect to randomness. The number of distinct X_i which are generated before repetition is called the period of the pseudo-random sequence. With an appropriate choices of b, X_0 and m, it is possible to obtain large periods and achieve also a desired degree of randomness of the sequence $\{X_i/m\}$. However, the multiplicative congruential algorithm does not guarantee that a period of length m can be achieved. Also the method has the inherent defect that the period decreases as one moves from the most significant to the least significant digits. These problems may be alleviated by using the mixed congruential algorithm

$$X_i \equiv [b X_{i-1} + c] \pmod{m}. \qquad (2.5)$$

If we choose, $b = (2^a+1)$ (where a is an integer ≥ 2) and c to be an odd integer, we obtain the maximum period m. The main effect of the addition of c is to ensure long periods in the least significant digits of X_i as well as in the most significant digits.

Finally, the third method of Fibonacci series does not involve multiplications. Given two numbers X_0 and X_1 one generates the sequence

$$X_{i+1} = X_i + X_{i-1}. \qquad (2.6)$$

If our word has n bits, the numbers we obtain are modulo 2^n.

Basic Features of MC

Suppose we have an infinite supply of random numbers. Buffon's needle problem can be treated by MC in the following way. We pick a pair of random numbers, ξ_1 and ξ_2 in the interval [0, 1] and derive the random numbers (d = 2ℓ for simplicity).

$$\eta_1 = 2\xi_1 \ell, \quad \eta_2 = (1 - 2\xi_2) \pi/2 \qquad (2.7)$$

as the numerical values for the random variables x and α. We then check whether the random numbers η_1 and η_2 satisfy the condition

$$\frac{\ell}{2} \cos \eta_2 < \eta_1 < 2\ell - \frac{\ell}{2} \cos \eta_2 \qquad (2.8)$$

and decide whether or not the needle intersects the field. The outcome of such an experiment is now described by a random variable which takes discrete values ζ_i. The interaction of the needle with the field of lines implies a success and $\zeta_i = 1$, otherwise $\zeta_i = 0$. Thus if the above process is repeated n times and that there are in all r intersections, then the MC experiment determines the mean

$$\bar{\zeta} = \frac{1}{n} \sum_{i=1}^{n} \zeta_i = \frac{r}{n} \qquad (2.9)$$

which is an estimate $\bar{p}$ for the probability p and also an estimate of the value of $1/\pi$ according to Eq.(2.2).

It is clear that the values ξ_i are sampled from a "parent" distribution which is binomial.

$$P(n,r) = {}^nC_r \, p^r (1-p)^{n-r}. \qquad (2.10)$$

Here P(n,r) is probability that r out of a total n trials ends in a success with p as the probability for a success in a given trial. p is also the ratio of successful trials to the total number of trials with a variance $\sigma^2 = p(1-p)$. In MC experiment we estimate p and σ^2 by $\bar{p}$ and s^2 respectively, which are given by

$$\bar{p} = \bar{\zeta}, \qquad s^2 = \frac{1}{(n-1)} \sum_{i=1}^{n} (\zeta_i - \bar{\zeta})^2 . \qquad (2.11)$$

The variance of $\bar{\xi}$ is then σ^2/n. In general, we do not know σ^2 and we use s^2/n as the variance of the estimator $\bar{\zeta}$. It is clear from this that the variance and hence the error in our estimate can be reduced either by taking larger samples (large n) or by sampling from a distribution with a reduced variance σ^2.

In the Buffon's experiment we may reduce effectively the variance σ^2 by increasing the probability p of intersection. This can be achieved by using, instead of a single needle, a cross made of two needles each of length ℓ joined at a right angles at their mid-points. A little computation shows that the probability of intersection of the cross with the parallel lines as $p_2 = 4\ell/\pi d$ (instead of $p_1 = 2\ell/\pi d$ for a single needle), which is doubled. The ratio of fractional errors in the two cases is easily obtained as (ℓ=2d)

$$\frac{(\sigma_2/P_2)}{(\sigma_1/P_1)} = \left[\frac{(1-P_2)P_1}{(1-P_1)P_2}\right]^{1/2} = \left[\frac{(\pi-2)}{2(\pi-1)}\right]^{1/2} \approx \frac{1}{2}. \quad (2.12)$$

Thus, one expects that for the same number of tosses n, the estimate of π, obtained by using a cross is much better (the fractional error is essentially reduced by about a factor of two).

This simple example brings out the three basic issues involved in all MC methods.
i) Selection of suitable probability process which can simulate the problem. (For example, relating π with the random throws of the needle).
ii) Generation of sample values of the random variables on a computer. (Picking x and α values leading to the sampling of ζ_1).
iii) Variance reduction techniques (Replacing the needle by a cross).

In a sense these three issues are interlinked. For example, the methods which are used for variance reduction are strongly dependent upon the probability model and on techniques used to generate values of the random variables. The greatest gains in variance reduction are often made by exploiting specific details of the problem rather than by routine application of general principles. An important point here is that in a MC problem, the experimenter has complete control of his sampling procedure. This fact is used in the idea of importance sampling and the ultimate of zero-variance sampling. Lastly, if a part of the problem can be done analytically it contributes to the reduction of the overall variance. Thus, if some degrees of freedom can be eliminated from the path integral, the subsequent MC evaluation reduces the variance.

Evaluation of integrals by MC

Consider a multiple integral of the form

$$\theta = \int_a^b f(\vec{q}) \, d\vec{q} \qquad (2.13)$$

where for brevity the notation $\vec{q} \equiv (q_1, q_2, \ldots, q_n)$; $d\vec{q} = \prod_{j=1}^{n} dq_j$ is used. The integral can be estimated by MC in a straightforward manner by repeatedly sampling n random numbers $\vec{\eta} \equiv (\eta_1, \eta_2 \ldots \eta_n)$ with $a \leq \eta_i \leq b$. The value of the integral is then

$$\bar{f} = \frac{1}{n} \sum_{k=1}^{n} f(\vec{\eta}_k) \qquad (2.14)$$

where n is sufficiently large. For finite n, this estimate will have the statistical uncertainty associated with incomplete sampling which may be shown to be proportional to $\sigma_f/\sqrt{n}$, where σ_f is the standard deviation of f defined theoretically by

$$\sigma_f^2 = \int (f(\vec{q}) - \theta)^2 \, d\vec{q} \quad . \qquad (2.15)$$

In practice, we estimate the standard deviation from the formula

$$\sigma_f^2 = \frac{1}{n-1} \sum_{i=1}^{N} (f_i - \bar{f})^2. \qquad (2.16)$$

Now, the time required to complete a one-dimensional integral using a simple quadrature formula involving r number of points would be proportional to r. A straightforward generalization of this would imply that the time required for evaluating a N-dimensional integral would be proportional to r^N. On the other hand, a simple MC method like the above which is based on a uniform sampling points from the integration space would involve a time which is only linear in N. A MC method may not be as efficient as an ordinary quadrature technique when integral of very few dimensions are involved; but as the dimensionality N of the integrals is increased, MC soon gains efficiency, and when N is very large it probably becomes the only method that is practical.

In physical problems, one is concerned with the evaluation of averages

$$\langle f \rangle = \frac{\int f(\vec{q}) W(\vec{q}) d\vec{q}}{\int W(\vec{q}) d\vec{q}} \qquad (2.17)$$

where $f(\vec{q})$ is the quantity to be averaged over a normalized probability distribution $W(\vec{q})$. The integration is necessarily over a large dimension and recourse to MC methods is warranted. A MC scheme based on a uniform sampling of points from the integration space would be grossly inadequate here; because, for most we may be picking the points where the weightage from $W(\vec{q})$ is small. In the early 1950's, Metropolis *et al* invented an ingenious algorithm for sampling points from a given distribution $W(\vec{q})$. Once n points $\vec{x}_k$ are sampled this way, the average $\langle f \rangle$ can be estimated as

$$\langle f \rangle = \frac{1}{n} \sum_{k=1}^{n} f(\vec{q}_k) \qquad (2.18)$$

with the variance given by $\sigma_f^2 = (\langle f^2 \rangle - \langle f \rangle^2)/n$. Crudely speaking the Metropolis method is a procedure of evolving from a given ensemble to the desired one. This is achieved by replacing old state by new ones using importance sampling in such a way that one gets the correct probability density distributions $W(\vec{q})$ in the limit of a large number of such replacements. Moreover, regardless of the initial state used, repeated applications of the sampling procedure always brings the system into the correct ensemble and once we reach the equilibrium ensemble, further applications of the method keeps one in the same ensemble.

Metropolis algorithm

In order to understand the basic structure of the Metropolis algorithm, it may be necessary to outline at first the theory of Markov processes. Let us consider a collection of states a forming an ensemble $[a]_Q$. The state a itself may in general be a vector $(a_1, a_2, \ldots a_N)$ in N dimensions. These states may be grouped into a vector with components $Q(a)$. $Q(a)$ is in fact, the density distribution of state a in the ensemble. Consider another collection of states b belonging to another ensemble $[b]_{Q'}$. These states may be grouped into another vector with components $Q'(b)$. A Markov process is defined by a transition probability $P(b \leftarrow a)$ which generates the ensemble $[b]_{Q'}$ from the ensemble $[a]_Q$. This means that if P is applied to every member of the ensemble $[a]_Q$, it generates the ensemble $[b]_{Q'}$. In the more transparent matrix notation we may write

$$Q'(b) = \sum_a P(b \leftarrow a) Q(a) \qquad (2.19)$$

or briefly as $Q' = P Q$.

The evolution is a Markov process if the transition probability $P(b \leftarrow a)$, apart from being strictly positive, has the additional property

$$\sum_a P(b \leftarrow a) = 1 \qquad (2.20)$$

and the vector $Q(a)$ satisfies

$$\sum_a Q(a) = 1 . \qquad (2.21)$$

The condition that the transition probability is strictly positive is also known as the "strong ergodicity condition". It implies that there is a finite non-zero probability for any initial state to go to every final state in one Markov step. Equation (2.20) merely states the conservation of probability, that is P must take every state a into some state b. The last condition (2.21) implies that at any given step in the evolution, the system must certainly be in one of the allowed states.

In numerical simulations, we would like to have an evolution scheme, defined by repeated applications of P on some initial ensemble Q_0, such that after a sufficiently large number of steps, we reach the given ensemble W. Moreover, once we arrive at W, we must remain there, that is, subsequent repeated application of P must also generate the same ensemble. Thus it is desirable that for arbitrary Q_0, P **must** satisfy

$$\lim_{n \to \infty} P^n Q_0 = W \quad . \qquad (2.22)$$

A necessary condition for this to be satisfied is that W is a fixed point of P, that is, P W = W . Markovion character of the process then guarantees that W is the only fixed point of P.

Metropolis algorithm also imposes the condition of detailed balance on $P(\text{\&} \leftarrow a)$, viz., the condition

$$P(\text{\&} \leftarrow a) \, W(a) = P(a \leftarrow \text{\&}) \, W(\text{\&}) \; . \qquad (2.23)$$

It turns out that this condition is sufficient to give P the desired properties although it is not a necessary condition. In fact, from the detailed balance (2.23) and normalization condition (2.20) it follows that

$$\sum_a P(\text{\&} \leftarrow a) \, W(a) = \sum_a P(a \leftarrow \text{\&}) \, W(\text{\&}) = W(\text{\&}) \qquad (2.24)$$

that is, W is an eigenvector of P with eigenvalue unity. Perron's theorem then guarantees that this eigenvalue of P is nondegenerate and exceeds all other eigenvalues in modulus . In other words, as long as, the initial vector Q_0 has a nonzero component along the desired vector W, repeated applications of P on Q_0 will result in the enhancement of the component along W.

The Metropolis algorithm may now be stated as :

$$P(\text{\&} \leftarrow a) = 1 \; , \qquad W(a) < W(\text{\&}) \qquad (2.25a)$$

$$P(\text{\&} \leftarrow a) = \frac{W(\text{\&})}{W(a)} \; , \qquad W(a) \geq W(\text{\&}). \qquad (2.25b)$$

It is clear that P is non-negative and also that $\sum_{\text{\&}} P(\text{\&} \leftarrow a) = 1$ since either $W(a) \geq W(\text{\&})$ or $W(a) < W(\text{\&})$. If P is modified to satisfy the strong ergodicity condition, it defines a Markov process. We now show that P satisfies the property of detailed balance

If $W(a) < W(\mathscr{b})$, then from (2.25a),

$$P(\mathscr{b} \leftarrow a) \, W(a) = W(a) \qquad (2.26a)$$

and from (2.25b)

$$P(a \leftarrow \mathscr{b}) \, W(\mathscr{b}) = \left(\frac{W(a)}{W(\mathscr{b})} \right) W(\mathscr{b}) = W(a). \qquad (2.26b)$$

On the other hand, if $W(a) \geq W(\mathscr{b})$, then from (2.25b)

$$P(\mathscr{b} \leftarrow a) \, W(a) = \left(\frac{W(\mathscr{b})}{W(a)} \right) W(a) = W(\mathscr{b}) \qquad (2.27a)$$

and from (2.25a)

$$P(a \leftarrow \mathscr{b}) \, W(\mathscr{b}) = W(\mathscr{b}). \qquad (2.27b)$$

Thus detailed balance condition is satisfied in every case.

If the states are updated one by one, the strong ergodicity condition is not satisfied. Rather, P satisfies the weaker condition, $P(a \leftarrow \mathscr{b}) \geq 0$. Consider such a transition probability which preserves W. We shall show that

$$P'(a \leftarrow c) = \sum_{\mathscr{b}} P(a \leftarrow \mathscr{b}) \, P(\mathscr{b} \leftarrow c) \qquad (2.28)$$

also preserves W. The idea here is to sequentially update every variable in the system by applying the Markov transition probability P to them one by one; the cumulative result of updating all the variables together is then considered as a single Markov step. It is clear that the cumulative Markov step is strongly ergodic and preserves W.

We first show that P' defines a Markov process. Since $P \geq 0$, it is clear that $P' > 0$. Moreover, since $0 \leq P(a \leftarrow \mathscr{b}) \leq 1$, we have

$$P'(a \leftarrow c) \leq \sum_{\mathscr{b}} P(\mathscr{b} \leftarrow c) = 1 \qquad (2.29)$$

where the second step follows from the conservation of probability. The latter property also allows us to prove

$$\sum_a P'(a \leftarrow c) = \sum_{a\,\&} P(a \leftarrow \&) P(\& \leftarrow c) = \sum_{\&} P(\& \leftarrow c) = 1. \quad (2.30)$$

Thus P' satisfies the conditions of a Markov process. Finally, since W is a fixed point of P,

$$\sum_c P'(a \leftarrow c)W(c) = \sum_{c\,\&} P(a \leftarrow \&)P(\& \leftarrow c)W(c) = \sum_{\&} P(a \leftarrow \&)W(\&) = W(a) \quad (2.31)$$

which means that W is also a fixed point P'.

Finally one can prove the convergence of the sampling scheme, that is, starting from any initial ensemble, the asymptotic ensemble is always W. The reader is advised to consult further literature cited at the end of the chapter for details of this proof.

In practical applications, for example, in the evaluation of a multiple integral by MC methods, the algorithm of (2.25) is applied step by step to each variable. As an illustration, we apply the Metropolis algorithm to generate a set of values for a single variable $x \in (-\infty, \infty)$ with a given probability distribution W(x). For this purpose, we start from any x and intend to affect a change to a new value $x' = x + \varepsilon_1$ or $x' = x - \varepsilon_2$. The decision about whether the change $x \rightarrow x'$ is accepted or not is now made using (2.25). If the initial value was chosen from a distribution Q(x), the final distribution is given by

$$Q'(x') = \frac{1}{2} \sum_{\substack{x=x'-\varepsilon_1 \\ x=x'+\varepsilon_2}} \left\{ \theta[W(x')-W(x)] + \frac{W(x')}{W(x)} \theta[W(x)-W(x')] \right\} Q(x)$$

$$+ \left[1 - \frac{1}{2} \sum_{\substack{x''=x'+\varepsilon_1 \\ x''=x'-\varepsilon_2}} \left\{ \theta[W(x'')-W(x')] + W(x'')/W(x') \theta[W(x')-W(x'')] \right\} \right] Q(x')$$

$$(2.32)$$

where $\theta(x)$ is the step function (that is, $\theta(x) = 1$, for $x \geq 0$, and zero otherwise. In Eq. (2.31), the first term represents the probability that x is changed to $x' = x + \varepsilon_1$ or $x' = x - \varepsilon_2$. This happens either directly if $W(x') > W(x)$ or else with probability $W(x')/W(x)$ if $W(x') < W(x)$. The second term in Eq. (2.32) is the probability that x was x' initially and was not changed in the updating step. It is now clear that $Q = W$ is a fixed point of (2.32), that is, $Q = Q' = W$ satisfies (2.32) if and only if $\varepsilon_1 = \varepsilon_2$. Thus, it is necessary to select a trial value x' for x from a distribution which is symmetric around x.

In the next section, we discuss an application of this algorithm to evaluate a path integral.

11.3. Path Integral by Monte Carlo

The simplest problem which illustrates summing paths on a computer is the calculation of ground state energy of a quantum problem. Consider the motion of a spinless particle in one dimension under the influence of a potential $V(x)$. The particle motion can be described by the propagator

$$K(x,T;x_0,0) = \int_{x(0)=0}^{x(T)=x} \exp\left\{\frac{i}{\hbar}\int_0^T \left[\frac{m\dot{x}^2}{2} - V(x)\right]dt\right\} \mathcal{D}[x(t)] . \tag{3.1}$$

Suppose that the potential $V(x)$ is such that the problem has a complete set of eigenfunctions ϕ_n with the corresponding eigenvalues E_n. Therefore the propagator admits an expansion

$$K(x,T;x_0,0) = \sum_{n=0}^{\infty} \phi_n^*(x_0) \phi_n(x) \exp[-iE_n T/\hbar]. \tag{3.2}$$

We now assume that the validity of this relation extends to imaginary values of time and let $T \rightarrow -iT$ and $x_0 = x$, so that,

$$K(x,-iT;x,0) = \sum_{n=0}^{\infty} |\phi_n(x)|^2 \exp(-E_n T/\hbar) . \tag{3.3}$$

In particular assuming that the ground state energy E_0 is non-degenerate then for sufficiently long T, that is, for, $T \gg \hbar/(E_1-E_0)$ where E_1 is the energy of first excited state, the first term dominates in the sum on the right-hand side of Eq.(3.3). Hence the ground state wavefunction can be isolated as

$$|\phi_0(x)|^2 = \lim_{T\to\infty} \exp(E_0 T/\hbar) \, K(x,-iT;x,0). \tag{3.4}$$

We may also write this as

$$|\phi_0(x)|^2 = \lim_{T\to\infty} \mathcal{J}(T) \, K(x,-iT;x,0) \tag{3.5a}$$

where the symbol $\mathcal{J}(T)$ is defined by the integral

$$\mathcal{J}(T) = \int_{-\infty}^{\infty} K(x,-iT;x,0)dx \tag{3.5b}$$

so that the factor $\exp(E_0 T/\hbar)$ is eliminated

Metropolis importance sampling (MIS) method

The idea now is to use discretized form of the propagator by dividing the imaginary time interval T in small steps of size $\tau = T/N$. Writing

$$t_k = k\tau, \quad x_k = x(t_k), \quad x_N = x(t_N) = x, \quad k = 0, 1, \ldots, N \tag{3.6a}$$

we may write

$$K(x,-iT;x_0,0) = \lim_{N\to\infty} A_N \int_{-\infty}^{\infty} \prod_{k=1}^{N-1} dx_k \, \exp\{-\tau \, E(\xi(\tau))/\hbar\} \tag{3.6b}$$

where the normalization constant, the discretized path $\xi(\tau)$ and $E[\xi(\tau)]$ are defined as

$$A_N = \left[\frac{m}{2\pi\hbar\tau}\right]^{N/2}, \quad \xi(\tau) \equiv (x_0, x_1, \ldots x_{N-1}, x) \tag{3.7}$$

$$E(\xi(\tau)) = \sum_{k=0}^{N-1} \left\{ \frac{m}{2}\left(\frac{x_{k+1}-x_k}{\tau}\right)^2 + V(x_k) \right\}. \tag{3.8}$$

Here $E[x(\tau)]$ is like the "energy" associated with the path $x(\tau)$. The propagator is now a $(N-1)$-dimensional Riemann integral. The multiplicity $(N-1)$ being very large the use of a Monte Carlo method is necessary. The problem of computing the sum in Eq.(3.1) is therefore reduced to the computation of a large dimensional integral. In particular, the expression for the ground state wavefunction takes the form

$$|\phi_0(x)|^2 = \frac{1}{Z}\int\cdots\int dx_0\, dx_1\ldots dx_{N-1}\, \delta(x-x_0)\, e^{-\tau E(\xi(\tau))/\hbar} \qquad (3.9)$$

where Z looks very similar to the partition function in classical statistical mechanics (with $\beta = \tau/\hbar$):

$$Z = \int\cdots\int dx_0 dx_1\ldots dx_{N-1}\, e^{-\tau E(\xi(\tau))/\hbar}. \qquad (3.10)$$

Equation (3.9) has a form which is similar to Eq.(2.17). It is appropriate for applying the metropolis algorithm with the probability distribution $\exp(-\tau E/\hbar)/Z$. We describe the procedure here. First choose an initial path $\xi^{(0)}$ which is a sequence of randomly chosen points $(x_0, x_1, \ldots x_N = x_0)$. A new path is to be generated by moving only one randomly selected point x_j to x_j'. The new path ξ' differs from $\xi^{(0)}$ only at jth component. Now select randomly an integer j between 0 and $N-1$ and a real number η in the interval $[0,1]$, and let

$$x_j' = x_j + \alpha\,(2\eta-1) \qquad (3.11)$$

where α is like τ and N is a parameter characterizing the numerical integration and discretization. Now let

$$\Delta E = E(\xi'(\tau)) - E(\xi^{(0)}(\tau)). \qquad (3.12)$$

If $\Delta E < 0$, the "energy" E is lowered and we take $\xi^{(1)} = \xi'$. If $\Delta E > 0$, we let $\xi^{(1)} = \xi'$ with probability $\exp(-\tau\Delta E/\hbar)$. This decision is implemented by picking another random number η' in $[0,1]$ such that if η' is less than $\exp(-\tau\Delta E/\hbar)$, we let $\xi^{(1)} = \xi'$, otherwise $\xi^{(1)} = \xi^{(0)}$. Next, $\xi^{(2)}$ is obtained from $\xi^{(1)}$ in the same manner. The sequence $\xi^{(k)}$ tends to relax towards a class of paths making most important contribution to the sum. Moreover, because of the rule allowing for the positive values of ΔE as well, the sequence does not get stuck at a local minima of E. Note that a crude MC calculation would involve choosing the sequence $\xi^{(n)}$ randomly, rather than as a member of a Markov chain (random walk) and then computing $\exp(-\tau E(\xi^{(k)}(\tau))/\hbar)$. This results in choosing with high probability sequences (paths) where $\exp(-\tau E/\hbar)$ is small and hence of relatively low weight. This has the consequence of making the wavefunction accurate near the center of the system (origin), but highly inaccurate at the tails. Instead of choosing paths randomly and then weighting them with $\exp(-\tau E/\hbar)$, it is therefore preferable to choose paths with probability $\exp(-\tau E/\hbar)$ and weigh then evenly. Note also that this introduces a normalization factor

$$\int K(x,-iT;x,0)\,dx \simeq e^{-E_0 T/\hbar} \qquad (3.13)$$

(for large T) which cancels in the final result. Note that the paths encountered in the present calculations are closed paths ($x = x_0$) so that any point along the path may be considered as both the "beginning" as well as the "end point" of the closed path. Therefore, each sampled path may be used in calculating the probability density at every point along the path, not just at one point.

An efficient way to evaluate $\phi_0(x)$ from $(\xi^{(k)})$, therefore, is as follows. The real axis is divided into a large number of bins $[m\Delta,(m+1)\Delta]$, $m = 0, 1, 2,$ etc. with the size Δ chosen suitably. Consider now a particular step of the above stochastic process that generates $(\xi^{(k)})$. At the (k+1)th step, for example, some coordinate x_j is allowed to assume a new value $x_j + \alpha(2\eta-1)$. The jth component of $(\xi^{(k+1)})$ is, therefore, either x_j or $x_j + \alpha(2\eta-1)$. In either case jth component of $\xi^{(k+1)}$ falls into some bin $[m\Delta,(m+1)\Delta]$, and a point is scored for that bin. As the stochastic process continues, each bin accumulates points with each successful entry or persistence in that bin. $|\phi_0|^2$ is then proportional to the total number of the points in the bin associated with a given x and by normalizing the point score by the total number of configurations, an estimate of the square of the wavefunction is obtained.

How does the method work? As stated before the Metropolis algorithm is a procedure of evolving from a given ensemble to the desired one. This is achieved by replacing old states (in our case "paths") by new ones using importance sampling in such a way that one gets the correct normalized probability distribution (1/Z) exp $\{-\tau E/\hbar\}$ in the limit of large number of such replacements. As shown in Sec. 11.2, regardless of initial states used, repeated application of the sampling procedure always brings the system into correct ensemble and once we reach the equilibrium ensemble, further application of the method keeps one in the same ensemble. From computational point of view, one has to disregard several of the initial configurations before counting of scores in the bin is undertaken. This is required in order to allow sufficient "time" (measured in terms of the number of steps) before the Markov chain might get stuck in a metastable state during the course of its evolution towards the equilibrium. This can be avoided by repeating the procedure using different initial states. A better estimate of the integral is often obtained by taking an average over the results obtained with different initial states. The choice of various parameters like the step size α, time step τ (or equivalently N) and the bin size Δ depends on the problem and is often guided by experience.

This scheme works reasonably well in various one-dimensional quantal systems: a square well, harmonic oscillator, Morse potential and hydrogen atom. The wavefunction may be computed with better than 5% accuracy within a reasonable computing time. Also, it is, in principle, straightforward to generalize the method to more than one dimension or to multi-particle systems.

Path integral MC based on MIS can be used to treat both equilibrium and non-equilibrium problems at finite temperature. The general approach here is to consider the quantum mechanical density matrix $\rho(x,x_0;\beta)$ defined in terms of the Hamiltonian H, as

$$\rho(x,x_0;\beta) = <x|\exp(-\beta H)|x_0>. \quad (3.14)$$

The path integral representation of $\rho(x,x_0;\beta)$ is discussed in Sec. 1.6 of chapter 1. Basically $\rho(x,x_0,\beta)$ is just like the propagator $K(x,x_0,-iT)$ with T replaced by $\hbar\beta$. The knowledge of ρ for real values of $\beta = 1/k_B T$ is enough to obtain equilibrium properties. If in addition, we have knowledge of ρ for complex $\beta = \beta_r + i\beta_i$, we can extract dynamical information. The Metropolis algorithm has been used in literature to treat both the equilibrium and non-equilibrium problems. In the latter case, since β is complex, the sampling of paths is to be done with the real part of the exponential leaving the imaginary part with the variables to be averaged. The oscillatory nature of integral limits the calculations to small times, that is, to near equilibrium situations. Similar problems arise in the applications of the MIS to tunneling problems and the calculations of the time correlated functions. In some of these problems the direct sampling path integral Monte Carlo approach may be used.

Direct sampling MC method (DSMC)

Direct sampling path integral MC is often useful in problems where potential changes are abrupt. This sort of situation is present in tunneling problems in condensed matter system and also in ultra small semiconductor devices where one is interested in accurate modeling of the electronic properties. In such problems, high resolution in path configurations is required. For this purpose, we rewrite the propagator (or the density matrix ρ) as

$$K(x,-iT;x_0,0) = K_0(x,-iT;x_0,0) <\{\exp[-\frac{1}{\hbar}\int_0^T V[x(t)]dt]\}> \quad (3.15)$$

where $<...>$ denotes the expectation value with respect to the paths generated with the free space probability distribution for paths given by

$$\rho_0[x(t)] = \exp\left\{-\frac{1}{2\hbar}\int_0^T m\,\dot{x}^2\,dt\right\}. \qquad (3.16)$$

The free particle propagator is written as

$$K_0(x,-iT;x_0,0) = \int_{x(0)=0}^{x(T)=x} \rho_0[x(t)]\,\mathcal{D}\,[x(t)]\,. \qquad (3.17)$$

The expectation value $<\ldots>$ is then written as

$$<\exp[-\frac{1}{\hbar}\,V\,]> \;=\; \frac{\int \exp\!\left[-\frac{1}{\hbar}\,V\right] \rho_0[x(t)]\mathcal{D}\,[x(t)]}{K_0\,(x,-iT;x_0,0)}\,, \qquad (3.18a)$$

where we have set for brevity

$$V = \int_0^T V[x(t)]dt\,. \qquad (3.18b)$$

These equations suggest a simple two-step MC procedure for evaluating K for any given pair (x_0,x). The first step is to generate stochastically a set of path configurations based only on their kinetic energy from the probability distributions $\rho_0[x(t)]$. The second step is to evaluate $\exp[-V/\hbar]$ for each corresponding path from x_0 to x, find the average value of this quantity and multiply by the free space propagator K_0. Note that the kinetic energy of the path is translationally invariant and varies for all paths by the same amount $m(x-x_0)^2/2T$ as the paths are varied between the fixed end points by addition of a constant velocity path. Hence the entire propagator (or the density matrix) can be uniformly sampled with the same set of path configurations. In short, while the configuration space of paths is sampled by MC, the coordinate space of the function involving the potential is sampled in a uniform manner.

It is convenient to generate the path configuration in Fourier space first and then convert them to coordinate space before storing for subsequent use. A path $x(t)$ satisfying the end-point conditions $x(0) = x_0$ and $x(T) = x$, may be expanded in a Fourier series as

$$x(t) = x_0 + (x-x_0)\frac{t}{T} + \sum_k b_k \sin(k\pi t/T). \qquad (3.19)$$

It should be noted here that the deviations from the direct classical path have been expanded in a Fourier sine series. This expansion conveniently reduces the free space probability distribution as the product of Gaussian distributions

$$\rho_0[x(t)] = \exp\left[-\frac{m}{2T}(x-x_0)^2\right] \prod_{k=1}^{\infty} \exp\left[-b_k^2/2\sigma_k^2\right] \qquad (3.20)$$

where

$$\sigma_k^2 = \frac{2\hbar}{mT}\left(\frac{T}{k\pi}\right)^2 . \qquad (3.21)$$

Thus, each Fourier coefficient b_k is sampled from appropriate Gaussian distribution independent of all other coefficients. The path configurations in coordinate space are then generated from these coefficients using (3.19). In practice, it is useful to rewrite

$$\langle \exp[-V]\rangle = \frac{\int \prod_k db_k \exp\left\{-\sum_k b_k^2/2\sigma_k^2 - V\right\}}{\int \prod_k db_k \exp\left\{-\sum_k b_k^2/2\sigma_k^2\right\}} = \langle \exp[-V]\rangle_b \qquad (3.22)$$

where $\langle \ \rangle_b$ represents an average over $\exp(-\sum_k b_k^2/2\sigma_k^2)$.

Applications to equilibrium problems ($T_c = \hbar\beta$) have shown a reasonable convergence of the results with respect to the number of Fourier coefficients included in the description of the paths. Moreover, it is sufficient to use ordinary Gauss quadrature methods of low order for approximating the integral involving the potential $V[x(t)]$ over the time integral $[0,T]$. Once the above expectation value is computed it is to be multiplied by the free particle propagator K_0 in order to obtain the complete propagator K.

One can extend this approach to complex time $T_c = T_r + iT_i$ which implies $\beta_c = \beta_r + \beta_i$ by noting that the theory sketched above is formally applicable. However, for MC evaluation it is convenient to rewrite the expectation value of (3.22) in the following manner by separating the real and the imaginary parts in the Gaussian. Thus,

$$\langle \exp[-V_c] \rangle = \frac{\int \prod_k db_k \exp\left\{-\sum_k b_k^2/2\tilde{\sigma}_k^2 + i\sum_k b_k^2 \alpha_k - V_c\right\}}{\int \prod_k db_k \exp\left(-\sum_k b_k^2/2\tilde{\sigma}_k^2 + i\sum_k b_k^2 \alpha_k\right)}$$

$$= \frac{\langle \exp\left\{i\sum_k b_k^2 \alpha_k - V_c\right\}\rangle_\&}{\langle \exp\left(i\sum_k b_k^2 \alpha_k\right)\rangle_\&} \qquad (3.23)$$

where $\tilde{\sigma}_k^2 = (2\hbar/mT_r)(|T_c|/k\pi)^2$, $\alpha_k = [T_i/(T_r \tilde{\sigma}_k^2)]$ and V_c is still given by (3.18b) with T replaced by T_c. The averages occurring in the last step are taken over the modified distribution $\exp\{-\sum_k b_k^2/2\tilde{\sigma}_k^2\}$. In general the MC evaluation of (3.23) presents difficulties for large values of T_i. This is due to highly oscillatory nature of the integral.

We might add here that the computation of diagonal propagator $K(x,-iT;x,0)$ offers some simplicities. The paths are now closed and each sample path may be used in calculating the propagator at every point along the path, not just at one point. The appropriate Fourier expansion for paths is

$$x(t) = a_0 + \sum_k \left\{a_k \cos(2\pi kt/T) + b_k \sin(2\pi kt/T)\right\} \qquad (3.24)$$

so that $x(0) = x(t) = x$. The free space probability distribution now takes the form

$$\rho_0[x(t)] = \prod_{k=1}^{\infty} \exp\left[-(a_k^2 + b_k^2)/2\sigma_k^2\right] \qquad (3.25)$$

where $\sigma_k^2 = (\hbar/2mT)(T/k\pi)^2$ and as before each Fourier coefficient is chosen from its Gaussian distribution independently of all other coefficients. For each closed path the 0th order coefficient a_0 is generated with a uniform probability distribution over the range of integration of the coordinate space. The path configurations in the coordinate space are then generated from these coefficients using (3.24). Once the paths have been generated it is easy to compute the expectation value (3.18) and hence the propagator (or density matrix) for $x = x_0$. As in the previously discussed MIS scheme for the ground state energy calculation, we take advantage of the fact that the paths are closed. Each sample point on the path is both the beginning and the end of the path and contributes to the information about K. For this purpose, we overlay the potential structure with a spatial grid of desired resolution. The centre of each previously generated path configuration is then located within each grid box successively, according to the value of a_0. This process moves every sample point along every path configuration to every grid box successively. Whenever a sample point along a path falls into a particular grid box, a contribution of $\exp[-V]$ for that path is added to the value of the probability density in that box. When this value is divided by the number of contributions to the box, we obtain an estimate of the expectation value of Eq.(3.18). This expectation value is to be multiplied by the known form of the free particle propagator (in this case, only a normalization constant) to obtain the value of $K(x,-iT;x,0)$. Incidentally, if this K is integrated over all x, we obtain

$$\int_{-\infty}^{\infty} K(x,-iT;x,0)\, dx = \sum_n \exp(-E_n T/\hbar) \qquad (3.26)$$

and it is possible to recover information about eigenenergies from this expression.

11.4. Deterministic techniques of path summation

Monte Carlo methods rely on the representation of a propagator as a large dimensional Riemann integral. Deterministic methods are based on the semi-group composition law

$$K(x,t;x_0,0) = \int K(x,t;x',t') K(x',t';x_0,0) \, dx'. \qquad (4.1)$$

This semi-group property may not be valid in general. In particular, it fails when a reduced description based on elimination of some degrees of freedom is used. The discrete short "time" propagator (we use imaginary time here and subsequently)

$$K(x_2,x_1;\varepsilon) = \left(\frac{m}{2\pi\hbar\varepsilon}\right)^{1/2} \exp\left[-\frac{1}{\hbar}\left\{\frac{(x_2-x_1)^2}{2\varepsilon} + \frac{\varepsilon[V(x_1)+V(x_2)]}{2}\right\}\right] \qquad (4.2)$$

also obeys the semi-group property. Thus starting from the short time propagator (4.2) one may apply the semi-group composition law (4.1) repeatedly to obtain the large time propagator.

In actual calculations the upper and lower limit of integration is replaced by L/2 and -L/2 where L is chosen to be large enough in order to ensure convergence. The entire range (L) of the coordinate space is uniformly divided into N points. The position coordinates of the kernel K are allowed to assume only these N discrete set of values. The kernel is now a matrix and the required computer storage is of the order of N^{2d} in d dimensions. The integration over the intermediate variable is carried out using the simple trapezoidal rule, that is,

$$K(x_3,x_1;2\varepsilon) = \Delta \sum_{i=1}^{N} K(x_3,x_i,\varepsilon) K(x_i,x_1,\varepsilon) \qquad (4.3)$$

where Δ is the spacing $(x_{i+1} - x_i)$ between two consecutive points. The process is repeated n times to get $K(x_{n+1},x_1,2^n\varepsilon)$ where $2^n\varepsilon = T$.

In order to achieve a reasonable accuracy, N has to be sufficiently large. This tends to increase the storage requirements (N^{2d}). Note also that (4.3) implies a simple matrix multiplication. The total number of arithmetic operations involved is $(2N^d - 1)N^{2d}$ in d dimensions. It is clear that for large N, the computer time required increases with the dimension d of the system. Thus the problem soon becomes prohibitively expensive both in terms of the storage requirements and the computational time.

It is possible to improve upon the simple numerical matrix multiplication scheme discussed above. We observe that the short time kernel is of the form $\exp(-f(x))$. The evaluation of the finite time kernel involves integration of such an exponential function. It is clear that the dominant contribution to the integral arises from a few points in the neighbourhood of the minima of $f(x)$. This is similar to the usual WKBJ strategy of approximating the Feynman path integral. The idea is to identify an appropriate set of few (ℓ) points rather than a large number (N) of points. The choice of this small set is guided by the Gauss quadrature formula. Thus the simple trapezoidal rule of (4.3) is to be replaced by a more refined quadrature formula :

$$K(x_3, x, 2\varepsilon) \simeq \sum_{i=1}^{\ell} W(x_i) g(x_i) \quad ; \quad g(x) = q(x) h(x) \qquad (4.4)$$

and $q(x)$ is a function whose moments

$$\mu_k = \int q(x) x^k \, dx \qquad (k = 0, 1, 2, 3, \ldots) \qquad (4.5)$$

are finite. Thus (4.4) may be replaced by

$$K(x_3, x, 2\varepsilon) \simeq \sum_{i=1}^{\ell} w(x_i) h(x_i) \qquad (4.6)$$

where $w(x_i)$ are new weights. The Gauss quadrature formulae are closely related to orthogonal polynomials. If $q(x)$ has a form similar to the weight function of any of the orthogonal polynomials, then the x_i's are chosen to be the zeroes of the polynomial concerned. The associated weights $w(x_i)$ and the points are usually listed in literature.

The precision of the ℓ-point Gauss quadrature formula is equivalent to the interpolation formula obtained using $(2\ell - 1)$ base points. This means that the ℓ-point formula is exact if $h(x)$ is a polynomial of degree $\leq (2\ell - 1)$. By the same token the trapezoidal rule technique is exact if $g(x)$ is linear in x ($\ell = 1$).

We also observe from (4.2) that the form of the kinetic energy term in the kernel implies that the weight function $q(x) \propto \exp(-x^2/\hbar\epsilon)$. This is the weight function for Hermite polynomials and the ℓ-points are to be chosen as the zeroes of the Hermite polynomial $H_\ell(x/\sqrt{\hbar\epsilon})$ with the corresponding weight factors $w(x_i)$. The quadrature formula (4.6) is to be used repeatedly to obtain the large time propagator $K(x,x_0;2^n\epsilon=T)$. Information about wavefunctions and eigenenergies may be obtained subsequently from the computed finite time propagator. This technique has been applied to study quantum mechanics of one-dimensional potentials. It has been shown that the Gauss-Hermite quadrature formula using $\ell = 8$ or 16 points yields results on ground state wavefunction and energy comparable to numerical matrix multiplication technique (trapezoidal rule) using $N = 100$ points.

Finally one may point out here the basic similarity in the various approximation methods of evaluating Feynman path integral (in imaginary time). The general idea is to identify and sample a finite number of points of the infinite dimensional integral. This is like sampling the paths which contribute most to the Feynman integral. In the MC methods, the important paths are generated either using the canonical Boltzmann distribution or by means of the random sampling of the Fourier expansion coefficients of the paths from a Gaussian probability distribution. The WKBJ method characterizes the important paths by their stationarity property. In the deterministic method treated here, the identification of the important paths is carried out by means of the zeroes of the appropriate orthogonal polynomial.

Notes and References

A good exposition on the random sampling methods is found in the following books.

J. M. Hammersley and D. C. Handscomb, "Monte Carlo Methods", John Wiley, New York (1964),

Y. A. Shreider (Ed.), "The Monte Carlo Method", (Pergamon, New York, 1967).

Applications of Monte Carlo methods to quantum problems are discussed in the following book.

M. H. Kalos (Ed.) "Monte Carlo Methods in Quantum Problems" D. Reidel, Dordrecht (1984).

For a more recent discussion the reader should consult the "Proceedings of The Conference on Frontiers of Quantum Monte Carlo" which appear as an issue of

J. Stat. Phys. **43**, 729 (1986).

Original reference to Metropolis algorithm is the following paper

N. Metropolis, A. W. Rosenbluth, M. N. Marshell and A. H. Teller, J. Chem. Phys. **21**, 1087 (1953).

For a readable account on Metropolis method and other alternative MC schemes see

G. Bhanot, Rep. Prog. Phys. **51**, 429 (1988).

Imaginary time PIMC based MIS was first used for obtaining ground state wavefunctions and energies of quantal systems in the following papers.

S. V. Lawande, C. A. Jensen and H. S. Sahlin, J. Comp. Phys. **3**, 416 (1969),

S. V. Lawande, C. A. Jensen and H. S. Sahlin, J. Comp. Phys. **4**, 451 (1969),

S. V. Lawande, C. A. Jensen and H. S. Sahlin, J. Chem. Phys. **54**, 445 (1971).

Another recent reference on this technique is

P. K. Mackeown, Am. J. Phys. **53**, 880 (1985)

Direct sampling PIMC based on Fourier expansion of paths treated in this chapter was first used for statistical mechanics calculations in the papers :

L. D. Fosdick, J. Math. Phys. **3**, 1251 (1962),

L. D. Fosdick and H. F. Jordan, Phys. Rev. **143**, 58 (1966).

A good discussion of MIS versus Direct sampling PIMC is given in a recent paper :

L. F. Register, M. A. Stroscio and M. A. Littlejohn, Super Lattices and Microstructures, **6**, 233 (1989)

As mentioned in the text real-time PIMC computations present more severe problems than imaginary-time PIMC computations because of the oscillatory integrals involved. However, there have been attempts to perform such computations in a variety of applications in chemical physics and other areas. Some recent references are listed below

G. Scher, M. Smith and M. Boranger, Ann. Phys. (NY) **130**, 290 (1980),

D. Thirumalai and B. J. Berne, J. Chem. Phys. **79**, 5029 (1980),

E. C. Behrman, G. A. Jongeward and P. G. Wolynes, J. Chem. Phys. **79**, 6277 (1983),

J. D. Doll, J. Chem. Phys. **81**, 3536 (1984),

H. B. Schüttler and D. J. Scalapino, Phys. Rev. **B34**, 4744 (1986),

B. Mason and K. Hess, Superlattices and Microstructures, **3**, 421 (1987),

J. D. Doll, R. D. Coalson and D. L. Freeman, J. Chem. Phys. **87**, 1641 (1987),

J. Chang and W. H. Miller, J. Chem. Phys. **87**, 1648 (1987),

L. F. Register, M. A. Stroscio and M. A. Littlejohn, Superlattices and Microstructures, **4**, 61 (1988),

J. D. Doll, T. L. Beck and D. L. Freeman, J. Chem. Phys. **89**, 5753 (1988).

Some references for deterministic techniques of path summation treated in section 10.4 are:

D. Thirumalai, E. J. Bruskin and B. J. Berne, J. Chem. Phys. **79**, 5065 (1983),

C. C. Gerry and J. Kiefer, Am. J. Phys. **56**, 1002 (1988),

A. Das and V. A. Singh "An efficient algorithm for Feynman propagator" in "Path-Integral Methods and Their Applications", Ed. D. C. Khandekar and S. V. Lawande, (IPA, Bombay 1990), p. 158.

Additional References :

For PIMC applications for tunneling problems see

M. Creatz and B. Freedman, Ann. Phys. **132**, 427 (1981),

E. L. Pollack and D. M. Ceperley, Phys. Rev. **B30**, 2555 (1984),

J. W. Negele, "Path integral Monte Carlo study of tunneling in quantum many-particle systems", in "Path integrals from meV to Mev", Ed. M. C. Gutzwiller, A. Inomata, J. R. Klauder, and L. Streit, (World Scientific, Singapore, 1986), p.63.

CHAPTER 12

MATHEMATICAL NATURE OF FEYNMAN PATH INTEGRAL

12.1. Introduction

In the last several chapters we discussed various applications of Feynman propagators by using the time-slicing or the polygonal approach to evaluate the desired propagator. While as an operational definition, the polygonal scheme works very well, it is yet to receive a rigorous mathematical justification. The difficulty in properly interpreting it lies in the oscillatory nature of the "integrand". It can be best understood by considering finite dimensional oscillatory integrals. To start with let us examine a one-dimensional oscillatory integral.

$$\mathcal{I}(\alpha) = \int_{-\infty}^{\infty} \exp\{i\alpha x\} \, dx . \tag{1.1}$$

Since, the integrand is a unimodular function, $\mathcal{I}$ does not have a meaning in the sense of ordinary integration. However, one of the ways in which the integral in Eq. (1.1) can be understood is through the following limiting procedure, $viz.$,

$$\mathcal{I}(\alpha) = \lim_{\varepsilon \to 0} \int_{-\infty}^{\infty} e^{-\varepsilon|x| + i\alpha x} \, dx . \tag{1.2}$$

It is immediately apparent that $\mathcal{I} = 0$ for every $\alpha \neq 0$ and is unbounded for $\alpha = 0$. Moreover, it is easy to verify that

$$\int_{-\infty}^{\infty} I(\alpha, \varepsilon) \, d\alpha = 2\pi . \tag{1.3}$$

It is clear from this discussion that the integral $\mathcal{I}(\alpha)$ possesses the familiar properties of the so called "δ-function" which in a strict sense does not belong to the class of ordinary functions. Next, consider, the oscillatory Gaussian integral $\mathcal{I}_1(\alpha)$ frequently encountered while dealing with path integrals :

$$\mathcal{I}_1(\alpha) = \int_{-\infty}^{\infty} e^{i\alpha x^2} dx = \lim_{\varepsilon \to 0} \int e^{-\varepsilon x^2 + i\alpha x^2} dx \ . \tag{1.4}$$

Thus we see that the oscillatory integrals can be interpreted through an appropriate limiting procedure. Here the integrand is multiplied by a suitable function, $g(x,\varepsilon)$. One finds that, while $\mathcal{I}(\alpha)$, or $\mathcal{I}_1(\alpha)$ needed reinterpretation, the function $J(\alpha,\varepsilon)$ defined as

$$J(\alpha,\varepsilon) = \int f(x,\alpha) \, g(x,\varepsilon) \, dx \tag{1.5}$$

exists for every g belonging to a particular class. For example, in Eq.(1.2), $f(x,\alpha) = e^{i\alpha x}$, $g(x,\varepsilon) = e^{-\varepsilon|x|}$. In other words, the integral must be understood in the sense of a distribution. Its value is defined over a class of test functions. In more than one-dimension, such integrals can be interpreted through test functions defined in an appropriate space. The Feynman propagator is usually interpreted as an "infinite dimensional" integral. One may attempt to give it a precise meaning through an appropriate limiting procedure. One such scheme appeals to the use the Trotter product formula.

12.2. Definition through Limiting Procedures

12.2.1. *Trotter product formula*

The wave function ψ_t at time t is related to its value at time t = 0, through the relation

$$\psi_t = \exp\left[-\frac{i}{\hbar} t \hat{H}\right] \psi_0 \tag{2.1}$$

where $\hat{H}$ is a well defined self-adjoint Hamiltonian operator associated with the quantum system under consideration. The wavefunctions ψ belong to a class of square integrable functions $L^2(\mathbb{R}^d)$. Then if the potential is bounded and continuous, the Trotter product formula asserts

$$e^{-(i/\hbar)t \hat{H}} = \text{S-lim}_{n \to \infty} \hat{R}_n, \quad \hat{R}_n = \left(e^{-(i/\hbar)\varepsilon V} e^{-(i/\hbar)\varepsilon \hat{H}_0}\right)^n \quad (2.2)$$

where $\hat{H} = \hat{H}_0 + \hat{V}$, $\hat{H}_0$ being the free particle Hamiltonian operator. The quantity V denotes the potential. The limit in Eq.(2.2) is taken in the strong sense. This means that $\|(\hat{R}_n - \hat{R}_m)\psi\|$ can be made arbitrarily small for every ψ belonging to $L^2(\mathbb{R}^d)$ for all n and m greater than a certain $N(\psi)$. Next, we know that

$$e^{-(i/\hbar)t \hat{H}_0} \psi_0(x) = \int K_0(x,t;y,0) \psi_0(y) \, dy, \quad t > 0 \quad (2.3)$$

where $K_0(x,t;y,0)$ denotes the free particle propagator. A repeated use of the result in Eq.(2.3) yields

$$\psi_t(x) = \lim_{n \to \infty} \left(\frac{m}{2\pi i \hbar \varepsilon}\right)^{n/2} \int \exp\left[\frac{i}{\hbar} S_n[x, x_1, \ldots, x_{n-1}, y]\right] \psi_0(y) \prod_{j=1}^{n-1} dx_j \, dy. \quad (2.4)$$

Here S_n corresponds to the discretized form of action occurring in the polygonal definition of the propagator. Equation (2.4) can be used to give a precise definition to Feynman integral. However, it still does not mean that Eq.(2.4) can be interpreted in the sense of usual integrals because one has still take a limit $n \to \infty$ every time. Hence it is desirable to generalize our notions of integrals where limits are inherently defined and the limiting integrals are properly handled. One such definition involves the generalization of the usual Gaussian integrals to infinite dimensions.

12.2.2. *Generalized Gaussian integrals*

The basic idea in defining Feynman integrals in this framework consists in reinterpreting the Gaussian integrals so that the definitions can be easily extended to infinite dimensions. For example, consider a finite dimensional Gaussian integral

$$I_n = \int \exp\left[\frac{i}{2} <\vec{q}, B\vec{q}> + i\vec{k} \cdot \vec{q}\right] d\vec{q} \quad (2.5)$$

where **B** is a n x n positive definite matrix. Since **B** is Hermitian, it can be diagonalized and hence I_n can be thought of as a product of n independent Gaussian integrals, viz.,

$$I_n = \prod_{p=1}^{n} I_p \quad ; \quad I_p = \int \exp\left[\frac{i}{2} \lambda_p q^2 + i\vec{k}_p \cdot \vec{q}\right] d\vec{q} \quad . \tag{2.6}$$

Each of I_p can be interpreted as a limiting integral (1.4), which leads to

$$I_n = \frac{(2\pi i)^{n/2}}{\sqrt{\det B}} \exp\left[-\frac{i}{2} < \vec{k}, B^{-1} \vec{k} >\right] \quad . \tag{2.7}$$

Now let us consider the quantity $\tilde{I}_n$,

$$\tilde{I}_n = \frac{\sqrt{\det B}}{(2\pi i)^{n/2}} I_n = \exp\left[-\frac{i}{2} < \vec{k}, B^{-1} \vec{k} >\right] \quad . \tag{2.8}$$

From Eq. (2.8) we see that while I_n does not have a well defined limit as $n \to \infty$, $\tilde{I}_n$ does possess a limit provided B^{-1} exists. We can immediately generalize this result to any function f(x) which is a Fourier transform of a bounded measure $d\mu_f(\vec{k})$. Thus we may write

$$\tilde{I}_n(f) = \int \exp\left[-\frac{i}{2} < \vec{k}, B^{-1} \vec{k} >\right] d\mu_f(\vec{k})$$

$$= \frac{\sqrt{\det B}}{(2\pi i)^{n/2}} \int \exp\left[\frac{i}{2} < \vec{q}, B \vec{q} >\right] f(\vec{q}) d\vec{q} \quad . \tag{2.9}$$

Moreover, for a sequence of functions $f_1, f_2, \ldots, f_k \ldots$ approaching in the limit $k \to \infty$, to a function f, one can establish

$$\lim_{k \to \infty} \tilde{I}_n(f_k) \to \tilde{I}_n(f) \quad . \tag{2.10}$$

We interpret the space of functions f_k as admissible for the integral $\tilde{I}_n$. Since the right-hand side of Eq. (2.9) exists for all functions belonging to the admissible space, we can use this as a definition of $\tilde{I}_n(f)$. We are now ready to generalize the notion of oscillatory integrals to infinite dimensions. For this purpose we introduce a separable Hilbert space $\mathcal{H}$. On this space we consider a bounded symmetric operator B with bounded symmetric inverse. Then the functional integral

$$\lim_{n \to \infty} \tilde{I}_n(f) = \int_{\mathcal{H}} \exp\left[\frac{i}{2} < \vec{x}(\tau), B \vec{x}(\tau) >\right] f(\vec{x}(\tau)) \mathcal{D} [\vec{x}(\tau)]$$

$$= \int_{\mathcal{H}} \exp\left[-\frac{i}{2} < \vec{k}, B^{-1} \vec{k} >\right] d\mu_f(\vec{k}) \tag{2.11}$$

is well defined for all f belonging to the space of admissible functions. We denote this limit as $\tilde{I}(f)$ and define this as the normalized integral of f with respect to B. We can now use these ideas to define Feynman propagator. For this purpose we split the action function $S[x(t)]$ as

$$S[x(t)] = S_0[x(t)] + S_1[x(t)], \tag{2.12}$$

S_0 being the quadratic part of S. Now, let us concentrate on Eq. (2.8). The splitting (2.12) implies that the discretized form S_N of the action S can be written as

$$S_N(x, x_1, \ldots, x_{N-1}, y) = < \vec{x}, B \vec{x} > + S_1(x, x_1, \ldots, x_{N-1}) \tag{2.13}$$

and hence Eq. (2.4) can be cast into a form

$$\psi_t(x) = \lim_{n \to \infty} \int \exp\left[\frac{i}{\hbar} <\vec{x}, B \vec{x}>\right] \exp\left[\frac{i}{\hbar} S_1\right] \psi_0(y) \prod_{j=1}^{N-1} dx_j \, dy. \tag{2.14}$$

At this point we identify $\exp[(i/\hbar)S_1]$ with $f(\vec{x})$. We see that the above prescription can be used as a rigorous definition for Feynman propagator for a restricted class of potentials.

12.3. Definition Through White Noise Calculus

12.3.1. White noise calculus

We begin our understanding of White noise calculus by recollecting the properties of a single Gaussian random variable y of mean zero and variance 1. The probability distribution for such a variable is given by

$$P(y)dy = \frac{1}{\sqrt{2\pi}} e^{-y^2/2} dy.. \tag{3.1}$$

Alternatively, a distribution function can be characterized completely by specifying all its moments. For a Gaussian random variable, it is well known that the $P(y)dy$ of (3.1) is uniquely characterized by the first two moments viz: $E(y) = 0$, and $E(y^2) = 1$, E denoting the expectation value.

A method of providing complete information about the moments and hence the distribution function is to define the associated characteristic function $C(k)$:

$$C(k) = E(e^{iky}) = \frac{1}{\sqrt{2\pi}} \int e^{iky} e^{-y^2/2} dy = e^{-k^2/2} . \tag{3.2}$$

Equation (3.2) is nothing but the usual Fourier transform of $P(y)$. One can obtain all the moments of the probability distribution (3.1) by expanding $C(k)$ of Eq. (3.2) as a power series in k. The coefficient of k^n is related to $E(x^n)$. Next, let us consider a function $f(y)$ of the Gaussian random variable y. In an analogous way, the function $f(y)$ is totally specified by its $\mathcal{J}$-transform defined as

$$(\mathcal{J} f)(k) = E(f(y) e^{iky}) . \tag{3.3}$$

With these notations the characteristic function $C(k)$ can be viewed as $\mathcal{J}$-transform of a "Unit function". Assume that the function $f(y)$ has a bounded Fourier transform, then it is easy to see that $(\mathcal{J} f)(k)$ admits an expansion

$$(\mathcal{J} f)(k) = e^{-k^2/2} \sum_n k^n F_n \tag{3.4a}$$

where the quantity F_n has the expression

$$F_n = \int \frac{(-k)^n}{n!} \tilde{f}(k) e^{-k^2/2} dk , \tag{3.4b}$$

$\tilde{f}(k)$ being the Fourier transform of $f(y)$. The relations (3.3)--(3.4) can be used as a definition for the function $f(y)$ in the sense that F_n's defined in Eq.(3.4b) are unique and completely characterize the function f(y). Moreover, if f and g are two functions of the Gaussian random variable y with expansion coefficients F_n and G_n, then it is easy to see

$$E(f \cdot g) = \sum_n n! \, F_n G_n . \tag{3.5}$$

At this point the advantage of using $\mathcal{J}$-transform should be clear. First, the $\mathcal{J}$-transform is a point function whereas the probability distribution is a set function. Moreover, in actual computations one very often needs to evaluate the expectation value of the product of functions of a random variable. In such a case the use of Eq.(3.5) facilitates the computations.

The results can be immediately generalized to any finite number of identically distributed independent random variables $y_1, y_2, \ldots, y_n$. The distribution function will have the form

$$P(y_1, y_2, \ldots, y_n) \, dy_j = \prod_{k=1}^{n} \frac{1}{\sqrt{2\pi}} e^{-y_k^2/2} \, dy_k \tag{3.6a}$$

which implies

$$E(y_k) = 0, \qquad E(y_k \cdot y_s) = \delta_{ks} . \tag{3.6b}$$

If the variables $\{y_k\}$ are viewed as the components of an n-dimensional vector $\vec{q}$, the results can be given the following interpretation. The different components of $\vec{q}$ are completely uncorrelated and have the same variance.

Physicists have been familiar with the continuum analogue of these properties through the concept of white noise. A random signal is called white noise if all frequencies are present with equal intensity which is what Eq.(3.6b) implies. Hence to generalize these statements to infinite dimensions we consider a one-parameter family of random variables y(t), indexed by t, $(0 < t < \infty)$ with the properties

$$E(y(t)) = 0 , \qquad E(y(t), y(s)) = \delta(t-s) \tag{3.7}$$

However, y(t) can no more be interpreted as ordinary random variable because the correlation function $E(y(t) \, y(s))$ ceases to be an ordinary function in the sense of analysis. Thus y(t) represents a generalized random variable and the family {y(t)} represents a generalized process.

In order to formally define the white noise we shall have to provide the sample space (the analogue of the points k). We will confine ourselves to test functions belonging to Schwartz space $\mathscr{S}$. The functions belonging to this space go to zero at infinity faster than any power of y. The probability space considered in the analysis is the dual space $\mathscr{S}^*$ of measures.

Given the probability space and the test function space, we can characterize the white noise distribution by its characteristic functional $C[\xi(\tau)]$

$$C(\xi) = E(e^{i<y,\xi>}) = \int_{\mathscr{S}^*} e^{i<y,\xi>} d\mu[y] = \exp\left[-\frac{1}{2}\|\xi\|^2\right] \qquad (3.8)$$

where $<y,\xi>$ is the bilinear form connecting $\mathscr{S}$ and $\mathscr{S}^*$. If the spaces $\mathscr{S}$ and $\mathscr{S}^*$ have the inner product defined then $<y,\xi>$ can be taken as the inner product. $\|\xi\|$ denotes the L^2 norm of ξ.

Having defined the white noise measure, one still has to define the functionals of white noise. Remember that the basic probability space considered is $L^2(\mathscr{S}^*)$, so that almost all y belonging to $\mathscr{S}^*$ can be viewed as a sample function of white noise : Any element of $L^2(\mathscr{S}^*)$ can be termed as the functional of white noise. However, to characterize these functionals completely one has to take recourse to the $\mathscr{T}$-transform. The definition of $\mathscr{T}$-transform of a functional φ now takes the form

$$(\mathscr{T}\,\varphi)\,(\xi) = \int_{\mathscr{S}^*} e^{i<y,\xi>} \varphi(y)\, d\mu[y]. \qquad (3.9)$$

Just to establish the correspondence with the definition of Eq.(2.3), one can map $\xi \to k$, $\varphi(y) \to f(y)$ and $d\mu(y) \to P(y)dy$. A powerful expedient technical tool of white noise analysis is the generalization of the result in Eq.(3.4a). Each random variable φ on $L^2(\mathscr{S}^*,\mu)$ with finite variance is uniquely determined by $\mathscr{T}$-transform which admits a unique expansion

$$(\mathcal{T}\varphi)(\xi) = C(\xi) \sum_{n=0}^{\infty} i^n \int d^n\tau\, F_n(\tau_1,\tau_2,\ldots,\tau_n)\, \xi(\tau_1)\ldots\xi(\tau_n) \qquad (3.10)$$

where F_n is a symmetric function in $L^2(\mathbb{R}^n)$. Moreover

$$\|\varphi\|^2_{L^2(\mathscr{S}^*)} = \sum_{n=0}^{\infty} n!\, \|F_n\|^2_{L^2(\mathbb{R}^n)} . \qquad (3.11)$$

Notice that Eq.(3.10) has enabled us to characterize the random variables (in infinite dimensions) through symmetric functions F_n defined on $\mathbb{R}^n$. With the above identification the space $(L^2) = L^2(\mathscr{S}^*,\mu)$ is just the Bosonic Fock space defined over $L^2(\mathbb{R})$.

In order to see the nature of the random variables more clearly, let us consider the subspace $\mathcal{H}_n$ consisting of the set of elements of (L^2) for which only the kernel F_n is nonzero. For example, the set of constant random variables will be characterized by $F_0 \neq 0$, while all other expansion kernels are zero. An element characterized by F_1 will be of the form $\langle y, F_1 \rangle$. In the context of path integrals random variables characterized by a nonzero F_2 are very often used. Hence it is desirable to have an explicit formula for their $\mathcal{T}$-transform. It reads as

$$\mathcal{T}\left(\exp\left[-\frac{1}{2}\lambda\langle y, Fy\rangle\right]\right)(\xi) = \{\det(I+\lambda F)\}^{-1/2} \exp\left[-\frac{1}{2}\langle \xi, (I+\lambda F)^{-1}\xi\rangle\right] \qquad (3.12)$$

It follows from the above formula that

$$\mathcal{T}(\langle y, Fy\rangle)(\xi) = C(\xi)\, [-\langle \xi, F\xi\rangle + \mathrm{Tr}(F)] . \qquad (3.13)$$

We are now ready to discuss the relation of the above mathematical machinery to physics. To begin with let us consider the characterization of Brownian motion within this framework. The Brownian motion $B(t)$ is one parameter family of Gaussian random variables possessing the properties

$$E(B(t)) = 0, \quad E(B(t)B(s)) = \min(t,s) . \qquad (3.14)$$

Now consider the random variable $B_1(t) = \langle y, \chi_{0,t}(\tau)\rangle$ where χ denotes the characteristic function for the interval. It assumes the value 1, for $0 < \tau < t$ and 0 otherwise. It is easy to see that the random variable $B_1(t)$ possesses all the properties of Brownian Motion. First, we observe from Eq.(3.8) that

$$E\left(e^{i\lambda\langle y,\chi_{0,t}\rangle}\right) = \exp\left[-\frac{\lambda^2}{2}\|\chi\|^2\right] = \exp\left[-\frac{\lambda^2}{2}t\right]. \quad (3.15)$$

It follows from the above equation that $E(B_1(t)) = 0$ and $E(B_1^2(t)) = t$. Similarly using the expectation value $E\left(e^{i\lambda\langle y,\chi_{0,t}\rangle} e^{i\eta\langle y,\chi_{0,s}\rangle}\right)$ one can easily derive $E(B_1(t) B_1(s)) = \min(t,s)$.

Thus within this framework the Brownian motion is a functional of white noise. Further, suppose heuristically that we interpret

$$B(t,y) = \langle y,\chi_{0,t}\rangle = \int_0^t y(\tau)\, d\tau \quad (3.16)$$

then one can identify y with $\dot{B}(t)$ (the velocity of a Brownian particle) as a generalized random variable. Mathematically, of course we know that since the sample paths of Brownian motion are nowhere differentiable, $\dot{B}(t)$ should not be construed as the physical velocity of a particle.

Levy's stochastic area

As an illustration of this technique let us consider the evaluation of Levy's stochastic area S_T. This is defined as the area spanned by a two-dimensional Brownian motion starting at time $t = 0$. It can be expressed as

$$S_T = \int_0^T [B_1(t)\, dB_2(t) - B_2(t)\, dB_1(t)]. \quad (3.17)$$

To evaluate the generating functional for Levy's area we have to first generalize our various formulae to two-dimensions. The two-dimensional Brownian motion can be realized on the probability space of one-dimensional white-noise as

$$B_1(t) : y \subset \mathscr{S}^*(\mathbb{R}) = B_1(t,y) = (y, \chi_{0,t}[\tau]), \quad (3.18a)$$

$$B_2(t) : y \subset \mathscr{S}^*(\mathbb{R}) = B_2(t,y) = (y, \chi_{0,-t}[\tau]). \quad (3.18b)$$

Orthogonality of two characteristic functions $\chi_{0,t}[\tau]$ and $\chi_{0,-t}[\tau]$ indeed takes care of the independence of two components. Using this representation of the Brownian motion, we can immediately write

$$S_T = \int_{L^2(\mathbb{R}^2)} y(\tau) \, \mathcal{L}(\tau,\sigma) \, y(\sigma) \, d\tau \, d\sigma \qquad (3.19)$$

where the kernel $\mathcal{L}(\tau,\sigma)$ has the form

$$\mathcal{L}(\tau,\sigma) = -[\chi_{0,T}(\tau)\,\chi_{-\tau,0}(\sigma) + \chi_{0,T}(\sigma)\,\chi_{-\sigma,0}(\tau)]/4$$

$$+ [\chi_{0,-\tau}(\sigma)\,\chi_{-T,0}(\tau) + \chi_{0,-\sigma}(\tau)\,\chi_{-T,0}(\sigma)]/4 \, . \qquad (3.20)$$

The kernel can be represented graphically as in Fig. 12.1 below.

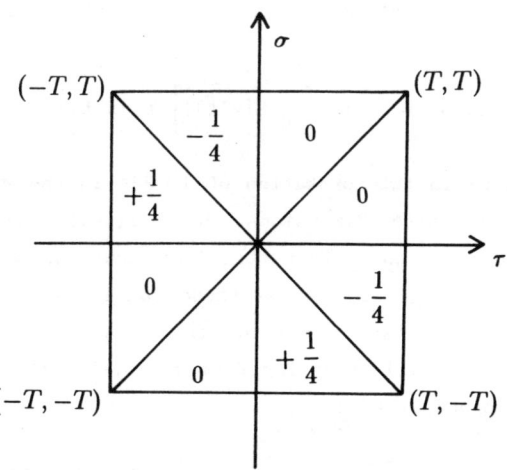

Fig. 12.1 Diagrammatic representation of the kernel $\mathcal{L}(\tau,\sigma)$

One can easily solve the eigenvalue problem associated with the kernel $\mathcal{L}(\tau,\sigma)$. The eigenvalues $\{\mu_k\}$ are given by

$$\mu_k = \frac{T}{2\pi(2k+1)} \, , \quad k = 0, \pm 1, \pm 2, \ldots ,$$

each eigenvalue being doubly degenerate. Hence the characteristic functional $Z(\lambda)$ of the area S_T can be evaluated by using the defining relation for the generating functional and the formula for $\mathcal{J}$-transform given in Eq. (3.12). This reads as

$$Z(\lambda) = E\left(e^{-i\lambda S_T}\right) = \mathcal{J}\left(e^{-i\lambda S_T}\right)(\xi = 0) = \left[\cosh\left(\frac{\lambda T}{2}\right)\right]^{-1} \quad (3.21)$$

a result derived earlier in chapter 6 by using the time slicing approach.

12.3.2. Feynman propagator

While developing his path integral formulation as a sum over trajectories, Feynman suggested that these trajectories were somewhat similar to the trajectories of a Brownian particle. The definition of Feynman propagator as an expectation over the sample paths of a Brownian particle makes this intuitive definition more precise.

Towards this goal, let us recall the heuristic definition of the propagator $K(x'',t'';x',t')$ as

$$K(x'',t'';x',t') = \int \exp\left[\frac{i}{\hbar} S[x(t)]\right] \mathcal{D}[x(t)] \quad . \quad (3.22)$$

The first difficulty in interpretation of (3.22) is the absence of any flat measure in infinite dimensions. However, if one defines the integral with respect to some other "true" measure, one must "formally" compensate for the change. Guided by these considerations, Streit and Hida suggested that one may choose the trajectories in (3.22) as the paths of a Brownian particle starting from x', i.e., $x(t) = x' + B(t)$ and interpret the propagator as

$$K = E\left(N \exp\left[\frac{i}{\hbar} S[x(t)]\right] \delta(x(T) - x'') \exp\left[\frac{1}{2}\int_0^T \dot{B}^2 d\tau\right]\right)$$

$$= E\left(I \exp\left[\frac{i}{\hbar}\int_0^T V(x(t))\right]\right) \quad (3.23)$$

where the symbol I formally denotes the quantity

$$I = N \exp\left[\frac{1}{2}\left(1 + \frac{i}{\hbar}\right)\int_0^T \dot{B}^2 d\tau\right] \delta(B(T) - Y) \quad (3.24)$$

with $Y = x'' - x'$.

One can immediately see that an extra factor $\exp[\frac{1}{2}\int_0^T \dot{B}^2 \, d\tau]$ has been introduced in Eqs. (3.23)--(3.24) to compensate for the "Gaussian" density of white noise measure. The δ-function ensures that the trajectories pass through the fixed point (x'', T). However, the interpretation (3.23) runs into difficulty due to the fact that I of Eq. (3.24) is not a functional of white noise because of the presence of $\int_0^T \dot{B}^2(\tau)d\tau$. To understand the difficulty recall that for well defined quadratic functionals represented by the kernel $F(\tau_1, \tau_2)$ the $\mathcal{J}$-transform is given by Eq. (3.13). However, for $\int_0^T \dot{B}^2(\tau)d\tau$, the kernel function $\delta(\tau_1 - \tau_2)$ is highly singular and does not have a bounded trace. The way to solve this difficulty is to use some additive renormalization so as to compensate for this divergence. The renormalized functional can be interpreted as a Brownian functional.

Problem 12.1

Consider the functional $\delta(B(T)-Y)$. Evaluate the expansion functions F_n. Further show that $\sum n! \, \|F_n\|^2$ diverges.

The way to get out of these difficulties as far as Feynman integral is concerned is to take recourse to the continuous analogue of the expansion (3.5). Let φ and ψ be two random variables of white noise with the corresponding expansion functions $\{F_n\}$ and $\{G_n\}$ respectively. Then

$$E(\varphi \, \psi) = \sum_{n=0}^{\infty} n! \int_{L^2[\mathbb{R}^n]} F_n(\{\tau_j\}) G_n(\{\tau_j\}) \prod_j d\tau_j \, . \qquad (3.25)$$

Next, notice that each of the terms on the right-hand side of Eq. (3.25) may have a well defined meaning for highly singular $F_n(G_n)$ provided G_n (F_n) is regular enough to compensate for the singularities of $F_n(G_n)$ or more generally, the integral in Eq. (3.25) can be interpreted as a canonical pairing between the members of a Gelfand triplet in which the space of symmetric functions is embedded. Similarly, the series (3.25) may have nice convergence properties even if the series containing only F_n does not. This intuition is made precise by embedding of (L^2) itself in a Gelfand triplet

$$(L^2)^+ \subset (L^2) \subset (L^{2-}) . \qquad (3.26)$$

The right-hand side of (3.25) then defines a canonical pairing between the generalized functional $\varphi \ \varepsilon (L^2)^-$ and test functional $\psi \ \varepsilon (L^2)^+$ for which even with the abuse of notation the symbol $E(\varphi, \psi)$ will be retained.

Nature of I

To gain further insight into the nature of I, we observe that the kinetic energy part of the action functional along with the formal factor $\frac{1}{2} \int_0^T \dot{B}^2 d\tau$ can be expressed as

$$\frac{1}{2} \int_0^T \dot{x}^2 \, d\tau + \frac{1}{2} \int_0^T \dot{B}^2 d\tau = -\frac{1}{2} \langle y, Fy \rangle \qquad (3.27)$$

where for notational convenience we have set $y = \dot{B}$. Further, the kernel $F(\tau_1, \tau_2)$ has the form

$$F = -(1+i) \, \delta(\tau_1 - \tau_2) \, \chi_{[0,T]} [\tau_2] . \qquad (3.28)$$

Next, we recall that for well behaved kernel F,

$$\mathcal{J} \left[\exp\left(-\frac{1}{2} \langle x, Fx \rangle\right) \right](\xi) = \frac{1}{\sqrt{\det(I+F)}} \exp\left[\frac{1}{2} \langle \xi, F(I+F)^{-1} \xi \rangle\right] . \qquad (3.29)$$

However, for F of Eq.(3.28) $\det(I+F)^{-1/2}$ is undefined. Therefore following the general strategy we use a multiplicative renormalization and set

$$(N \exp[-\frac{1}{2} \langle x, Fx \rangle])(\xi) = \mathcal{J}^{-1}[C(\xi) \exp[\frac{1}{2} \langle \xi, F(I+F)^{-1} \xi \rangle]] . \qquad (3.30)$$

Note that the left-hand side of Eq.(3.30) belongs to the space $\mathcal{H}_2$ of random variables. If F happens to be a Hilbert-Schmidt operator then the variable of Eq.(3.30) differs from the renormalized kernel by a finite factor, the modified Fredholm determinant. In Eq.(3.30) N can be formally related to the exponential by an infinite renormalization. Hence the right-hand side of Eq.(3.30) can be used as a definition for the left-hand side. Next, the quantity $\delta(B(T)-x''+x')$ can be interpreted in the sense of a distribution as

$$\delta(B(T)-x''+x') = \frac{1}{2\pi} \int_{-\infty}^{\infty} dp \; e^{-ip(x''-x')} \; e^{-i<y,p \; \chi_{[0,T]}(\tau)>} . \quad (3.31)$$

With these interpretations, we now focus our attention on the free particle propagator which can be written $K(x'',x';T) = E(I,1)$ where 1 is a unit functional. It is easy to verify from Eq.(3.10) that the unit functional can be characterized by $G_0 = 1$, $G_k = 0$ for all $k \geq 1$. Hence, it follows from (3.23) that $K(x'',x';T) = F_0$. Further using the definition of $\mathcal{J}$-transform it is easy to see that

$$F_0 = \mathcal{J}(I) \; (\xi = 0)$$

$$= \int_{-\infty}^{\infty} dp \; \mathcal{J} \; [N \; \exp(-\frac{1}{2} <y,Fy>) \; e^{i<y,p \; \chi_{[0,T]}>} e^{-ip(x''-x')}](\xi = 0) \quad (3.32)$$

where F is as given in Eq.(3.28), which can be readily evaluated using Eq.(3.30) and yields the well known free particle propagator.

Thus we see that the unit functional is an admissible test functional for the generalized functional I. The next question is what class of interaction terms form the admissible test functionals? A general answer to this question is not available at present, though attempts in this direction are continuing and we provide the reference for this work at the end of this chapter.

As far as specific cases are concerned, all quadratic actions (both local and non-local) can be handled within this framework. For this purpose, we would need the $\mathcal{J}$-transform of the expression

$$J = \mathcal{J}(N \; \exp[-\frac{1}{2}<y,Fy>] \; \exp[-\frac{1}{2}<y,Ly> + i<y,f>] \; \delta(B(T)-x''+x'))(\eta), \quad (3.33)$$

where F is as defined earlier and the kernel L is Hilbert-Schmidt type. The expression in Eq.(3.33) can be cast as before in the form

$$J = \int_{-\infty}^{\infty} dp \; \mathcal{J} \; [N \; \exp(-\frac{1}{2} (y, (I+F+L)y)) \; e^{-ip(x''-x')}] \; (\xi = \xi_0) , \quad (3.34)$$

where the function ξ has the value $\xi = f + \chi_{[0,T]} + \eta$. Next we observe that

$$\mathcal{J}\ [N\ \exp\ [-\tfrac{1}{2}\ (y,Fy)]\ \exp[-\tfrac{1}{2}\ \langle y,Ly\rangle])\ (\xi)$$

$$= \det\ [I+L(I+K)^{-1}]^{-1/2}\ \exp(-\tfrac{1}{2}\ \langle \xi,(I+K+L)^{-1}\xi\rangle)\ . \qquad (3.35)$$

The various quantities on the right-hand side of Eq(3.35) can be easily evaluated and the expression for J takes the form

$$J = \{\ 2\pi T\ (e, M^{-1}e)\ \det(I + L(I+F)^{-1})\ \}^{-1/2}\ \exp\left[-\tfrac{1}{2}\ (\eta,\ M^{-1}\eta)\ -\right.$$

$$\left. - \frac{1}{2\ (e,M^{-1}e)}\left\{\frac{Y}{\sqrt{T}} - i((\eta+f)\ M^{-1}e)\right\}\left\{\frac{Y}{\sqrt{T}} - i\ (e,\ M^{-1}(\eta+f))\right\}\right]. \qquad (3.36)$$

As an illustrative example of a quadratic action let us consider the evaluation of the propagator for a particle moving in an uniform magnetic field of strength B along z direction. In this case we can write

$$\exp\left[\frac{i}{\hbar}\int_0^T \vec{A}(\vec{q}).\vec{\dot{q}}\ dt\right] = \exp\ [-\tfrac{1}{2}\ \langle y,\ Ly\rangle]\ , \qquad (3.37)$$

where $\vec{A}$ is the associated vector potential and the kernel L is given by $L = 4i\gamma\mathcal{L}$, $\gamma = eB/\hbar c$ is the cyclotron frequency. The integral operator $\mathcal{L}$ has already been evaluated while obtaining Levy's stochastic area in Eq.(3.20). The propagator therefore can be written as

$$K(q'',q';T) = E(I\ \exp[-\tfrac{1}{2}(y,\ Ly)]\exp[-i\gamma y'\chi_{[0,T]} + i\gamma x'\chi_{[-T,0]}]). \qquad (3.38)$$

Using the result (3.36) the expectation value in Eq.(3.38) can be evaluated and one recovers the familiar expression for the propagator. Similar results can be obtained for the propagators corresponding to both local and nonlocal quadratic actions. However, we shall not pursue the explicit calculations for these cases but only quote the structure of the kernel L.

Harmonic oscillator of frequency ω

$$\int_0^T dt \; \frac{m\omega^2}{2} x^2 = (y, Ly) \quad : \quad L = i\hbar\omega \; (T - \max(\tau_1, \tau_2)) \tag{3.39a}$$

Bezak action

$$\frac{\Omega^2}{4T} \int_0^T dt \int_0^T ds \; [x(t) - x(s)]^2 = (y, Ly) \; :$$

$$L = i \; \Omega^2 [T - \max(\tau_1, \tau_2) - (T-\tau_1)(T-\tau_2)/T]. \tag{3.39b}$$

12.3.3. *A new expansion for the propagator*

The results of the white noise calculus can be exploited to generate an expansion for the propagator which is different from the conventional perturbation and eigenfunction expansions. The crucial step in deriving the expansion is the relation (3.25) in which we identify the variables φ as I and ψ as $\exp(-i \int_0^T V[x(\tau)] \, d\tau)$. In order to use the relation (3.25) effectively we must find out the expansion coefficients F_n and G_n for I and ψ respectively. First let us consider the quantity I. The $\mathcal{T}$-transform of I is explicitly given as

$$\mathcal{T}(I) \; (\xi) = \mathcal{T}(N \exp(-\tfrac{1}{2} \langle y, Fy \rangle)) \; \delta(B(T) - (x''-x')) \; (\xi), \tag{3.40}$$

which can be re-expressed as

$$\mathcal{T}(I) \; (\xi) = \frac{1}{2\pi} \int_{-\infty}^{\infty} e^{-ip(x''-x')} \; \mathcal{T}(N \exp(-\tfrac{1}{2} \langle y, Fy \rangle)) \; (\xi + p \, \chi_{[0,T]}(\tau)) \; dp \; . \tag{3.41}$$

The $\mathcal{T}$-transform in Eq.(3.41) can be readily evaluated using the result in Eq.(3.30). Subsequently performing integration over p, we obtain

$$\mathcal{J}(I)(\xi) = \frac{1}{\sqrt{2\pi i T}} \exp(-\frac{1}{2}\|\xi\|^2) \exp\left[\frac{i}{2T}\left\{Y^2 + \left(\int_0^T \xi(\tau)d\tau\right)^2 + 2Y\int_0^T \xi(\tau)d\tau\right\}\right].$$

(3.42)

A straightforward expansion of $\mathcal{J}$-transform in power series of $\int_0^T \xi(\tau)d\tau$ yields the coefficient functions F_n. They are given by

$$F_n = F_0 \sum_{p=0}^{[n/2]} \left[\frac{1-i}{2}\right]^p \frac{(2iT)^{(n-2p)/2}}{p!(n-2p)!} H_{n-2p}\left(-\frac{2Y}{\sqrt{(2iT)}}\right) \prod_{k=2p+1}^n \chi_{[0,T]}(\tau_k)$$

$$\times \delta(\tau_1-\tau_2)\delta(\tau_3-\tau_4)\ldots\delta(\tau_{2p-1}-\tau_{2p}), \qquad (3.43)$$

where $H_n(x)$ is the Hermite polynomial. The quantity F_0 is given by Eq.(3.32) and can be identified as the free particle propagator.

We now turn our attention to the potential term. Consider the quantity $\mathcal{J}\left(\exp\left(-\frac{i}{\hbar}\int_0^T V(x(t))dt\right)\right)$. Setting $x(t) = x' + B(t)$, we can write

$$\mathcal{J}\left(\exp\left(-\frac{i}{\hbar}\int_0^T V(x(t)dt)\right)\right) = C(\xi)\sum_{n=0}^\infty i^n \int_R V_n(\tau_1,\tau_2\ldots\tau_n) \prod_{k=1}^n \xi(\tau_k)\,d\tau_k,$$

(3.44)

where V_n are the corresponding expansion coefficients of the argument of $\mathcal{J}$-transform in Eq.(3.31). Now using Eq.(3.25), it follows

$$K(x'',x';T) = E(I\exp(-\frac{i}{\hbar}\int_0^T V(x'+B(\tau)d\tau)) = \sum_{n=0}^\infty n! \int_{-\infty}^\infty F_n V_n\,d^n\tau. \quad (3.45)$$

The expansion has several peculiar features. First, note that in Eq.(3.44) the potential term does not contain any x''. Hence the expansion coefficients V_n are independent of x''. In other words the entire dependence of the propagator on x'' is contained in the coefficients F_n. Recall that in chapter 1 we had established that for

Brownian motion $\langle \Delta x^2 \rangle \simeq T$. The argument of Hermite polynomials in the expansion coefficients F_n is proportional to $Y/\sqrt{T}$. This is directly related to our assertion that the paths entering in the "functional integration" are Brownian motion trajectories superimposed over a fixed path. In this picture we can reinterpret the potential term as providing some kind of bias to "free Brownian motion".

Problem 12.2 :

Consider the time-dependent potential $V = \alpha(t)\, x(t)$. Show that

$$\left(\exp\left(-\frac{i}{\hbar}\int_0^T V\, d\tau\right)\right)(\xi) = \exp(ix'f(0)/\hbar)\, \exp\left(-\frac{1}{2}\,\|(f,\chi_{[0,T]}) + \xi\|^2\right)$$

(3.46)

and hence obtain the expansion coefficient V_n

$$V_n = \frac{i^n}{n!}\, e^{ix'f(0)/\hbar}\, \exp\left(-\frac{1}{2}\int_0^T f^2(\sigma)\, d\sigma\right) \prod_{k=1}^n f(\tau_k)\, \chi_{[0,T]}(\tau_k)\, . \quad (3.47)$$

Use the expansion for the propagator (3.45) and the expression (3.43) for F_n's to derive an expression for the propagator corresponding to the linear potential.

The result stated in the above mentioned problem can be easily generalized to establish that all potential terms having a bounded Fourier transform provide admissible interactions to the generalized functional I. To see this clearly let $V(x)$ be the potential with a bounded Fourier transform. Therefore, we can write

$$\exp\left[-\frac{i}{\hbar}\int_0^T V(x(\tau))d\tau\right] = \sum_{n=0}^\infty \frac{(-1)^n}{n!} \int d^n\tau \int \prod_{\nu=1}^n d\mu(k_\nu)\, \exp(k_\nu x(\tau_\nu)). \quad (3.48)$$

The $\mathcal{J}$-transform of the potential term in Eq.(3.48) has now been related to the $\mathcal{J}$-transform of the linear potential of strength $\alpha(\tau) = k_\nu \delta(\tau-\tau_\nu)$ and can be evaluated using the result (3.47). Now using the properties of the boundedness of the measure $d\mu(k_\nu)$ it can be easily shown that the expansion (3.45) corresponds to the well known Dyson series.

Notes and References

The mathematical nature and precise meaning of Feynman integrals through Lie-Trotter formula has been nicely discussed by

S. A. Albeverio and R. Hoed-Krohn, "Mathematical Theory of Feynman Path Integrals" in Lect. Notes Maths 523, (Springer Verlag, Berlin, 1976).

Our treatment is entirely based on Albeverio's work. However, the conditions on V can be relaxed. This can be found in

E. Nelson, J. Math. Phys. I, 332, (1984),

P. Exner, "Open Quantum Systems and Feynman Integrals", D. Reidel, Dordrecht (1985).

The method of redefining oscillatory Gaussian integrals in infinite dimensions has been considered by K. Ito and is very extensively developed by Albeverio et al. This is discussed in a review article by

S. Albeverio "Some recent developments and applications of path integrals" in "Path Integrals from meV to MeV", Eds. M. C. Gutzwiller, A. Inomata, J. R. Klauder and L. Streit (World Scientific, Singapore, 1986)

This definition of path integrals has been generalized in various directions. For example see :

V. Mandrekar, "Some remarks on various definitions of Feynman Integrals", Lect. Notes. Math. (Ed. Beck and K. Jacob 1983).

An excellent treatment of the mathematical theory of Brownian Motion and the generalized stochastic process can be found in

T. Hida, "Brownian Motion" (Translated jointly by T. Hida and T. P. Speed) (Applications of Mathematics 11) (Springer Verlag, New York, 1980).

The material on Functionals of Brownian Motion can be found in

T. Hida, Ricerche di Mathematica XXXIV, 183, (1985).

The formulation of the Feynman Path Integral in terms of the expectation value with respect to White noise measure is contained in

L. Streit and T. Hida, Stochastic Processes and Their Applications, 16, 55 (1983).

Later de Falco and Khandekar applied this technique to compute Feynman propagators for some specific problems. They are contained in

Diago de Falco and Dinkar C. Khandekar, Stochastic Processes and Their Application 29, 257 (1988).

A general description regarding the use of white-noise calculus in the computation of Feynman Propagator can be found in a review article :

D. C. Khandekar, "White Noise Analysis as a tool in Computing Feynman Integrals" in "White Noise Analysis; Mathematics and Applications", Eds. T. Hida, H.H. Kuo, J. Potthoff, and L. Streit (World Scientific, 1990).

This review also contains the derivation of the new expansion of Feynman propagator.

The question regarding what potentials are admissible within this formulation has been discussed in terms of the characterization theorems and can be found in

J. Potthoff and L. Streit, "The Feynman integrand as a Hida distribution", BiBoS preprint No. 439 (1990),

and in article, titled as

D. C. Khandekar and L. Streit, "Constructing the Feynman Integrand" Ann Physik 1, 49 (1992).

There have been several other attempts to give Feynman Integrals a proper definition. All of these cannot be accommodated here. The notable amongst the omissions are the approach due to Klauder based on Coherent States and is discussed in

J. R. Klauder and Bo-Sture Skagerstam, "Coherent States" (World Scientific, 1985)

J. R. Klauder and I. Daubechies, Phys. Rev. Lett. 52, 1161 (1984)

Another approach based on the theory of pro-distributions has been developed by C. Dewitt-Morette and can be found in

C. DeWitt-Morette, Comm. Math. Phys. 28, 47 (1972),

C. DeWitt-Morette, A. Maheshwari and B. Nelson, Phys. Reps. 50, 255 (1979),

C. DeWitt-Morette, Ann. Phys. 97, 367 (1976).